A Course in Cosmology

This new graduate textbook adopts a pedagogical approach to contemporary cosmology that enables readers to build an intuitive understanding of theory and data, and of how they interact, which is where the greatest advances are currently being made. Using analogies, intuitive explanations of complex topics, worked examples, and computational problems, the book begins with the physics of the early universe, and goes on to cover key concepts such as inflation, dark matter and dark energy, large-scale structure, and cosmic microwave background. Computational and data analysis techniques, and statistics, are integrated throughout the text, particularly in the chapters on late-universe cosmology, while another chapter is entirely devoted to the basics of statistical methods. A solutions manual for end-of-chapter problems is available to instructors, and suggested syllabi, based on different course lengths and emphasis, can be found in the Preface. Online computer codes and datasets enhance the student learning experience.

Dragan Huterer is a professor of physics at the University of Michigan, where he studies cosmology at the interface of theory and data. He is a leading expert in developing data-driven techniques to understand the nature and origin of "dark energy," a component that dominates the dynamics of the universe and causes its accelerated expansion. A fellow of the American Physical Society, he is also a recipient of the University of Michigan's Henry Russel Award and the Humboldt Foundation's Friedrich Wilhelm Bessel Award.

"An incredibly practical approach to introducing pivotal concepts in cosmology, with a preciously rare focus on preparing its readers to engage with the science of cosmology in a real-life manner."

Vera Gluscevic, University of Southern California

"This is a welcome addition to the growing number of textbooks on cosmology. The subject is concisely but exceptionally clearly presented, suitable for a one-semester course. The book stands out by covering the full range of material needed for a beginning graduate student contemplating entering research in cosmology: from the foundations all the way to active research topics, such as neutrinos, dark energy, modern large-scale structure and cosmic microwave background experiments, as well as analysis techniques. I am looking forward to adopting this book for my own cosmology course."

Zoltan Haiman, Columbia University

"A complete guide to form the next generation of well-rounded cosmologists. This book is a unique combination of theoretical, observational, and statistical aspects, with pedagogical and extensive descriptions of modern statistical methods in cosmology."

Alessandra Silvestri, Leiden University

"A clearly written introduction to astrophysical cosmology emphasizing data/theory connections. This text is unique in introducing the relevant techniques for those new to the field."

Curtis Struck, Iowa State University

"This superb first introduction to cosmology for advanced undergraduate or graduate students is sure to inspire and teach young inquisitive minds for years to come. Huterer weaves the history of the field with elegant and lucid explanations of the most salient topics in modern cosmology."

Balša Terzić, Old Dominion University

A Course in Cosmology

From Theory to Practice

DRAGAN HUTERER

University of Michigan, Ann Arbor

CAMBRIDGE
UNIVERSITY PRESS

Shaftesbury Road, Cambridge CB2 8EA, United Kingdom

One Liberty Plaza, 20th Floor, New York, NY 10006, USA

477 Williamstown Road, Port Melbourne, VIC 3207, Australia

314–321, 3rd Floor, Plot 3, Splendor Forum, Jasola District Centre, New Delhi – 110025, India

103 Penang Road, #05–06/07, Visioncrest Commercial, Singapore 238467

Cambridge University Press is part of Cambridge University Press & Assessment,
a department of the University of Cambridge.

We share the University's mission to contribute to society through the pursuit of
education, learning and research at the highest international levels of excellence.

www.cambridge.org
Information on this title: www.cambridge.org/highereducation/isbn/9781316513590

DOI: 10.1017/9781009070232

First published 2023

Printed in the United Kingdom by TJ Books Limited, Padstow, Cornwall, 2023

A catalogue record for this publication is available from the British Library

A Cataloging-in-Publication data record for this book is available from the Library of Congress

ISBN 978-1-316-51359-0 Hardback

Additional resources for this publication at www.cambridge.org/huterer

Contents

Part II The Early Universe

Part III The Later Universe

The plate section can be found between pp 210 and 211

Preface

A few decades ago, cosmology slowly made a shift from being a small, speculative corner at the intersection of physics and astronomy pursued by several dozen physicists and astronomers to a thriving, visible, highly quantitative field which nowadays occupies thousands of scientists worldwide. This shift happened as an avalanche of new and better data from an impressive variety of probes became available and were used to test theory. This field's achievements are widely recognized, perhaps most visibly by seven physics Nobel Prizes that have been awarded in the twenty-first century for discoveries in cosmology and astrophysics. Cosmology is indeed in its golden age (as cosmologist David Schramm already argued in the 1990s), and we cosmologists are lucky to be part of such a vibrant and active adventure to understand the universe.

The greatest advances in cosmology have undoubtedly happened at the interface of theory and data. We need look no further than the current cosmological analyses from the galaxy clustering, gravitational lensing, or cosmic microwave background data, which have produced percent-level constraints on some of the key physical quantities from the early and late universe. In these analyses, we would typically start with millions of observations (say, pixelized temperature measurements of the cosmic microwave background on the sky, or galaxy shapes measured in a lensing survey), and from there determine ten or so fundamental parameters of the cosmological model. The many intermediate steps in this procedure are non-trivial and involve sophisticated mathematics and statistics. Yet these techniques are, with a few exceptions, not covered in the cosmology textbooks currently available, which instead concentrate on theoretical background and concepts. These data-oriented topics are rather covered in specialized texts that focus on mathematical rigor and do not make the specific connections to cosmology transparent. This is where the present book comes in.

Motivation and Aim

This book was born out of my lecture notes for a semester-long course in cosmology at the University of Michigan. The idea was to provide a text that both covers basic cosmology and introduces the student to key methods of statistics, data analysis, and numerical modeling – what we will refer to as computational cosmology. The organic inclusion of computation, particularly some of the most commonly used statistical and numerical techniques and concepts, in a cosmology text was my first major motivation for writing this book.

While there were few, if any, cosmology-specific textbooks at the turn of the millennium, the situation is now very different, with a number of good textbooks covering either all of cosmology or its specific sub-areas. Some of these existing books provide very basic introductions to the field, while others are quite technical and cater to advanced graduate students or practicing researchers working in cosmology. What is missing is a book that starts with basic concepts and connects them to sufficiently quantitative results so that students are prepared for graduate-level research in the field. This intermediate-level coverage of cosmology was my second major motivation for putting pen to paper.

This book is aimed at advanced undergraduate and graduate students. Its primary use is envisioned as the main self-contained textbook for an advanced undergraduate or introductory graduate course in cosmology.

A secondary use of the book might also be that it serves as a reference for practicing cosmologists. Because the text has been written at an intermediate level, it is likely to be most readily used as a reference by research students and cosmologists who wish to consult a topic that is *not* in their own area of expertise.

Organization and Approach

The book is organized in three parts. Part I (Foundations; Chapters 1–3) provides a general background and introduces basic quantities in cosmology. The next two parts review the processes in the universe roughly chronologically. Part II (The Early Universe; Chapters 4–8) covers the thermodynamical history of the universe, Big Bang nucleosynthesis, neutrinos, and inflation, along with some other topics relevant for the early universe. Part III (The Later Universe; Chapters 9–14) covers structure formation, cosmic microwave background, dark matter and dark energy, and gravitational lensing.

There are three main thrusts of emphasis to the book, namely that it:

1. Be pedagogical and intuitive, and emphasize key concepts and useful tools starting from the basics, thus complementing other textbooks that tend to be either very elementary or very technical.
2. Include key modern developments in the field, especially at the data/experiment/ theory interface, thus complementing the traditional emphasis on theoretical fundamentals.
3. Emphasize computation and statistical and data analysis techniques, which is typically lacking in extant textbooks.

There are two principal ways in which I have integrated the computational/stats/ data aspect of cosmology. First, Chapter 10 introduces statistical and data methods, specifically tailored for work in cosmology and astrophysics. Second, at the end of most chapters I include one or more problems labeled [**Computational**], which require computer programming to be solved. These problems are tailored to cover a broad range of computational skills typical for analyzing data or comparing data to theory in cosmological applications.

I attempted to derive all the main results "from scratch" as far as possible. [To strike a balance and keep the book from getting overly technical or long, there are certainly exceptions, particularly when it comes to the adoption of some results from general relativity that are motivated by Newtonian derivations or simply quoted, rather than derived from scratch.] The idea was to remove the veil of mystery that sometimes conceals how results are obtained. I have adopted the same philosophy for the figures in the book, most of which I produced using Python codes that will be made available to instructors online (and which they could share with their students, if desired). Finally, I made an effort to provide intuitive explanations for some of the most common concepts, where I could come up with such analogies; a good example is an intuitive yet quantitative picture illustrating the meaning of the power spectrum in Fig. 9.5.

Key Features

I have included the following features to aid students and instructors in getting the most out of this text:

- Shaded key equations and takeaway points, to highlight their importance.

- Boxes (on average two per chapter) that include either a simple, worked example to illustrate how to obtain basic results, or else more advanced, optional material that can be skipped on first reading.

- Bibliographical notes at the end of each chapter pointing students to recommended further readings.

- Computational problems, to help students master the connection between data and theory.

- Appendices listing the quantities required for analytical and numerical problem-solving in cosmology, for easy reference.

- Solutions to all end-of-chapter exercises for instructors, available online along with computer codes and datasets.

Teaching with this Book

There are 14 chapters in the book and 14 weeks in a typical semester. Therefore, at first sight it may seem that, covering one chapter per week, the instructor could cover the whole book in a single semester. However, a few of the chapters – Inflation (8), Large-Scale Structure in the Universe (9), and Cosmic Microwave Background (13) – require at least two weeks' teaching, based on my experience. Moreover, most of the chapters are now somewhat longer than my original set of lecture notes (which I *would* usually get through during a semester-long course). Therefore, instructors of one-semester courses will have to make their own judgment call, and select a few topics to either skip or assign as optional reading, depending on the focus of their course. For example, it might make sense for a course emphasizing the late universe and observations to go light on Chapters 5 (Neutrinos in Cosmology), 6 (The Boltzmann Equation and Baryogenesis), and 7 (Big Bang Nucleosynthesis). Whereas, for a course concentrating on early-universe cosmology, some of the later chapters (9, 13, and 14, for example) could be covered in a condensed form. Finally, a two-semester course should be able to comfortably cover all of the topics in the book at the level at which they are introduced.

Chapter 10 covers a range of statistical methods used in cosmology, starting from the basics and working up to fairly advanced topics. This chapter could be covered earlier in the semester, if desired. It could also be split into sections which are covered throughout the semester, accompanying individual statistical concepts as they first appear in the text. Clearly, there are multiple ways to introduce this material; what is important is not precisely which of them is followed, but rather that students become introduced to the statistical techniques at some point during their study of cosmology.

Each chapter ends with a number of select exercises for students, to test their assessment of the material and understanding of key concepts. Some problems are of the standard, pencil-and-paper variety and, as mentioned above, most chapters also contain computational problems. The level of difficulty of the theory problems ranges from easy to moderately difficult; some of the more difficult problems are accompanied by hints to help students work through various steps. Many of the computational problems are fairly straightforward, while others (e.g., analysis of type Ia supernova data, or simulation of the evolution of the inflationary scalar field) contain multiple parts and may be time-consuming, and are thus perhaps better suited for an end-of-course project rather than a weekly homework. The range of difficulties and flavors to the end-of-chapter problems is intended to give a broad experience and physical insight to students of cosmology, and to cater for a range of abilities.

The "from-scratch" logic adopted in the text extends to the end-of-chapter problems, both pencil-and-paper and computational. A good example is Problem 9.9 in Chapter 9, where students are asked to code the matter power spectrum and then mass function from scratch, without using any publicly available tools. While the students definitely *can*, and *should*, learn to use and adopt such tools (starting with, e.g., `astropy`, and continuing on to `CAMB` and `CLASS`), there is great value, in my opinion, in also being able to understand the connection between physics and computation at a very basic level, before any numerical black box has been made available and used. This logic has been applied to most of the computational problems assigned in this book.

Acknowledgments

I would like to thank a small army of cosmology colleagues and friends who commented on the individual chapters and other scientific aspects of this book: Chris Blake, Gus Evrard, Arya Farahi, Elisa Ferreira, Cameron Gibelyou, Vera Gluščević, Yacine Ali-Haïmoud, Eiichiro Komatsu, Andrew Long, Luisa Lucie-Smith, Marta Monelli, Jessie Muir, Ken Nollett, Fabian Schmidt, Lucas Secco, Evangelos Sfakianakis, Blake Sherwin, Josh Spitz, Sherry Suyu, Scott Watson, Noah Weaverdyck, and Mijin Yoon. Their input significantly improved both the presentation and the correctness of the material. I also gratefully acknowledge help from three supremely talented students, Otávio Alves, Dhayaa Anbajagane, and Andrija Kostić, who worked through most of the end-of-chapter problems in the book. I truly feel privileged to have received help, advice, and encouragement from all these excellent cosmologists and physicists.

I thank my PhD advisor Michael Turner, and my postdoc advisors Wayne Hu, Lawrence Krauss, and Glenn Starkman. All of them significantly contributed to my "worldview" of cosmology, and helped me become a cosmologist in the first place.

It has been a pleasure to work with Cambridge University Press, particularly my contacts there, Melissa Shivers and Arya Thampi.

I am thankful to the Physics Department and the Leinweber Center for Theoretical Physics at the University of Michigan, both of which have been my home during the years I taught graduate cosmology and developed some of this material. I am fortunate to have mentored and worked with a number of talented students and postdocs. This experience contributed to the philosophy and methodology of exposition in this book.

My work has also been supported by the Department of Energy, National Science Foundation, and NASA. The book effort has specifically been supported by the Friedrich Wilhelm Bessel Research Award, given to me by the Humboldt Foundation, which supported my year-long sabbatical stay at the Max Planck Institute for Astrophysics (MPA) where I carried out a big chunk of this work.

Finally, I gratefully thank my family – Gin, Miriam, and Marcel – for their support and love.

PART I

FOUNDATIONS

1 Introduction to Cosmology

1.1 Cosmology: Science of the Universe

Humanity has looked out to the sky and wondered about our past, present, and future for a long time. The history of observing and contemplating the cosmos is undoubtedly thousands of years old. Probably the first known representation of objects in the cosmos outside of the solar system – the Nebra sky disk (see Fig. 1.1) – dates from the Bronze age, c. 1600 BCE. Early observations were limited to what could be seen by the naked eye, as the telescope was only invented much later and first used to observe celestial objects by Galileo in 1609.

The goal of cosmology is to understand the universe as a whole: how it started, how it came to be what we see today, what particles and fields are in it, and what is its future. In this regard, cosmology subtly differs from astrophysics, which is concerned with physical processes of *objects* in the universe (galaxies, neutron stars, black holes). For a cosmologist, galaxies are often test particles used to infer properties of space and time. However, areas of cosmology also study individual objects, so that the overlap between cosmology and astrophysics is nevertheless significant.

The birth date of modern cosmology, the quantitative science that compares data with theory to infer the physical laws of the cosmos, could arguably be traced to the time and work of Isaac Newton, as his *Principia* revolutionized all of physics with its highly quantitative, groundbreaking results. Perhaps a more appropriate start date that is specific to cosmology could be assigned to the early twentieth century, when Einstein's special and general theories of relativity were quickly adopted to build the first modern cosmological models, and Edwin Hubble discovered that the universe is expanding. Finally, if we are talking about the beginning of modern *observational* cosmology, where data are compared to theory using sophisticated statistical methods and where the parameters are measured at percent-level (rather than factors of a few) accuracy, then COBE experiment's discovery of the fluctuations in the cosmic microwave background, made in 1992, deserves to be highlighted. COBE provided stunning confirmation of the hot Big Bang model, and ushered in a flood of high-quality data in cosmology from which we have been benefitting ever since.

1.2 Observing the Universe

We are in a unique situation in that we observe the universe from a single vantage point in space and time. We rely on the measurements of photons from stars and galaxies. Because these reach us with the speed of light, which is large but finite, we see nearby objects as they were recently, and faraway objects as they were a long time ago. While this effect is small for nearby, solar-system observations (we see the Moon as it appeared 1.25 seconds ago and the Sun as it appeared 8 minutes ago), for cosmological observations we typically observe galaxies and their properties – their shapes and luminosities, their mutual clustering – as they were billions of years ago. By collecting information from objects at different distances, we therefore automatically map out the *temporal* history of the universe as well. For example, a population of galaxies that is observed at a large distance typically features a lower clustering signal than a population observed at a smaller distance, as clustering of galaxies grows over time (as we will discuss in Chapter 9).

How can we learn about the universe, given that we can neither travel back in time nor to faraway distances to inspect physical conditions at those epochs and locations? The answer lies in the remarkable generality of the laws of physics, which govern the structure and evolution of the universe starting from an unimaginably shorter period after the Big Bang until the present day. The laws of physics, including classical and quantum mechanics, statistical physics, relativity, and particle physics, dictate how nuclei and atoms form, particles interact, stars and galaxies form, and the universe expands. These physical laws allow us to interpret observations coming from different epochs in one single framework, completing the theoretical missing pieces where necessary. For example, observations of motions of stars around galaxies, interpreted via Newton's law of gravity, indicate the existence of dark matter (see Chapter 11). As importantly, the physics framework allows us to reject scenarios that are ruled out by observations interpreted in the context of physical laws. For example, the steady-state theory of the universe, which states that the universe is infinitely old and unchanging, can be ruled out based on several pieces of observational evidence – the "three pillars of the Big Bang" discussed just below, in Sec. 1.4.

1.3 The Expanding Universe

The universe is expanding. This fundamental fact was discovered by Edwin Hubble around 100 years ago, and is a cornerstone of our understanding of the universe. The expansion is referring to the fact that individual galaxies recede from each other. It is *not* the case that everyday objects expand (e.g., the book that you are reading, the radius of Earth, or the distance between Earth and the Moon). The reason for that is that these objects are being held by forces that resist the

Fig. 1.1   Nebra sky disk, a $\sim$3500-year-old bronze artifact depicting the Sun or full Moon, a lunar crescent, and stars. This is possibly the oldest known representation of cosmic objects. Credit: Schellhorn/ullstein bild/Getty Images.

expansion, for example the electromagnetic force that governs chemistry and holds most everyday objects together, or else the force of gravity that keeps Earth and the Moon near each other. One notable object that is still refusing to participate in the expansion is the Andromeda galaxy, our nearby companion, which is scheduled to collide with our Milky Way galaxy in a few billion years. In more rigorous language, the peculiar (own) velocity of Andromeda is greater than its expansion velocity and has a roughly opposite sign. But, apart from Andromeda (and some other, very nearby dwarf galaxies), all galaxies are receding away from us, following the Hubble law that will be discussed at length in Chapter 2.

Moreover, the universe has no center. Every point in space is as important as any other. It should therefore not be surprising to know that the universe is not expanding away from any one point in space, but rather that every point is receding from every other (according again to the Hubble law). A two-dimensional analogy often used to illustrate this is the surface of an expanding balloon, where every point recedes from every other, and there is no center of expansion on the surface itself.

Because the universe is expanding, an immediate question may be how it started – what happens when you run the film back in time all the way to the beginning? Since the late 1960s, we have been fairly certain that the universe originated some time ago – now known to be 13.8 billion years ago – in an event called the **Big Bang**. Right at the time of the Big Bang ($t = 0$ by convention), the universe was, formally, infinitely small, infinitely dense, and infinitely hot. We say "formally" since the details of the physics *right at* the Big Bang are unknown, and have thus far evaded the possibility of observational confirmation. In contrast, an important physical process called inflation that occurred a tiny moment thereafter (perhaps at $t \sim 10^{-35}$ s) – and that most cosmologists consider as the de facto beginning of the universe – is rather well understood and can be probed with present-day data.

1.4 Three Pillars of the Big Bang Model

As far as a general model of the universe is concerned, the Big Bang model[1] is the only game in town. The model stipulates that the universe was small and hot after its birth, and has been expanding and cooling ever since. The only serious competitor to the Big Bang model was the steady-state model, which proposed that the universe is infinitely old and, on average, unchanging – so, not expanding.

The steady-state model was actually quite appealing on account of its simplicity, but the Big Bang model dealt it a crushing defeat with a number of discoveries made in the decades spanning the mid-twentieth century. The reasons for the dominance of the latter are often referred to as the "three pillars of the Big Bang" (model); they are

1. The discovery of the **expansion of the universe**. The expansion implies that the universe was smaller and hotter in the past, hence favoring the Big Bang model.
2. The existence of the **cosmic microwave background (CMB)** radiation. We observe a nearly uniform bath of photons, which is expected in the Big Bang model due to its early, hot beginnings, but would otherwise be difficult to explain.
3. The observed cosmic **abundances of the lightest elements** from the periodic table. These elements were created during Big Bang nucleosynthesis (BBN), about 3 minutes after the Big Bang. To correctly predict the observed abundances requires high temperatures, as well as a suitable scaling of temperature with the expansion rate, both of which are available in the Big Bang model.

The expanding universe will be discussed further in Chapter 2, the CMB in Chapter 13, and BBN in Chapter 7. Note that no serious competitors to the Big Bang model have emerged since the ultimate demise of the steady-state model in the 1960s.

1.5 Timeline of Key Events

The timeline of the evolution of the universe is shown in Fig. 1.2. During the early inflationary epoch, the universe expands very rapidly. The process of reheating after

---

[1] A more complete name is the *hot* Big Bang model, to differentiate it from models where the universe starts out small but is cold. For brevity, we will just refer to the hot model as the Big Bang.

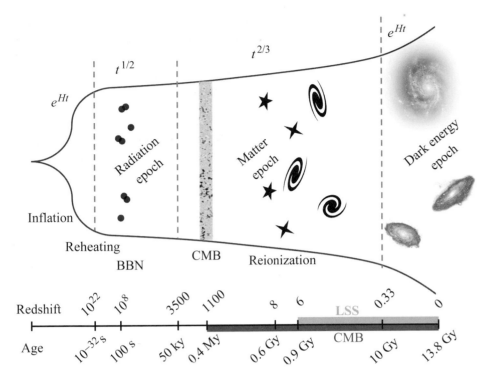

Fig. 1.2   Timeline of principal stages in the evolution of the universe. The expressions near the top show how the scale factor $a(t)$ (to be introduced in Chapter 3) scales with time t. The scale on the bottom shows redshift (to be introduced in Chapter 2) and time since the Big Bang. Figure courtesy of Noah Weaverdyck.

inflation generates particles of the Standard Model of particle physics – quarks and gluons (which a bit later combine into three-quark particles called baryons, such as the proton or the neutron), leptons (such as the electron or the neutrinos), and photons. It also presumably generates the dark-matter particle.

About a minute after the Big Bang, protons and neutrons combine into hydrogen and helium nuclei, along with small quantities of a few more lightest elements. This is the epoch during which energy in radiation (photons and relativistic neutrinos) dominates. Later, around 380,000 years after the Big Bang, the protons and electrons combine into atoms in a process called recombination. At that time, the CMB photons stop scattering with the electrons (which are now bound in atoms), and stream freely. The universe is now matter dominated as the energy density is dominated by a combination of baryonic ("normal") matter and dark matter. Following the dark ages when the only light around is that from the CMB, the largely neutral universe ionizes thanks to photons originating from the first stars and galaxies. Following this, the objects familiar to us (stars and galaxies) form. These structures are baryonic but thought to be enveloped by "halos," and often connected by filaments, of a ubiquitous web of dark matter.

Finally, just when we thought that the most important events in the history of the universe have all taken place, there is a surprise. About 10 billion years after the Big Bang, a mysterious component called dark energy causes the accelerated expansion of the universe, counteracting the tendency of dark and baryonic matter to slow down the expansion. The universe again starts to expand very rapidly.

Some of the most important discoveries in cosmology, made over the past century, are listed in Box 1.1.

Box 1.1   **Cosmology: Top Hits**

Author's list of best-of-the-best discoveries in cosmology made over the past century.

- 1915: Albert Einstein completes his general theory of relativity.
- 1922: Alexander Friedmann discovers an expanding-universe solution to Einstein's equations of general relativity.
- 1927 and 1929: Georges Lemaître theoretically predicts, and Edwin Hubble (partly thanks to recession velocity data from Vesto Slipher) discovers, the expansion of the universe.
- 1933: Fritz Zwicky finds the first evidence for dark matter.
- 1940s: Ralph Alpher, George Gamow, and Robert Herman outline how the lightest nuclei are produced in the hot early stage of the Big Bang model. They also predict existence of the relic CMB radiation.
- 1965: Arno Penzias and Robert Wilson accidentally discover the CMB, confirmed shortly by Bob Dicke and David Wilkinson's group.
- 1970s: Vera Rubin and collaborators confirm dark matter by establishing "flat rotation curves" of galaxies.
- 1970s: Jim Peebles, Rashid Sunyaev, Yakov Zeldovich, and others work out the physical foundations of modern cosmology.
- 1981: Alan Guth (followed shortly by Andrei Linde, Andreas Albrecht, and Paul Steinhardt) proposes inflation.
- 1992: NASA's COBE experiment discovers fluctuations in the CMB temperature.
- 1998: Two teams simultaneously discover that the expansion of the universe is accelerating using type Ia supernovae; the source of the acceleration is named dark energy.
- 2001–present: NASA's WMAP and ESA/NASA's Planck CMB experiments, along with ground-based and balloon-borne CMB experiments and large-scale structure surveys, determine many of the cosmological parameters to unprecedented accuracy.
- 2016: LIGO experiment detects gravitational waves from merging black holes and neutron stars, confirming a key prediction of Einstein's general theory of relativity, and ushering in an era of gravitational-wave cosmology.

1.6 Homogeneity and Isotropy

We observe the universe from a single vantage point – our own. While this makes inferences a bit difficult, we can still learn a lot by looking at the sky. As far as we can tell, on large scales – that is, on scales of hundreds of megaparsecs or larger (see Box 1.2 for the definition of a parsec) – our universe is homogeneous and isotropic. These words mean:

> **Homogeneous** ⇔ looks the same from every point in space.
> **Isotropic** ⇔ looks the same in every direction from our vantage point.

Note that it didn't have to be that way. The universe could be isotropic and not homogeneous, homogeneous and not isotropic (in the so-called Bianchi models), or

---

Box 1.2   **Parsec**

A parsec is a natural unit of distance in that it is defined in terms of Earth's orbit around the Sun. At a distance of 1 parsec, the parallax – the apparent shift of an object on the sky due to Earth's motion over the period of the year – is one arc second. In other words, a parsec is equal to the distance at which 1 AU (astronomical unit – average distance between Earth and the Sun) is seen at an angle of one arc second. It just happens that the parsec is not far off from another popular unit of distance, the light-year (the distance light travels in a year), $1\,\mathrm{pc} \simeq 3.26\,\mathrm{lyr}$.

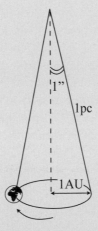

In cosmology, the usage of kiloparsecs ($1\,\mathrm{kpc} = 10^3\,\mathrm{pc}$), megaparsecs ($1\,\mathrm{Mpc} = 10^6\,\mathrm{pc}$), and gigaparsecs ($1\,\mathrm{Gpc} = 10^9\,\mathrm{pc}$) is omnipresent. Note that the parallax of objects that are much more than a kiloparsesc away is typically too small to be observed. The important part is that parsec furnishes a unit of distance that cosmologists have agreed to use.

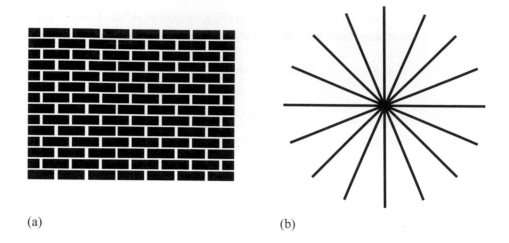

(a)   (b)

Fig. 1.3 A sketch illustrating homogeneity and isotropy of space. (a) A homogeneous but not isotropic space. (b) An isotropic (about the center) but not homogeneous space. [Inspired by a similar illustration on Ned Wright's cosmology tutorial, `www.astro.ucla.edu/~wright/cosmo_01.htm`.]

simply neither. See Fig. 1.3 for visual examples. Interestingly, one can prove that a space that is isotropic around every point in space is automatically homogeneous.

The notion of combined homogeneity and isotropy goes under the name of **cosmological principle** and is an essential ingredient in modern cosmology.[2] Although initially only a mathematical convenience, the cosmological principle has now been empirically confirmed; isotropy has been established by observations of the CMB to a high precision, while large-scale galaxy surveys confirmed that the distribution of matter is homogeneous at scales larger than about 100 Mpc. On the other hand, small deviations from homogeneity and isotropy are still possible and could be signatures of new physics, and in fact some hints of such departures from the expectation have been seen at large angular scales in the CMB. This makes testing the validity of the cosmological principle against observations an important topic on the forefront of cosmological research.

Aside from providing useful knowledge about the universe and presumably theories of its origin, homogeneity and isotropy have a very concrete benefit for the everyday work of a cosmologist. Assuming these two concepts leads to tremendous simplifications when connecting data to theory. For example, isotropy implies that one does not need to worry about the direction in which some pattern or phenomenon is observed, as things are, on average, the same in every direction. As another example, isotropy and homogeneity allow us to average counting pairs of

---

[2] There is also an even stronger concept than the cosmological principle called the *perfect cosmological principle*, stating that the universe is homogeneous, isotropic, and unchanging in time! The perfect principle describes the steady-state cosmological model, which we now know is wrong (in favor of the Big Bang). The universe *is* changing; for example, its density and temperature are both decreasing with time.

galaxies vs. their mutual distance, regardless of both the orientation of each pair in space and where on the sky they are located, in order to obtain the signal for galaxy clustering. Such averaging leads to an overall smaller error on key observed quantities, and also mathematically simplifies the analysis.

1.7 Units in Cosmology

We now shift gear to very practical matters and discuss the topic of units.

As far as the basics are concerned, we will follow any responsible physicist and stick to MKS (meter–kilogram–second) units; readers in some countries may be familiar with this as the SI (Système International) choice. So yes on the kilograms and meters, no on the pounds and feet.

We will also use **natural units** in this book. This means that we set some constants of nature to unity, and express quantities in terms of those constants (and their combinations). In particular, we set

$$c = \hbar = k_B = 1 \tag{1.1}$$

where c is the speed of light, $\hbar = h/(2\pi)$ the reduced Planck constant, and k_B the Boltzmann constant. Moreover, distance is equivalent to time since $L = vt$, and v is effectively dimensionless. Similarly, when we talk about temperature T, we can use units of energy E since $E = k_B T$, and $k_B = 1$. Also $E = mc^2$, so mass has the same units as energy as well. In fact, not only are there quantities that are equivalent to distance, and others that are equivalent to energy, but these two are also related to each other, since

$$[E] = [\hbar\nu] = [\nu] = [t]^{-1} = [L]^{-1} \tag{1.2}$$

where ν is frequency, t is time, and L is distance. Hence energy is also proportional to inverse time (or distance). In fact, as far as the vast majority of quantities discussed in this book is concerned, there is only one physical quantity because

$$[E] = [m] = [T] = [L]^{-1} = [t]^{-1}. \tag{1.3}$$

Other commonly used units can be derived from this. A particularly useful example is the energy density

$$[\rho] = \frac{[E]}{[V]} = \frac{[E]}{[L^3]} = [E]^4 \tag{1.4}$$

so that the energy densities are quoted in $(\text{eV})^4$ when the masses are quoted in eV, for example.

A trivial example of using natural units is to quote speed as a dimensionless number. For example, $v = 0.8$ is to be interpreted as $v = 0.8c$, or $240,000\,\text{km/s}$.

Another, less trivial example is the momentum of a ball of mass $1\,\mathrm{kg}$ moving with speed $v = 3\mathrm{m/s}$. We can express this in electronvolts as

$$p = mv = 3\,\mathrm{kg m/s} = 3\,\mathrm{kg} \times \frac{c}{3 \times 10^8} = 1 \times 10^{-8}\,\mathrm{kg} = 5.61 \times 10^{27}\,\mathrm{eV} \qquad (1.5)$$

using the conversion $1\,\mathrm{eV} = 1.782 \times 10^{-36}\,\mathrm{kg}$ (see Appendix B).

We will be making a lot of use of the Planck mass, which is the indispensable combination of fundamental constants of nature, $m_{\mathrm{Pl}} = (\hbar c/G)^{1/2} = 1.22 \times 10^{19}\,\mathrm{GeV}$, where G is Newton's gravitational constant. Since $\hbar = c = 1$, we have[3]

$$m_{\mathrm{Pl}} \equiv \left(\frac{\hbar c}{G}\right)^{1/2} = G^{-1/2}. \qquad (1.6)$$

This important relation furnishes a direct correspondence between Newton's gravitational constant and the Planck mass, $G = 1/m_{\mathrm{Pl}}^2$. Note also that we do *not* set Newton's gravitational constant to unity, a practice in some more aggressive applications of natural units.

With this result in hand, the dimensional analysis in natural units becomes quite straightforward. For example, Friedmann's first equation, which roughly reads as (see Chapter 2)

$$H^2 \simeq \frac{8\pi G}{3}\rho \equiv \frac{8\pi}{3m_{\mathrm{Pl}}^2}\rho \qquad (1.7)$$

where H is the Hubble parameter and ρ is energy density, implies that the units of H are

$$[H] = \sqrt{[G] \times [\rho]} = \sqrt{[E]^{-2} \times [E]^4} = [E] \equiv [t]^{-1}. \qquad (1.8)$$

These are precisely the expected units for the Hubble parameter which is usually quoted in $\mathrm{km/s/Mpc}$, so (distance/time)/distance, or just $1/\mathrm{time}$.

More examples about converting units are given in Box 1.3 and in Appendix A.

Box 1.3   **Worked Example: Natural Units**

Here we develop some practice working within the system of natural units, where $c = \hbar = k_B = 1$. The SI values of these parameters are

$$c = 3 \times 10^8\,\mathrm{m/s}$$

$$\hbar = 1.1 \times 10^{-34}\,\mathrm{J\,s} = 6.6 \times 10^{-16}\,\mathrm{eV\,s}$$

$$k_B = 1.38 \times 10^{-23}\mathrm{J/K},$$

where the $\hbar$ equation above contains two expressions that are related by the standard conversion $1\,\mathrm{eV} = 1.60 \times 10^{-19}\,\mathrm{J}$. One way to convert to alternate units is: start with the unit you would like to convert (into some other unit), and multiply it with the suitable powers (of SI values) of c, $\hbar$, and k_B – which are all equal to unity in natural units! – so as to get the desired new unit.

---

[3] Note that cosmologists sometimes use the *reduced* Planck mass, $m_{\mathrm{Pl}}^{\mathrm{reduced}} = (\hbar c/8\pi G)^{1/2}$, which is a factor of $\sqrt{8\pi}$ smaller than in our definition.

| Box 1.3 | Worked Example: Natural Units (continued) |

Let us convert a gigaelectronvolt (GeV), which characterizes the temperature or energy, into several alternative units. A GeV can be converted to joules as follows:

$$1\,\mathrm{GeV} = 10^9\,\mathrm{eV}\left(\frac{\hbar}{\hbar}\right) = 10^9\,\mathrm{eV}\,\frac{1.1 \times 10^{-34}\,\mathrm{J\,s}}{6.6 \times 10^{-16}\,\mathrm{eV\,s}} = 1.67 \times 10^{-10}\,\mathrm{J},$$

or to Kelvins:

$$1\,\mathrm{GeV} = 1\,\mathrm{GeV}\left(\frac{1}{k_B}\right) = 1.67 \times 10^{-10}\,\mathrm{J}\,\frac{1}{1.38 \times 10^{-23}\,\mathrm{J/K}} = 1.21 \times 10^{13}\,\mathrm{K},$$

or to kilograms:

$$1\,\mathrm{GeV} = 1\,\mathrm{GeV}\left(\frac{1}{c^2}\right) = 1.67 \times 10^{-10}\,\frac{\mathrm{kg\,m^2}}{\mathrm{s^2}}\,\frac{1}{(3 \times 10^8\,\mathrm{m/s})^2} = 1.85 \times 10^{-27}\,\mathrm{kg}.$$

Similarly, we can express an *inverse* GeV in units of distance – meters:

$$1\,\mathrm{GeV}^{-1} = 10^{-9}\,\mathrm{eV}^{-1}(\hbar c) = 10^{-9}\,\mathrm{eV}^{-1}(6.6 \times 10^{-16}\,\mathrm{eV\,s})(3 \times 10^8\,\mathrm{m/s})$$

$$= 1.98 \times 10^{-16}\,\mathrm{m},$$

or, similarly, to seconds. It is also useful to express a unit of mass or energy *density*, GeV^4, in kilograms per meter cubed:

$$1\,\mathrm{GeV}^4 = (1.67 \times 10^{-10}\,\mathrm{J})^4\left(\frac{1}{c^5\hbar^3}\right) = 7.72 \times 10^{-40}\left(\frac{\mathrm{kg\,m^2}}{\mathrm{s^2}}\right)^4$$

$$\times \frac{1}{(3 \times 10^8\,\mathrm{m/s})^5 \times (1.1 \times 10^{-34}\,\mathrm{J\,s})^3} = 2.39 \times 10^{20}\,\frac{\mathrm{kg}}{\mathrm{m^3}}.$$

Bibliographical Notes

The field of cosmology is (fortunately for us cosmologists) very appealing to the general public, and there are many popular books on the subject; we do not list them here. On a slightly more-technical-than-popular level, the textbook by Duncan and Tyler (2008) provides a very gentle and visually appealing introduction to the field. At a still slightly more technical level, the very clear textbook by Schneider (2010) is also recommended. This is also a good opportunity to recommend the books about the *history* of cosmology; Kragh (1999), Longair (2006), and Peebles (2020) are particularly good. A very interesting selection of articles about the history of cosmology is available in Kragh and Longair (2019), while Turner (2022) recounts how modern cosmology became a precision science. We will point out the historical articles about specific subfields of cosmology in the bibliographical notes to some of the upcoming chapters in this book.

2 The Hubble Law and Geometry of Space

One of the goals of this book is to mathematically describe the expansion of the universe. To do this, we first introduce the concept of redshift. We then move on to discuss the Hubble law, which governs the expansion of the universe, and introduce probably the most famous cosmological parameter – the Hubble constant. We next discuss the geometry of space, and introduce the Friedmann–Lemaître–Robertson–Walker metric. We end the chapter by deriving the first of the two Friedmann equations, which govern the expansion rate of the universe.

2.1 Cosmological Redshift

When we observe distant objects, the wavelength of light that we see is different from the emitted wavelength. This is because the stretching of space (i.e., the expansion) takes the wavelength along for the ride, and stretches it too. **Redshift** z is defined as the relative shift of the wavelength toward the red end of the spectrum:

$$z \equiv \frac{\lambda_{\text{observed}} - \lambda_{\text{emitted}}}{\lambda_{\text{emitted}}}. \tag{2.1}$$

An object nearby has (nearly) zero redshift because there is little space between us and it, and the light wavelength stretches negligibly during its travel toward us so that $\lambda_{\text{observed}} \simeq \lambda_{\text{emitted}}$. The more distant an object is from us, the higher its redshift, though the relation between distance and redshift is generally not linear as we will see below.

Redshift as defined in Eq. (2.1) clearly looks like the familiar Doppler shift from elementary physics. However, the shift in spectral lines is best interpreted not as a Doppler shift, but rather as the result of expanding space: the farther a galaxy is from us, the more space there is between it and us, and the greater the redshift we observe. As we will see below, at *low* redshift ($z \ll 1$) the Doppler analogy is quite appropriate, but at high redshift it is not useful any more. For example, we will see later in this chapter that objects may appear to be moving away from us at a speed greater than the speed of light, which expanding space allows, but could never be inferred from a Doppler-effect measurement.

Note also that redshift is not only a measure of distance, but also a measure of time. This is because light travels at a finite speed, so it takes longer to get to us from distant (high-redshift) objects than from nearby (low-redshift) ones. Observing the more distant objects therefore gives us a picture of what the universe was like at the time the light was emitted from them. So, for example, $z = 0$ means "here, now" while $z = $ (large) means "far away, distant past." A typical galaxy is at $z \sim 1$ (coinciding roughly with half the age of the universe); the most distant objects observed are at $z \sim 10$; the earliest objects probably formed at $z \sim 10 - 20$; cosmic microwave background radiation comes from $z \sim 1000$; the synthesis of the lightest elements (hydrogen and helium; $\sim$3 minutes after the Big Bang) happened at $z \sim 10^{10}$; and the Big Bang corresponds to $z = \infty$.

2.2 The Hubble Law

In 1929, Edwin Hubble made a groundbreaking contribution in discovering the expansion of the universe, and thus effectively ushered in an era of observational cosmology. Using the 100-inch Mt. Wilson telescope, then the largest in the world, he observed "nebulae" (now called galaxies). At the time these objects were the farthest things ever observed; nowadays the galaxies that Hubble observed are considered to be, from the cosmological perspective, so close to be practically in our backyard.

Hubble measured distances to these galaxies by utilizing **Cepheids**, pulsating stars whose pulsation period can be related to the Cepheid's luminosity. The latter result had been established a few years earlier by another great American astronomer, Henrietta Leavitt, who found the so-called period–luminosity relation, or Leavitt law, for Cepheids: the longer the pulsation period of a Cepheid, the larger its luminosity. Thanks to the Leavitt law, one can infer the luminosity of a Cepheid from its observed period, then combine this with the measurement of how bright the Cepheid appears to be (its flux) to estimate the distance between us and the galaxy hosting the Cepheid. [We will return to Cepheids in Chapter 12.]

Cepheids in observed host galaxies provided Hubble with distances, but Hubble still needed the recession velocities of these galaxies. These he obtained by using data collected by Vesto Slipher. Slipher was a spectroscopist and specialized in measurements of recession velocities; he deserves much credit for determining that most galaxies are *receding* from us. Slipher, however, used very nearby galaxies which have non-negligible random velocities relative to their recession velocities due to the expansion of the universe, so he never established the law that Hubble surmised.

Hubble found that nearly all galaxies recede from us. And the *farther* away they are, the *faster* they recede. He proposed what is now known as the Hubble law,

$$v = H_0 d \qquad \text{(Hubble law)}, \qquad (2.2)$$

where v is recession velocity, d is distance – to the galaxy (*which* distance? – we will resolve that a bit further below), and H_0 is the slope of that relation, now called the **Hubble constant**, traditionally measured in km/s/Mpc. The modern value of the Hubble constant is

$$H_0 \simeq 70 \, \text{km/s/Mpc} \quad (\text{modern value for } H_0). \tag{2.3}$$

That is, for each megaparsec that the galaxy is farther away, it is receding from us about 70km/s faster. Incredibly, Hubble got the value about seven times larger than what we measure today, inferring $H_0 \simeq 500 \, \text{km/s/Mpc}$ rather than the currently accepted $H_0 \simeq 70 \, \text{km/s/Mpc}$. Moreover, his velocity vs. distance data were not well fit by a line (see Fig. 2.1), and it is in fact remarkable that Hubble inferred that there should be a linear relationship between the two.

The Hubble constant's units are inverse time and, as we will show in Chapter 3, the *age* of the universe (time since the Big Bang) is approximately equal to $1/H_0$, where the value of the prefactor to this quantity depends on the expansion history of the universe. Because Hubble hugely overestimated H_0, he therefore massively underestimated the age of the universe, effectively deriving the latter to be about 2 billion years! The fact that the universe was inferred to be younger than Earth ($t_{\text{Earth}} \simeq 4.5 \, \text{Gyr}$) was the reason why cosmology as a whole did not have a very good reputation in the 1930s and 1940s, up until the Hubble constant's value was drastically revised and the inferred age of the universe correspondingly extended.

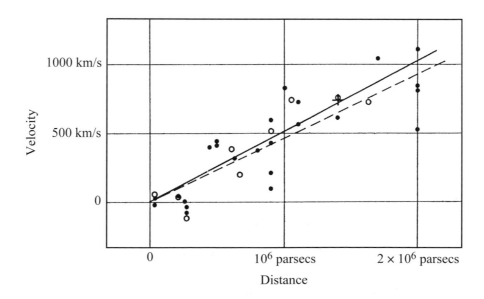

Fig. 2.1 Hubble's (1929) original measurements of the velocity of galaxies (vertical axis) vs. their distance (horizontal axis). Note the poor fit – how did he infer it was a line? The modern value for the slope (Hubble constant), $H_0 \simeq 70 \, \text{km/s/Mpc}$, is about seven times smaller than his original measurement shown here, $H_0 \simeq 500 \, \text{km/s/Mpc}$.

For many decades thereafter (roughly 1950–1990), the favored H_0 values varied between 50 km/s/Mpc (favored by astronomer Allan Sandage) and 100 km/s/Mpc (favored by Gérard de Vaucouleurs). The large variation between various measurements was due to the presence of systematic errors in these measurements and inferences. This impasse was mercifully resolved starting in the 1990s, when new, more precise measurements using the Hubble Space Telescope obtained $H_0 \simeq$ 70 km/s/Mpc, roughly halfway between the two hotly contested values. However, recent measurements of the Hubble constant revealed a new discrepancy between measurements obtained by two methods, leading to the **Hubble tension**. Measurements by the so-called distance ladder (where distances to nearby galaxies are measured and compared to their recession velocities) reveal $H_0 = (73.3 \pm 1.0)$ km/s/Mpc, while inferences from the cosmic microwave background give $H_0 = (67.4 \pm 0.5)$ km/s/Mpc. While they look similar in the historical context, the two measurements are discrepant due to their small corresponding measurement errors. It remains to be seen whether this Hubble tension will be resolved with better accounting of the systematic errors in the corresponding measurements or, perhaps, with new physics.

The remarkable importance of Hubble's 1929 paper comes from the fact that he correctly surmised that the relation between recession velocity and distance is linear – a hallmark of expanding space. In fact, every expanding space, not just our universe, can be characterized by a "Hubble constant" that encodes the rate of its expansion (see Problem 2.1).

To reiterate: the Hubble constant H_0 tells us how rapidly the universe is expanding *today*. A full expansion history of the universe is described by the **Hubble parameter**, $H(t)$. As we will see in Chapter 3, the Hubble parameter is decreasing in time for all cosmological models of interest. Conversely, as we turn back the clock far enough, the distance between galaxies goes to zero and, to preserve finite recession velocities, the Hubble law [Eq. (2.2)] shows that the Hubble parameter goes to infinity as we approach the Big Bang.

Note finally that the Hubble constant is sometimes written as

$$H_0 = 100\, h \,\text{km/s/Mpc}, \tag{2.4}$$

where h parametrizes our lack of knowledge of the exact value, and is now known to be $h \approx 0.7$. While the use of h as a "buffer" in equations of cosmology has abated since the Hubble constant measurement has stabilized around the value in Eq. (2.3), it remains omnipresent in the definition of units of distance (e.g., h^{-1}Mpc), as we will explain in Chapter 3.

2.2.1 Scale Factor

Consider two galaxies located far away from each other, with no individual ("peculiar") velocities of their own relative to their respective local rest frames. Due to the expansion of the universe, the distance between them increases with time, and goes as some function of time, $a(t)$. This function is called the **scale factor**.

As long as we assume a homogeneous and isotropic universe, the scale factor $a(t)$ contains complete information about its expansion history. A scale factor that is increasing with time corresponds to an expanding universe, while a decreasing scale factor corresponds to a contracting universe.

By convention, the scale factor today (at time t_0 after the Big Bang) is set to unity, $a_0 \equiv a(t_0) = 1$. At the Big Bang itself, the scale factor is zero.

2.2.2 Comoving Coordinates

Let us carry on with our example featuring two galaxies with no peculiar velocities, receding from each other due to the expansion of the universe. The distance between them increases with time. However, the distance is fixed in coordinates that stretch with the expansion. The distance in these expanding coordinates is called the **comoving distance**.

To help us understand the comoving distance, we can think about it as a distance in new units, which we will call *notches*.[1] We will imagine that the rulers used to measure notches expand at the same rate everywhere in the universe. A distance between objects is a fixed number of notches at all times, provided again that there are no peculiar velocities, which would slightly change the notches number. Of course, the *physical* distance between galaxies increases with time, because galaxies reside in space that is expanding. This then motivates why we use the comoving coordinates: we do so when we wish to study the changes in distance *beyond* those due to the expansion of the universe.

In particular, the relation between the physical and comoving coordinates is

$$d_{\mathrm{phys}} = a(t)d_{\mathrm{comov}}. \tag{2.5}$$

In other words,

> The comoving distance "takes out the expansion," which is often useful when discussing distances in cosmology.

Figure 2.2 illustrates the comoving coordinates.

When referring to comoving distances, we do not use notches or any such new unit, but rather adopt the standard units of distance with the word "comoving" appended. For example, consider the distance corresponding to the so-called sound horizon, a geometrical feature in the distribution of galaxies that we will discuss in Chapter 13. This distance as measured today corresponds to about $150\,\mathrm{Mpc}$. However, the sound horizon has been expanding along with the universe. For example, measured around the time of recombination ($a \simeq 0.001$, see Chapter 13), the physical distance corresponding to the sound horizon is $d_{\mathrm{sh}} = ad_{\mathrm{SH,comov}} \simeq$

[1] Hat tip to Alan Guth for this argument.

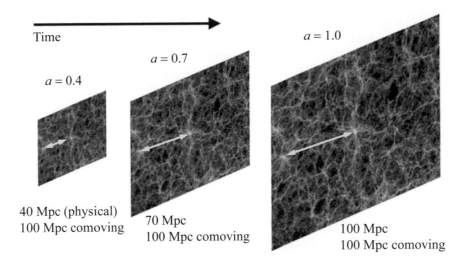

Fig. 2.2   Illustration of comoving coordinates. The plot shows a part of the universe (from the Millennium computer simulation of structure in the universe; Springel *et al.*, 2005) at three epochs in time: scale factor 0.4, 0.7, and today ($a = 1$). Assuming no evolution except for the expansion of space, the feature in the map shown with an arrow corresponds to a different physical distance at the three epochs, but the same comoving distance (100 Mpc comoving, in this example).

$150\,\mathrm{kpc}$ (note, *kilo*parsecs). Nevertheless we often refer to the comoving distance of the sound horizon as "150 megaparsecs, comoving" even when referring to it at this earlier time. In the absence of velocities not associated with the expansion of the universe, the comoving distance remains unchanged, and we quote it in units corresponding to the physical distance *as it would appear today*.

2.2.3 Hubble Parameter

Consider an arbitrary time t in the history of the universe, not necessarily equal to time t_0 today. We would like to know whether an observer living at that time would also measure the Hubble law. Taking the time derivative of Eq. (2.5), we get the recession velocity, $\mathbf{v}(t)$, of an object distance $\mathbf{d}(t)$ from the observer to be

$$\mathbf{v}(t) = \frac{d\mathbf{d}(t)}{dt} = \frac{da(t)}{dt}\mathbf{d}_{\mathrm{comov}} \equiv H(t)\mathbf{d}(t), \qquad (2.6)$$

where we have defined

$$H(t) \equiv H \equiv \frac{\dot{a}(t)}{a(t)} \qquad \text{(Hubble parameter)}, \qquad (2.7)$$

where the overdot is a derivative with respect to physical time. Indeed, the observer at time t also measures the Hubble law, but now featuring the **Hubble parameter** $H(t)$, which describes the expansion *rate* of the universe at some point in time t. Note that the Hubble parameter evaluated at $t = t_0$ is just the Hubble constant H_0. In Problem 2.2, you will prove that there is nothing special about our point of view when observing the Hubble law, and that *all* observers at a given cosmic time t see the same Hubble law featuring the Hubble parameter $H(t)$.

Now that we have related the Hubble parameter to the scale factor, we can provide another way to think about the Hubble law. Consider Taylor-expanding the scale factor at some time t close the present time t_0,

$$a(t) = a(t_0) + \dot{a}|_{t=t_0} (t - t_0) + \cdots = a(t_0) \left[1 + H_0(t - t_0) + \cdots \right], \tag{2.8}$$

where $H_0 \equiv (\dot{a}/a)|_{t=t_0}$. Note that $H_0 > 0$ corresponds to the expanding universe where galaxies are redshifted, and $H_0 < 0$ corresponds to a contracting universe where galaxies are blueshifted. Then, Eq. (2.8) shows that the Hubble constant can also be thought of as the rate of change of the present-day scale factor with time.

2.3 The Geometry of Space

We now move on to studying the geometry of space. This is a necessary ingredient in our understanding of modern cosmological observations. The geometry also provides for a fascinating application of mathematics to the real world – in fact, to the universe as a whole.

Note first that the number of dimensions of space refers to the number of coordinates required to describe a position in it. Clearly, a sheet of paper is an example of a two-dimensional (2D) space, as two coordinates are required to describe a location in it, say the Cartesian x and y coordinates. The surface of a sphere is also 2D, since two coordinates are required (say angles θ and ϕ). Note that even though the sphere may be embedded in our everyday three-dimensional (3D) space, its surface is 2D.

The space that describes our universe is of course 3D, and ultimately we will be interested in 3D geometry results. To warm up, however, we start by considering the mathematics of the geometry of 2D spaces. We start by considering the simplest case of a flat metric.

2.3.1 Flat-Space Metric

Consider the surface of a flat, non-expanding, 2D sheet. Let us work in cylindrical coordinates (r, θ) and denote the distance element as ds. Then the distance ds between two points separated by infinitesimal radial distance dr and angle $d\theta$ is given by a quantity known as the metric,

$$ds^2 = dr^2 + r^2 d\theta^2 \quad \text{(2D flat, static sheet)}. \tag{2.9}$$

How about the distance element in a 3D flat space? Clearly $ds^2 = dx^2 + dy^2 + dz^2$ in Cartesian coordinates. In spherical coordinates (r, θ, ϕ), and this becomes a bit more interesting:

$$ds^2 = dr^2 + r^2(d\theta^2 + \sin^2\theta\, d\phi^2) \quad \text{(3D flat, static space)}$$
$$= dr^2 + r^2 d\Omega^2,$$

(2.10)

where $d\Omega^2 \equiv d\theta^2 + \sin^2\theta\, d\phi^2$, and Ω is the solid angle.

2.3.2 Non-Euclidean Geometries

Euclid's famous fifth postulate says that two lines that are parallel at one location in space remain parallel everywhere. Not necessarily so Mr Euclid! Euclidean geometry corresponds to spatially flat space, which is only one of the three possibilities for the geometry of space. In particular:

- In Euclidean geometry, two parallel lines remain parallel (i.e., a constant distance from each other) even if extended to infinity.
- In hyperbolic geometry, two lines that are parallel at some location in space curve away from each other, increasing their mutual distance as one moves farther away from that location.
- In elliptic geometry, two lines that are parallel at some location curve toward each other, and eventually intersect.

The mathematicians who first studied curved space were German genius Carl Friedrich Gauss, Hungarian mathematician János Bolyai, and Russian mathematician Nikolai Lobachevski. Their work provides a foundation for the geometric description of the universe.

2D curved spaces. It is interesting to consider the sum of angles of a triangle on curved 2D spaces. In flat space, the sum is the familiar $180°$, or π radians. In a positively curved (elliptic) space, the sum is however *greater* than π – for example, the sum of angles in the triangle whose one side is along the equator of a sphere while the other two sides are along the $0°$ and $90°$ meridians and go to the north pole is $270°$. And in a negatively curved (hyperbolic) space, the sum of angles in a triangle is *smaller* than π. In fact, given a triangle with angles α, β, and γ, their sum is

$$\alpha + \beta + \gamma = \begin{cases} \pi & \text{(flat)} \\ \pi + \dfrac{A}{R^2} & \text{(positive curvature)} \\ \pi - \dfrac{A}{R^2} & \text{(negative curvature),} \end{cases}$$

(2.11)

where A is the area of the triangle and R is the radius of the elliptical or parabolic space. The corresponding metric is, for the three respective cases,

$$ds^2 = \begin{cases} dr^2 + r^2 d\theta^2 & \text{(flat)} \\ dr^2 + R^2 \sin^2\left(\dfrac{r}{R}\right) d\theta^2 & \text{(positive curvature)} \\ dr^2 + R^2 \sinh^2\left(\dfrac{r}{R}\right) d\theta^2 & \text{(negative curvature).} \end{cases} \quad (2.12)$$

Note the appearance of sine (for the positively curved case) and hyperbolic sine (for the negatively curved case). These two functions are ubiquitous in mathematical considerations of non-flat geometry. Intuitively, the flat case separates the "slower" elliptical from the "faster" hyperbolic radial metric, as $\sin x \leq x \leq \sinh x$. Note also that, in all cases, in the limit $R \gg r$ both elliptical and hyperbolic metrics become equal to the flat one. This makes sense for the same reason that the surface of Earth appears flat to us: typical distances r are much smaller than the radius of curvature of that space ($R_E = 6400\,\text{km}$).

3D curved spaces. The 2D results generalize quite straightforwardly to three dimensions; the only difference is that we now have another coordinate, ϕ. The metric is now given by

$$ds^2 = \begin{cases} dr^2 + r^2 \left[d\theta^2 + \sin^2(\theta)d\phi^2\right] & \text{(flat)} \\ dr^2 + R^2 \sin^2(r/R) \left[d\theta^2 + \sin^2(\theta)d\phi^2\right] & \text{(positive curvature)} \\ dr^2 + R^2 \sinh^2(r/R) \left[d\theta^2 + \sin^2(\theta)d\phi^2\right] & \text{(negative curvature).} \end{cases} \quad (2.13)$$

It is useful to define κ, a dimensionless constant that signifies whether the space is positively or negatively curved, or not curved at all. Kappa takes one of three discrete values:

$$\kappa = \begin{cases} 0 & \text{zero curvature} \\ +1 & \text{positive curvature} \\ -1 & \text{negative curvature.} \end{cases} \quad (2.14)$$

We will use κ just below when we introduce the Friedmann–Lemaître–Robertson–Walker metric. For the moment, however, we use it simply to label the different geometry scenarios. Then we can write the 3D metric in an arbitrary geometry as

$$ds^2 = dr^2 + S_\kappa(r)^2 d\Omega^2, \quad (2.15)$$

where

$$S_\kappa(r) \equiv \begin{cases} r & (\kappa = 0, \text{ zero curvature}) \\ R\sin(r/R) & (\kappa = +1, \text{ positive curvature}) \\ R\sinh(r/R) & (\kappa = -1, \text{ negative curvature}), \end{cases} \quad (2.16)$$

and $d\Omega^2 = d\theta^2 + \sin^2(\theta)d\phi^2$, as before. An example of 2D spaces with negative, positive, and zero curvature is shown in Fig. 2.3.

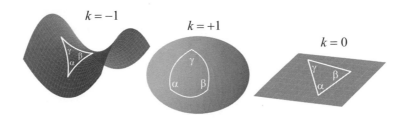

Fig. 2.3   Sketch of a negatively curved, positively curved, and flat space in two dimensions.

2.4 Friedmann–Lemaître–Robertson–Walker Metric

In relativity, space and time are intertwined, and need to be considered together when studying motions of particles and other physical phenomena. A 3D space with one-dimensional (1D) time dimension is therefore described by a four-dimensional (4D) *spacetime* metric. Let us now consider spacetime metrics in cosmology.

For the Minkowski (flat) space, one can ask what is the spacetime separation between two events, one with four-position (t, r, θ, ϕ) and the other with $(t + dt, r + dr, \theta + d\theta, \phi + d\phi)$. The result is the so-called Minkowski metric

$$ds^2 = -dt^2 + dr^2 + r^2 d\Omega^2, \tag{2.17}$$

with the opposite sign of the time term compared to signs of the space terms.

Fine – but what about an expanding space, that also happens to be homogeneous and isotropic? The torch was passed from Alexander Friedmann, who first found the expanding, homogeneous, and isotropic solutions of Einstein's equations of general relativity, to three scientists working independently in the 1930s: Belgian priest and cosmologist Georges Lemaître, American physicist Howard P. Robertson, and British mathematician Arthur Walker. They found that the unique metric consistent with Einstein's equations that describes the expanding homogeneous and isotropic space is[2]

$$ds^2 = -dt^2 + a^2(t) \left[\frac{dr^2}{1 - \kappa r^2} + r^2 d\Omega^2 \right]. \tag{2.18}$$

Here $a(t)$ is the scale factor which multiplies the spatial part of the metric indicating the isotropic expansion of space. See Box 2.1 for a derivation of the FLRW metric from first principles.

---

[2] Lemaître's contribution is specifically not so much coming up with the FLRW metric, but rather giving an exceptionally clear physical interpretation of the expanding-universe solutions, something that Friedmann didn't do in his pioneering work.

It is a straightforward exercise to show that this metric can be rewritten by replacing the coordinate r with another, **radial coordinate** χ via[3]

$$ds^2 = -dt^2 + a^2(t)\left[d\chi^2 + S_\kappa(\chi)^2 d\Omega^2\right],\qquad (2.19)$$

where the radial coordinate χ is given in terms of r via

$$\chi = \begin{cases} \sin^{-1}(r) & (\kappa = +1) \\ r & (\kappa = 0) \\ \sinh^{-1}(r) & (\kappa = -1). \end{cases}\qquad (2.20)$$

Note that the radial distance is multiplied by the scale factor to give the radial displacement, $ds \simeq \chi a$. Therefore, χ too is a kind of comoving distance.

So far we have discussed uniform, homogeneous spaces. In detail, our universe is inhomogeneous because the cosmic structure is made up of galaxies and dark matter, both of which cluster (or clump). Similar mathematics but with more complicated metrics can describe non-uniform spaces as well. When observed on spatial scales of about 100 Mpc or larger, our universe appears sufficiently uniform so that the FLRW metric describes it accurately.

2.5 From Coordinates to Distances

In standard Newtonian physics, there is just one distance – *the* distance that all people on Earth agree on (even though they may use different *units* to measure distances). In cosmology, however, things are more complicated. Because of the non-trivial structure of spacetime, there are several different measures of distance. We now discuss those distance measures that are useful in cosmology. We finish the job in Chapter 3, where we discuss the two distance measures that are actually observable – the angular-diameter distance and the luminosity distance.

2.5.1 Proper Distance

It turns out we can easily get distance from the (spacetime) metric. As a warm-up, let us start with the Minkowski metric from Eq. (2.17). Light propagates on null

---

[3] The notation – both name and symbol – for the comoving distance r and the radial distance χ is notoriously variable across various textbooks and papers. For example, some authors call our r their x, our χ may be their r, and sometimes our comoving distance r is called the "comoving angular-diameter distance." Here we stick to the notation that we deem simplest and is the most commonly used.

Box 2.1   **Derivation of FLRW à la Weinberg**

Weinberg (2008) gives a concise derivation of the FLRW metric, which we recapitulate here. Consider a 3D hypersphere (3-sphere) embedded in a 4D space (for an analogy, remember that a circle is fundamentally a 1D object embedded in a 2D x–y space). Let the three coordinates be $\mathbf{x} = (r, \theta, \phi)$ and the fourth coordinate is z; we will use the latter only as a helping device, but we wish to express our results in terms of the former three coordinates. The metric and the equation of that hypersphere are given by, respectively,

$$ds^2 = d\mathbf{x}^2 + dz^2 \qquad z^2 + \mathbf{x}^2 = a^2, \tag{B1}$$

where a is the radius. Similarly, for a negatively curved space and a 3D hyperbola embedded in a 4D space,

$$ds^2 = d\mathbf{x}^2 - dz^2 \qquad z^2 - \mathbf{x}^2 = a^2. \tag{B2}$$

We can rescale the coordinates $\mathbf{x} = a\mathbf{x}'$ and $z = az'$, and then drop the primes, to get

$$ds^2 = a^2 \left[d\mathbf{x}^2 \pm dz^2 \right] \qquad z^2 \pm \mathbf{x}^2 = 1. \tag{B3}$$

The differential of the equation $z^2 \pm \mathbf{x}^2 = 1$ gives $zdz = \mp \mathbf{x}d\mathbf{x}$, so therefore

$$ds^2 = a^2 \left[d\mathbf{x}^2 \pm \frac{(\mathbf{x}d\mathbf{x})^2}{1 \mp \mathbf{x}^2} \right]$$

$$= a^2 \left[d\mathbf{x}^2 + \kappa \frac{(\mathbf{x}d\mathbf{x})^2}{1 - \kappa\mathbf{x}^2} \right] \tag{B4}$$

$$= a^2 \left[d\mathbf{x}^2 + \kappa \frac{(rdr)^2}{1 - \kappa r^2} \right],$$

where $\kappa = +1$ for elliptical, -1 for hyperbolic, and 0 for the flat case. In the last line, we remembered that $\mathbf{x}^2 = r^2$, and also that $\mathbf{x}$ is parallel to the dr direction and perpendicular to $d\theta$ and $d\phi$ (so that $\mathbf{x}d\mathbf{x} = rdr$).

Finally, using $d\mathbf{x}^2 = dr^2 + r^2 d\Omega^2$, we have

$$ds^2 = a^2 \left[r^2 d\Omega^2 + \frac{dr^2(1 - \kappa r^2) + \kappa(rdr)^2}{1 - \kappa r^2} \right]$$

$$= a^2 \left[r^2 d\Omega^2 + \frac{dr^2}{1 - \kappa r^2} \right] \qquad \text{QED.} \tag{B5}$$

geodesics, that is, trajectories on which $ds^2 = 0$. Assuming for simplicity that light propagates on a radial geodesic in our coordinate system (so $d\theta = d\phi = 0$), we have

$$\frac{dr}{dt} = \pm 1, \tag{2.21}$$

that is, $\pm c$ if we wish to reintroduce the speed of light.

Likewise, for the FLRW metric, the condition for a null geodesic implies

$$ds^2 = -dt^2 + a^2(t)\left[d\chi^2 + S_\kappa(\chi)^2 d\Omega^2\right] = 0. \tag{2.22}$$

Consider a galaxy somewhere in the universe. We define the **proper distance** d_p as the length of the geodesic between us and the galaxy – that is, between spatial coordinates $(\chi, \theta, \phi) = (0, 0, 0)$ and $(\chi, 0, 0)$ – at a *fixed time* (or fixed scale factor $a(t)$). In other words, and in the language of relativity, proper distance is the distance between two events in an inertial frame where those events are simultaneous.

For a radial geodesic with $d\theta = d\phi = 0$, the geodesic distance element is

$$ds = a(t)d\chi. \tag{2.23}$$

We can integrate this in order to get the proper distance

$$d_p = \int ds = a(t)\chi = a(t) \int_0^r \frac{dr'}{\sqrt{1 - \kappa r'^2}}. \tag{2.24}$$

Proper distance is therefore the distance measured at a fixed cosmic time. In fact, to "measure" the proper distance, you can imagine freezing the expansion of the universe, taking out a meter stick, and counting the number of meters between the locations of the two events. [This is why the proper distance is sometimes called the "instantaneous physical distance."] A measurement as described is impossible to do in practice, but proper distance is nevertheless a useful concept.

In particular, proper distance is the one featured in the Hubble law. To see that, consider the time derivative of Eq. (2.24),

$$\dot{d}_p = \dot{a}\chi = \frac{\dot{a}}{a}a\chi = H(t)d_p, \tag{2.25}$$

where $d_p = a\chi$ is the proper distance.

2.5.2 The Hubble Distance and Time

When is the recession velocity of a galaxy apparently equal to the speed of light? In other words, we are asking about the case when

$$\dot{d}_p \equiv v_p = H_0 d_p = c. \tag{2.26}$$

This happens when

$$d_p = \frac{c}{H_0} \equiv \frac{1}{H_0} \equiv d_H(t_0), \tag{2.27}$$

where in the second equality we remind the reader about the natural-units convention used in this book where $c = 1$.

The quantity $d_H(t_0)$ is called the **Hubble distance**. [The Hubble distance is confusingly often called the horizon distance, or simply the **horizon**. Strictly speaking, the latter is defined slightly differently – see Eq. (3.37) in Chapter 3 – though it is admittedly a close cousin to the Hubble distance that we are now discussing.] For $H_0 \approx 70\,\mathrm{km/s/Mpc}$, the Hubble distance is equal to

$$d_H(t_0) \approx 4500\,\mathrm{Mpc}, \tag{2.28}$$

or, more generally in terms of $h = H_0/(100\,\mathrm{km/s/Mpc})$,

$$d_H(t_0) \equiv \frac{c}{H_0} \equiv \frac{1}{H_0} = 2997.9\,h^{-1}\mathrm{Mpc} \qquad \text{(Hubble distance)}, \tag{2.29}$$

where the prefactor 2997.9 is just 1/100th of the speed of light in units of km/s. Likewise, one can define the **Hubble time**

$$t_H \equiv \frac{1}{H_0} \simeq 14.5\,\mathrm{Gyr} \qquad \text{(Hubble time)}. \tag{2.30}$$

Somewhat accidentally, the Hubble time just happens to be very close to the actual age of the universe today; we discuss this further in Chapter 3.

Are objects at large distances, $d_p > 1/H_0$, really moving away from us faster than the speed of light as Eq. (2.26) seems to imply? In special relativity, relative velocity and distance are measured with clocks and rulers in a fixed inertial frame. It turns out that it is possible to suitably define such a frame only on scales much smaller than the curvature of the (4D) spacetime, which is H^{-1}. Therefore, a simple, special-relativistic interpretation of objects' recession velocity cannot be made. More intuitively, the speed-of-light limit applies to velocities of objects made in their local reference frames. A distant galaxy may appear to us to exceed that limit, but only because the space between us and the galaxy is expanding.

2.5.3 Coordinate Distance and Conformal Time

We have already introduced the coordinate distance χ in the FLRW metric in Eq. (2.19), where we referred to it as the radial distance. We now discuss it in some more detail. Special relativity teaches us that light propagates on trajectories with $ds^2 = 0$. We make use of this to define the distance that would be covered by light traveling from time t to the present time t_0. This distance is called the **coordinate distance,** but is also known as the **conformal distance,** or else radial distance.[4] The condition $ds^2 = 0$ applied to Eq. (2.19) gives

$$\chi(t) = c \int_t^{t_0} \frac{dt'}{a(t')} \qquad \text{(coordinate distance)}, \tag{2.31}$$

---

[4] We solemnly promise to the reader that *most* other quantities in cosmology do not suffer from such multiplicity of naming conventions.

where we have temporarily restored the speed of light c for clarity. A very closely related quantity is **conformal time** η (sometimes labeled as τ, and sometimes called coordinate time), defined as

$$\eta(t) = \int_0^t \frac{dt'}{a(t')}. \tag{2.32}$$

Modulo the factor of c, coordinate distance and conformal time are the same thing, as they both measure the distance/time that a photon covers. The only difference is that comoving distance typically measures the distance from the present day to some event in our past, while conformal time typically measures time from the Big Bang to some event thereafter.

Coordinate distance χ is widely used in cosmology, and we will be repeatedly coming back to it in this book. Conformal time η is favored by theorists as it often makes perturbation-theory equations simpler than when using time t; we will however not use the conformal time very much in what follows.

2.5.4 Relation between Redshift and Scale Factor

We have already introduced the redshift z and scale factor a; we now show that they have a one-to-one correspondence. Consider again the FLRW metric, and let us consider the propagation of light which, as we recall from the Minkowski-space example, travels on null geodesics. For light therefore $ds^2 = 0$ so that (restoring c again in this section),

$$cdt = a(t)d\chi. \tag{2.33}$$

Therefore, we can integrate from the emission time t_e to the present time t_0 to get

$$c \int_{t_e}^{t_0} \frac{dt}{a(t)} = \int_0^{\chi} d\chi' \tag{2.34}$$

$$= \chi.$$

Consider now the next crest of light. It is emitted at time $t_e + \lambda_e/c$, and received at time $t_0 + \lambda_0/c$. Thus

$$c \int_{t_e+\lambda_e/c}^{t_0+\lambda_0/c} \frac{dt}{a(t)} = \chi. \tag{2.35}$$

Comparing Eqs. (2.34) and (2.35), we have

$$\int_{t_e}^{t_0} \frac{dt}{a(t)} = \int_{t_e+\lambda_e/c}^{t_0+\lambda_0/c} \frac{dt}{a(t)}$$

$$= \left[\int_{t_e}^{t_0} + \int_{t_0}^{t_0+\lambda_0/c} - \int_{t_e}^{t_e+\lambda_e/c} \right] \frac{dt}{a(t)} \tag{2.36}$$

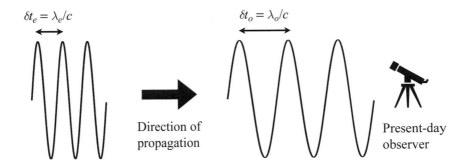

Fig. 2.4 Illustration accompanying the derivation of the relation between the scale factor and redshift. The time between two peaks of the photon wave is δt_e at emission, but becomes δt_o at observation today. The coordinate distance χ between the two peaks, however, remains unchanged.

or

$$\int_{t_e}^{t_e+\lambda_e/c} \frac{dt}{a(t)} = \int_{t_0}^{t_0+\lambda_0/c} \frac{dt}{a(t)} \tag{2.37}$$

$$\frac{1}{a(t_e)} \frac{\lambda_e}{c} = \frac{1}{a(t_0)} \frac{\lambda_0}{c}, \tag{2.38}$$

where we used $\lambda \ll ct$. Equation (2.37) is illustrated in Fig. 2.4. Recalling the definition of redshift $1 + z = \lambda_0/\lambda_e$, we get $1 + z = a(t_0)/a(t_e)$, or

$$1 + z = \frac{1}{a}, \tag{2.39}$$

where we assume the standard convention used in cosmology that $a(t_0) = 1$.

Equation (2.39) is a fundamental relation that shows that the redshift observed from a distant object directly tells us the size of the universe (relative to present) at the time when the light from that object left it. Clearly, redshift z and scale factor a are completely interchangeable. We will be using both of these fundamental quantities in cosmology in the remainder of this book.

2.6 The Friedmann I Equation

Now we introduce one of the most important equations in cosmology – the first of the two Friedmann equations, henceforth Friedmann I (we will discuss the second

Friedmann equation in Chapter 3). Both Friedmann equations are derived from Einstein's equations of general relativity for the case of a smooth, homogeneous, and isotropic universe that is expanding (or contracting) according to some scale factor $a(t)$. The general-relativistic derivation is "tedious but straightforward"; here we rely on the fact that Friedmann I just happens to be derivable from simple Newtonian physics. We now present this simplified calculation.

2.6.1 Newtonian Derivation

Assume a sphere of pressureless matter, and let $R_s(t)$ be the time-dependent radius of the sphere, $\rho(t)$ be its energy density, and $M_s = (4\pi/3)\rho(t)R_s(t)^3$ be its mass. The acceleration of a test particle on the surface of this sphere is, by Newton's second law,

$$\frac{d^2 R_s}{dt^2} = -\frac{GM_s}{R_s^2}. \tag{2.40}$$

By multiplying both sides with dR_s/dt and integrating, we get

$$\frac{1}{2}\left(\frac{dR_s}{dt}\right)^2 = \frac{GM_s}{R_s} + E, \tag{2.41}$$

where E is the constant of integration. The term on the left-hand side in this equation is essentially the test particle's kinetic energy, E_k; the *negative* of the first term on the right-hand side is its potential energy, $-E_p$. Then $E = E_k + E_p$ can be identified with the total energy of the test particle.

Since the expansion is assumed isotropic around the sphere's center, the radius of the sphere just scales with the scale factor

$$R_s(t) = a(t)r_s, \tag{2.42}$$

where we define r_s to be the comoving radius of the sphere. Using $M_s = (4\pi/3)\rho R_s^3$ (and dropping the implicit time dependence of R_s, a, and ρ for conciseness from here on), we can rewrite Eq. (2.41) as

$$\frac{1}{2}r_s^2\dot{a}^2 = \frac{4\pi}{3}Gr_s^2\rho a^2 + E, \tag{2.43}$$

or

$$\left(\frac{\dot{a}}{a}\right)^2 = \frac{8\pi G}{3}\rho + \frac{2E}{r_s^2}\frac{1}{a^2}. \tag{2.44}$$

This is the **Friedmann I equation** derived from Newtonian physics. It determines the time evolution of the scale factor given the time evolution of the energy density.

Note that the evolution depends on the sign of E:

- For $E > 0$, the right-hand side is always positive, and thus $\dot{a}^2 > 0$ always – the expansion never stops.

- If $E < 0$, the right-hand side starts out positive; however, it reaches zero at the finite value of the scale factor $a_{\max} = -GM_s/(Er_s)$. At that point the expansion stops, and turns around – the sphere starts shrinking.
- If $E = 0$, then we have a boundary case where the expansion only stops in the infinite future: $\dot{a} \to 0$ as $\rho(t) \to 0$, which happens at $t \to \infty$.

The Newtonian derivation above is necessarily approximate, and in particular does not include the effects of spatial curvature.[5] The full Friedmann I equation, derived using general relativity, is

$$\left(\frac{\dot{a}}{a}\right)^2 = \frac{8\pi G}{3}\rho - \frac{\kappa}{R_0^2 a^2}, \tag{2.45}$$

where κ is the curvature parameter that we already introduced in Eq. (2.14), and R_0 is the **radius of curvature** of space – that is, of the universe. Here ρ does not have to refer to matter; it can be *any* component with energy density (say, radiation). Note we are firmly setting $c = 1$ from here on; therefore, for us the energy density $\rho \equiv E/V$ and the mass density $m/V = (E/c^2)/V$ are identical – see Appendix A.

We often rewrite the Friedmann I equation using the dimensionful curvature parameter defined as $k = \kappa/R_0^2$, then

$$\left(\frac{\dot{a}}{a}\right)^2 = \frac{8\pi G}{3}\rho - \frac{k}{a^2} \qquad \text{(Friedmann I equation)}. \tag{2.46}$$

Note, each term in this equation has (natural) units of inverse distance squared. Note also that ρ is the energy density, and is equivalent to mass density given that $mc^2 = m$ in natural units.

Box 2.2 has a worked example on how to solve the Friedmann I equation for a flat, matter-dominated universe, while Box 2.3 and Problem 2.9 discuss the solutions for a positively or negatively curved, matter-dominated model.

2.6.2 Critical Density

Given the measured expansion rate $H \equiv \dot{a}/a$, there is a certain value of the density ρ for which the universe *would* be flat ($k = 0$ in Eq. (2.46)). This value is called the **critical density** of the universe; from the Friedmann I equation, it is equal to

$$\rho_{\text{crit}} = \frac{3H^2}{8\pi G}. \tag{2.47}$$

Note that critical density varies with time because the Hubble parameter does.

---

[5] The reason that the Newtonian derivation gets us very close to the exact general-relativistic answer is the existence of the **Birkhoff theorem** in general relativity, whose corollaries lead to a number of simple results for spherical mass distributions.

Of particular interest to us is the critical density evaluated *today*,

$$\rho_{\text{crit},0} = \frac{3H_0^2}{8\pi G} \qquad \text{(present-day critical density)}, \qquad (2.48)$$

which can be evaluated to be

$$\rho_{\text{crit},0} = 1.877 \times 10^{-26} h^2 \, \text{kg m}^{-3}$$

$$= 2.775 \times 10^{11} h^2 M_\odot \, \text{Mpc}^{-3} \qquad (2.49)$$

$$= 1.053 \times 10^{10} \, \text{eV}^4.$$

In other words, critical density is, very roughly, equal to one atom per cubic meter.

The critical density today, $\rho_{\text{crit},0}$, leads us to the definition of one of the most useful quantities in cosmology: by dividing the energy density at any given time by the critical density of that time, we obtain the **density relative to critical density**, omega:[6]

$$\Omega(t) \equiv \frac{\rho(t)}{\rho_{\text{crit}}(t)} \qquad \text{(density relative to critical – "omega")}. \qquad (2.50)$$

We can now rewrite the Friedmann I equation as

$$1 - \Omega(t) = -\frac{\kappa}{R_0^2 a(t)^2 H(t)^2}. \qquad (2.51)$$

Note that the right-hand side of Eq. (2.51) cannot change sign as the universe expands![7] Thus, if $\Omega(t) > 1$ at some time, $\Omega(t) > 1$ at any time. Likewise, if $\Omega(t) < 1$ at some point in time, $\Omega(t) < 1$ always. Finally, $\Omega(t) = 1$ (flat universe case) also holds forever if it holds at some point in the history of the universe.

Of particular interest is the density parameter *today*, often denoted Ω[8] – so plain omega, without the "of time" label, (t). We have

$$1 - \Omega = -\frac{\kappa}{R_0^2 H_0^2} = -\frac{\kappa}{H_0^2}. \qquad (2.52)$$

Thus it follows that:

---

[6] It does not have a nicer, more compact name, sorry. Cosmologists just call it "omega."

[7] A sign change in Eq. (2.51) would require that the sign of spatial curvature changes during the evolution of the universe. While not explicitly forbidden, such a feature would be *extremely* unusual, and is not predicted in any of the mainstream models of evolution of the universe.

[8] Conventionally, the present-day values of this and other Ω parameters defined in Chapter 3, evaluated at the present time, do *not* require the 0 subscript. Moreover, note that the energy density relative to critical and solid angle both have the same symbol, Ω.

| Box 2.2 | Worked Example: Flat, Matter-Dominated Universe |

We practice solving the Friedmann I equation for the simplest cosmological model: one that is spatially flat and matter dominated – the so-called Einstein–de Sitter universe. While we discuss cosmological models at length in Chapter 3, here we only need to know that the matter density scales with the inverse of the cubed scale factor,

$$\rho \propto a^{-3} \quad \text{(for matter).} \tag{B1}$$

For our flat model with $k = 0$, the Friedmann I equation becomes

$$\left(\frac{\dot{a}}{a}\right)^2 = \frac{8\pi G}{3}\rho = \frac{8\pi G}{3}\rho_0 a^{-3} \equiv H_0^2 a^{-3}, \tag{B2}$$

where $\rho = \rho_{\text{crit}}$ in a flat universe, and we identified the constant $(8\pi G/3)\rho_0 \equiv (8\pi G/3)\rho_{\text{crit},0}$ as the Hubble constant squared (since, at $a = 1$, the left-hand side of the Friedmann I equation is H_0^2). Solving this is easy:

$$\left(\frac{\dot{a}}{a}\right) = H_0 a^{-3/2}$$

$$a^{1/2}da = H_0 dt \tag{B3}$$

$$a = \left(\frac{3}{2}H_0 t\right)^{2/3} \equiv \left(\frac{t}{t_0}\right)^{2/3},$$

where the additive constant of integration is zero (since $a = 0$ when $t = 0$), and in the last line we used the boundary condition $a(t = t_0) = 1$ to identify the present-day age of the universe as $(2/3)H_0^{-1} \equiv t_0$. This gives us the well-known result that a flat, matter-dominated universe – not quite our universe, as we shall soon see! – has the scale factor that increases as time to the power of two-thirds.

Solutions of Friedmann I for the *non-flat* matter-dominated universe are discussed in Box 2.3 and Problem 2.9.

> If you know the density parameter today, Ω, you know the sign of curvature (closed, flat, or open).

Equation (2.52) can be rewritten as

$$R_0^2 = \frac{\kappa}{H_0^2(\Omega - 1)}. \tag{2.53}$$

Therefore, if you know both Ω and H_0, you can infer the radius of curvature R_0.

Finally, it is worth rewriting the Friedmann I equation purely in terms of potentially observable quantities Ω and H_0. Combining Eqs. (2.46), (2.48), (2.50), and (2.52), this becomes

$$H^2 = H_0^2\,\Omega + \frac{H_0^2(1-\Omega)}{a^2}, \tag{2.54}$$

where H, Ω, and a are all evaluated at an arbitrary time t.

Box 2.3   **It's a... Cycloid!**

The Friedmann I equation can be rewritten in a mathematically interesting, parametric form. For a closed universe, this is

$$t \quad = R(\theta - \sin\theta)$$
$$\frac{a}{\sqrt{k}} \quad = R(1 - \cos\theta) \qquad \text{(closed-universe parametric solution),}$$

where θ is a development angle which takes values $\theta = 0$ at the Big Bang and $\theta = 2\pi$ at the Big Crunch. Here R is some constant with units of length; recall that k has units of $(\text{length})^{-2}$, so the units work out.

The equations above describe a **cycloid**, a mathematical shape that a point on the circumference of a wheel describes as the wheel rolls. As time evolves, the scale factor increases to a maximum, then decreases to zero in a Big Crunch. This is illustrated in the figure below; here the axes are $x \equiv t$ and $y \equiv a/\sqrt{k}$.

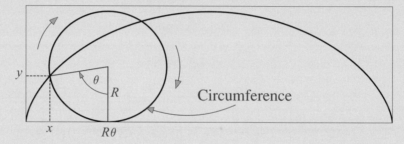

A similar parametric form can be obtained for an open universe; the equations are

$$t \quad = R(\sinh\theta - \theta)$$
$$\frac{a}{\sqrt{k}} \quad = R(\cosh\theta - 1) \qquad \text{(open-universe parametric solution).}$$

Therefore, the closed-universe equations use trigonometric functions, while the open-universe equations contain hyperbolic functions. In Problems 2.9 and 2.10 you will further explore these parametric solutions.

2.6.3 So... What is the Curvature of the Universe??

It's zero, and the geometry is flat.

Here we are revealing the answer to a particularly exciting detective story. Curvature measurements go back to the legendary mathematician Carl Friedrich Gauss, who climbed three peaks in Germany, measuring angles between two of them as seen from the third one, with the goal to verify whether the angles of the triangle sum to 180° or some other value. We now know that, given a reasonable curvature expected in cosmology (even before it was accurately measured), Gauss had no chance to detect it (see Problem 2.8).

In the 1970s and 1980s, many cosmological measurements were indicating a low value, $\Omega \sim 0.3$. A spectacular new discovery made near the end of the last millennium found the presence of **dark energy**, a new component with properties very different from matter that nevertheless contributes to energy density ρ, and hence to Ω. These details will be discussed further in Chapter 12. Here, we merely quote the modern measurement of the density parameter,

$$\Omega \xrightarrow{\text{now called}} \Omega_{\text{TOT}} = 0.999 \pm 0.002, \tag{2.55}$$

where we adopt the modern naming convention for the present-day density parameter (Ω_{TOT}, to be properly introduced in the next chapter).

The number in Eq. (2.55) reads as if the universe is desperately trying to tell us that the spatial geometry is flat, that is, $\Omega_{\text{TOT}} = 1$. This is indeed the prevailing interpretation that cosmologists have adopted. Nevertheless, it is possible that small but statistically significant deviations in either the positively or negatively curved direction will be found with future measurements. Such deviations are in fact predicted in some models of inflationary theory, to be discussed in Chapter 8. The quest to measure the geometry of the universe is not finished yet.

Bibliographical Notes

The material covered in this chapter is also covered by Ryden (2016), the book that most influenced the author in his teaching of this material. Other excellent books that cover the subject are Linder (1997) and Liddle (2015). The discovery paper of the Hubble law, and thus the expanding universe, is Hubble (1929). We adopted the derivation in Box 2.1 from Weinberg (2008). Finally, a useful discussion of the various misconceptions about the expanding universe, specifically when interpreting the recession velocities, is given in Davis and Lineweaver (2004).

Problems

2.1 **Hubble constant of raisin bread.** The Hubble law applies to any uniformly expanding space. Consider a very common example of the expanding universe in popular talks on cosmology – that of a raisin bread baked in an oven. The bread is expanding, and each raisin inside the bread is receding from all the other raisins – according to the Hubble law, in fact.

Estimate the Hubble constant for a typical raisin bread. Estimate also the "time to the Big Bang," H_0^{-1}. Use units of distance and time appropriate for a kitchen, not the universe!

2.2 Hubble law for another observer. At a given time in the history of the universe, we observe the Hubble law featuring the Hubble parameter $H(t)$. Demonstrate that *another* observer in the universe also measures the same Hubble law (provided that the universe is homogeneous and isotropic). That is, if $\mathbf{d}_O$, $\mathbf{d}_A$, and $\mathbf{d}_B$ are position vectors corresponding to us (the observer on Earth), and our friends who live in galaxies A and B in space, then demonstrate the following: our observation of the Hubble law toward the observers A and B implies that A and B also measure the same Hubble law when observing each other.

2.3 A circle in a non-Euclidean geometry. [Adapted from Guth (2013).] Consider a universe described by the Friedmann–Lemaître–Robertson–Walker metric which describes an open, closed, or flat universe, depending on the value of κ:

$$ds^2 = a^2(t) \left[\frac{dr^2}{1 - \kappa r^2} + r^2 \left(d\theta^2 + \sin^2\theta \, d\phi^2 \right) \right].$$

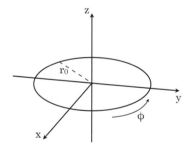

This problem will involve only the geometry of space at some fixed time, so we can ignore the dependence of a on t, and think of it as a constant. Consider a circle described by the equations

$$z = 0$$
$$x^2 + y^2 = r_0^2$$

or equivalently by the angular coordinates

$$r = r_0$$
$$\theta = \pi/2.$$

(a) Find the circumference S of this circle. *Hint:* Break the circle into infinitesimal segments of angular size $d\phi$, calculate the arc length of such a segment, and integrate.

(b) Find the radius R_c of this circle. Note that R_c is the length of a line which runs from the origin to the circle $(r = r_0)$, along a trajectory of $\theta = \pi/2$ and $\phi = $ const. Consider the case of open and closed universes separately, and take $\kappa = \pm 1$. *Hint:* Break the line into infinitesimal segments of coordinate length dr, calculate the length of such a segment, and integrate.

(c) Express the circumference S in terms of the radius R_c. This result is independent of the coordinate system which was used for the calculation, since S and R_c are both measurable quantities. Since the space described by this metric is homogeneous and isotropic, the answer does not depend on where the circle is located or on how it is oriented. For the two cases of open and closed universes, state whether S is larger or smaller than the value it would have for a Euclidean circle of radius R_c.

2.4 Volume of a closed universe. [Adapted from Alan Guth.] Calculate the total volume of a closed universe. It will be easiest to use the FLRW metric in the form

$$ds^2 = a^2(t) \left[dr^2 + R^2 \sin^2\left(\frac{r}{R}\right) \left(d\theta^2 + \sin^2\theta\, d\phi^2\right) \right].$$

Note that this form is a different writing of the same metric as in Problem 2.3.

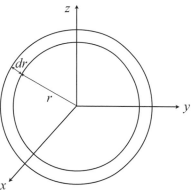

Break the volume up into spherical shells of infinitesimal thickness, extending from r to $r + dr$. As long as r is held fixed, the metric for varying θ and ϕ is the same as that for a spherical surface of radius $aR\sin(r/R)$ (what is it?). Multiply this area by the thickness of the shell (which you can read off from the metric!), and then integrate over the full range of r. Be careful to use the correct maximum value of r/R; this should correspond to reaching the antipodal point to the origin, which is when $S_\kappa(r) = 0$ again.

2.5 Redshift of very nearby objects. Familiarize yourself with nearby-universe calculations:

(a) Starting from Eq. (2.8), derive the first-order (i.e., valid for $z \ll 1$) relation between redshift z and distance d. [Do you need to worry about *which* distance?]

(b) The Andromeda galaxy is "in front of our nose" at $d = 780$ kpc away. What is the theoretically expected redshift of the Andromeda galaxy? Does this necessarily mean that its light is actually redshifted by this (small) amount? Why or why not – explain.

2.6 **Redshift drift.** In cosmology, usually we need to wait billions of years for some observable quantity to change visibly. Believe it or not, however, it is possible that we will soon observe the change in redshift of an object over the course of just a few years! Investigate as follows.

(a) Derive the expression for dz/dt_0, the change in redshift of an object as a function of elapsed time today, dt_0. Assume that the object is at redshift z and the proper time at its emission is t_1. Feel free to leave the answer in terms of $H_0 \equiv H(t_0)$ and $H(t_1)$, and obviously z. *Hint:* Start from $1+z = a(t_0)/a(t_1)$. At some point in the derivation you might need dt_1/dt_0, and for this Eq. (2.37) can help.

(b) For an object at some redshift $z = O(1)$, what is *approximately* – within an order of magnitude – the change in redshift dz if we take two measurements of the spectrum separated by $dt_0 = 10$ years? What is an experimental requirement for measuring such a dz (think spectrograph resolution $R \equiv \lambda/\Delta\lambda$ and be quantitative)? Looking at the right-hand side of your expression for dz/dt_0, which interesting quantity can you determine if you can measure dz/dt_0?

An important science goal of several ~ 30 m monster telescopes currently under construction (Extremely Large Telescope, Giant Magellan Telescope, Thirty Meter Telescope) is precisely detecting and measuring the redshift drift. They will do so using specially designed, ultra-high-resolution spectrographs.

2.7 **c-Field and steady-state cosmology.** The steady-state theory was proposed in 1948 by Fred Hoyle, Thomas Gold, Hermann Bondi, and others as an alternative to the hot Big Bang model. According to the steady-state model, new matter is continuously created as the universe expands, so that the **perfect cosmological principle** – that the universe looks on average the same everywhere in space *and at all times* – is satisfied. The steady-state model faced an uphill battle in explaining early observations that indicate evolution in the universe, such as the increased number density of quasars with distance, and was dealt a death knell with the discovery of the cosmic microwave background.

The steady-state model postulates a "c-field" (creation field) to create new matter and thus keep the large-scale matter density approximately constant. Estimate the rate of energy density production, $d\rho/dt$, that the c-field needs to achieve. Give your answer in both $\mathrm{kg}/(\mathrm{m}^3\,\mathrm{yr})$ and in $M_\odot/(\mathrm{Mpc}^3\,\mathrm{yr})$, and do not worry about factors of a few.

2.8 How big radius of curvature? Get a better feel for the curvature of the universe as follows.

(a) Estimate the radius of curvature R_0 of the universe in the formerly realistic, but as of about year 2000 ruled out, scenario that $\Omega \equiv \Omega_{\mathrm{TOT}} = 1.2$.

(b) Then estimate the radius of curvature in a hypothetical universe in which Gauss *would have* measured a one degree ($1°$) departure from the sum of the angles of a triangle connecting peaks of three mountains in Germany. Make reasonable assumptions about the distance between the mountain peaks (recall, Gauss could observe one while standing on another) but otherwise do not worry about factors of a few. By what factor is this radius smaller than that found in part (a)?

2.9 The cycloid solution. [Adapted from Alan Guth.] We now investigate the cycloid solution we discussed in Box 2.3. Consider a closed, matter-dominated universe with

$$t = R(\theta - \sin\theta)$$

$$\frac{a}{\sqrt{k}} = R(1 - \cos\theta),$$

where θ is a development angle, taking values $\theta = 0$ at the Big Bang and $\theta = 2\pi$ at the Big Crunch, and R is some constant with units of length.

(a) Find the Hubble parameter H as a function of R and θ.

(b) Find the matter density ρ as a function of R and θ.

(c) Find the matter density parameter Ω as a function of R and θ.

(d) For early times (i.e., suitably small t), use the first nonzero term of a power-series expansion to express θ as a function of t, and then a as a function of t.

(e) Even though these equations describe a closed universe, one still finds that $\Omega(t)$ approaches unity for very early times. In the same early-time limit as in part (d), the quantity $1 - \Omega(t)$ behaves as a power law in t. Find the expression for $1 - \Omega(t)$ in this limit, and identify the power law in t.

2.10 [Computational] Warm-up to computational cosmology: plot the cycloid. You will be assigned some computational problems in the back-of-chapter problems in this book. As a warm-up, make a simple, well-labeled plot in a programming language or plotting app of your own choosing (likely Python, but this is up to you).

Plot both the closed- and open-universe parametric solutions discussed in Box 2.3. Your x-axis will be t, and your y-axis $a/\sqrt{k}$. You can further assume that $R = 1$, so that you will be plotting simple functions of the development angle θ, where $\theta \in [0, 2\pi]$. It is probably simplest to generate a uniform array of values in θ, then calculate and plot the corresponding arrays in t and $a/\sqrt{k}$. When making your plot, and in all future plots that you make, please follow these good practices:

- make sure that the plotted curves are thick enough to be easily readable and mutually distinguishable (either with different colors or different line styles);
- make sure the plot is well-labeled, including x- and y-axis, as well as a legend briefly describing individual curves;
- make sure that all labeling text is LARGE and easily readable;
- make sure that the axis ranges are reasonable – in this problem, the y-axis range should not be high enough to encompass all of the values of the open-universe curve, but should cut it off at, say, $a/\sqrt{k} \simeq 5$ (assuming $R = 1$);
- make sure you can save the plot as a `pdf` or `jpg` file.

3 Contents of the Universe

In this chapter, we discuss the mass/energy contents of the universe, and how they govern the expansion rate. We also make the connection between the geometry of space and cosmological distances. We learn that there are several kinds of distance, in addition to those defined in Chapter 2, which are adopted depending on what precisely is being observed. Along the way, we establish the fiducial cosmological model – a set of cosmological parameters that we will use for all our results in the remainder of this book.

3.1 Continuity and Friedmann II Equations

In Chapter 2 we derived the first Friedmann equation. Now we introduce two more equations: the continuity equation and the second Friedmann (or acceleration) equation. Two of these three equations turn out to be independent; the third can always be derived from the other two.

3.1.1 Continuity Equation

To keep things simple, we again perform the derivation using classical, Newtonian arguments. We start from the first law of thermodynamics,

$$dQ = dE + PdV, \tag{3.1}$$

where dQ is the flow of energy in or out of the system, dE is change in internal energy, and P and dV are pressure of the gas (or, more generally, fluid) and volume change of the box enclosing the system, respectively. Note here, "system" can be considered to be any finite volume in our universe – we take it to be a sphere of comoving radius r. As before, we assume that the stuff filling the universe is smooth – no clustering.

In parallel with our Friedmann I equation derivation, we again assume a sphere of volume V filled with a component with energy density ρ and pressure P. We will be considering **adiabatic** expansion (or contraction) of the sphere, where there is no net heat flow in and out of the system, so $dQ = 0$. [Because there is no "outside" to the universe, it makes sense that there is no heat exchange with its surroundings.] We can now rewrite the first law of thermodynamics as

$$\dot{E} + P\dot{V} = 0, \tag{3.2}$$

where derivatives are with respect to (proper) time. The sphere's volume is

$$V(t) = \frac{4\pi}{3} r^3 a(t)^3, \tag{3.3}$$

where r is the comoving radius of the sphere, and $ra(t)$ is its physical radius. Then

$$\dot{V} = \frac{4\pi}{3} r^3 3a^2 \dot{a} = 3HV, \tag{3.4}$$

where, recall, $\dot{a}/a = H(t)$. Moreover, $E(t) = \rho(t)V(t)$, where $\rho(t)$ is the energy density of stuff in the box, so Eq. (3.2) implies that

$$\dot{\rho}V + \rho(3HV) + P(3HV) = 0, \tag{3.5}$$

or

$$\dot{\rho} + 3H(\rho + P) = 0 \qquad \text{(continuity eq.).} \tag{3.6}$$

This is the **continuity equation**. It tells us how the energy density of the stuff changes, given the energy density, pressure, and expansion rate at any given point in time. As with Friedmann I, the continuity equation is fully relativistic even though we simplified matters by sketching out a classical derivation that happens to give the correct result.

3.1.2 Friedmann II: The Acceleration Equation

There are two Friedmann equations – the one we already introduced in Eq. (2.46), and the one we will talk about now. We motivated Friedmann's first equation by a Newtonian derivation, while we will just quote the second Friedmann's equation – or the **Friedmann II equation**, or the **acceleration equation** – as

$$\frac{\ddot{a}}{a} = -\frac{4\pi G}{3}(\rho + 3P) \qquad \text{(Friedmann II eq.).} \tag{3.7}$$

Here, as before, ρ and P are the energy density and pressure of stuff in the universe. If there are multiple components (e.g., matter *and* radiation), then ρ and P refer to the *sum* of individual energy densities and pressures of the two components.

To repeat, the two Friedmann equations, (2.46) and (3.7), and the continuity equation, (3.6), are not all independent – knowing any two of them, we can derive the third.

Finally, let us also note that the first Friedmann equation can be written in a different way, by treating curvature just as another density component. We accomplish this by writing

$$\left(\frac{\dot{a}}{a}\right)^2 = \frac{8\pi G}{3}\rho(t) - \frac{k}{a(t)^2}$$

$$\equiv \frac{8\pi G}{3}(\rho(t) + \rho_k(t)), \tag{3.8}$$

where $\rho_k = -3k/(8\pi G a^2)$. Basically, curvature can just be treated as a contribution to energy density that scales as a^{-2}, so that ρ_k is referred to as the **"energy density in curvature."**

3.1.3 Equation of State

In order to solve the Friedmann equations and/or the continuity equation, we need to know about the ratio of pressure to energy density of any given component. Following standard practice from classical physics, we introduce the **equation of state** parameter w ("dubya") as

$$w = \frac{P}{\rho}. \tag{3.9}$$

Restoring the speed of light c, this reads $w = P/(\rho c^2)$.

Let us assume that we have a single-component universe, that is, a universe filled with only one fluid (e.g., matter only, or radiation only). Let us also assume that the equation of state is constant in time. Then, from the continuity equation, we have

$$\dot{\rho} = -3H\rho(1 + w). \tag{3.10}$$

Moreover, from $H = \dot{a}/a$ note that

$$\frac{d}{dt} = H\frac{d}{d\ln a}. \tag{3.11}$$

We boxed this equation as it is extremely useful in cosmology. Identity (3.11) is typically used in situations where we have derivatives with respect to time, and we wish to replace those with derivatives with respect to $\ln a$. The latter quantity is dimensionless, and also "steps" through the history of the universe in very sensible equal logarithmic intervals in the scale factor a, making it a favorite for implementation in computer programs. Notably, the second-order differential equations for the evolution of the inflaton field (Chapter 8) and the linear evolution of density perturbations (Chapter 9) can both be rewritten in a computer-friendly form by making use of Eq. (3.11).

Applying Eq. (3.11) to Eq. (3.10), we get

$$\frac{d\ln\rho}{d\ln a} = -3(1+w), \tag{3.12}$$

which can be easily solved to give

$$\rho(a) = \rho_0 a^{-3(1+w)} \qquad (\text{for } w = \text{const.}), \tag{3.13}$$

where ρ_0 is the energy density today and, recall, the present-day value of the scale factor is $a_0 = 1$ by convention. This means that scaling of the energy density with scale factor is given directly in terms of the equation of state of the given mass/energy component. This makes w a useful parameter.

With multiple components of stuff in the universe, we can define the *total* equation of state as

$$w_{\text{TOT}} = \frac{\sum_i P_i}{\sum_i \rho_i}, \tag{3.14}$$

where i runs over all components that fill the universe. It is this total equation of state that determines how rapidly the total energy density of the universe falls off. Note however that, in a multiple-component scenario (discussed in more detail in Sec. 3.3 below), w_{TOT} is typically not constant in time, so Eq. (3.13) does not hold.

3.2 Single-Component Universes

We now apply our results to a special case where the universe is dominated by a single component of mass/energy – the so-called **single-component universe**.

3.2.1 Matter-Dominated Universe

Let us first consider a matter-dominated universe. [This scenario is sometimes also called the dust-filled universe, because relativists traditionally refer to matter as dust.] Matter – dark *or* luminous – has the property that

$$w \approx 0 \qquad (\text{matter}). \tag{3.15}$$

To demonstrate the negligible impact of pressure for matter (dark or baryonic), that is $w \approx 0$, consider for example a box of gas molecules. The ideal gas equation is $P = nk_BT$, where n is the mean mass of gas particles, and T is the temperature of the gas, and where for clarity we hold on to the Boltzmann constant k_B. Moreover, for the gas we have two expressions for the energy per molecule:

$$E \simeq k_B T \simeq m \langle v^2 \rangle, \tag{3.16}$$

where $\langle v^2 \rangle$ is the root mean square (rms) velocity of gas particles. It follows that, for gas molecules,

$$w = \frac{P}{\rho} = \frac{n k_B T}{mn} = \frac{k_B T}{m} \simeq \langle v^2 \rangle \ll 1, \tag{3.17}$$

where, as usual, the velocity is in units of the speed of light. For a typical velocity of gas particles of ~ 500 m/s, $w \simeq 10^{-12}$. Clearly, the equation of state of matter can just be taken as zero; note in any case that it is strictly non-negative.

Since $w = 0$ for matter, the scaling of the energy density, from Eq. (3.13), is clear:

$$\rho_M(a) = \rho_0 a^{-3}. \tag{3.18}$$

Therefore, matter density dilutes with the volume of the expanding universe. This is the most intuitive of all single-component cases: the number of matter particles dilutes with the volume, and so does the total energy density. The total energy (energy density times the volume) is constant in the matter-dominated universe. Our classical, Newtonian, pre-relativistic intuition works – but not for long.

In parallel with the total density parameter Ω introduced in Eq. (2.50), one can define "omega in matter" evaluated today, or

$$\Omega_M \equiv \frac{\rho_M(t_0)}{\rho_{\mathrm{crit}}(t_0)}. \tag{3.19}$$

A single-component, flat, matter-dominated universe – that is, the one with $\Omega_M = 1$ – is called the **Einstein–de Sitter (EdS)** universe after two famous scientists, Albert Einstein and Willem de Sitter, who studied it early on. The EdS universe was for a long time a leading candidate model to describe our present-day universe, up until this was definitively overturned in the 1990s. More on that in Chapter 12.

3.2.2 Radiation-Dominated Universe

The next most important case is radiation, by which we refer to fully relativistic particles: photons and gravitons; massless neutrinos would also fall in this category. As we will see further below in this chapter, radiation was the dominant component over an important span in the history of the universe, starting just after inflation (ballpark 10^{-35} s after the Big Bang), until when matter took over, which was at redshift of about 3500 ($\sim$50,000 years after the Big Bang). For the moment, we will assume that the radiation is in photons.

Radiation is different from matter: while the photons dilute with the expansion exactly as matter particles do, their wavelength also *stretches* and each photon loses energy. Recall that the energy of the photon is given in terms of its frequency ν or wavelength λ as

$$E_{\text{photon}} = h\nu = \frac{hc}{\lambda}, \tag{3.20}$$

where h is Planck's constant. Since the wavelength stretches with the expansion, $\lambda \propto a$ and

$$E_{\text{photon}} \propto a^{-1}. \tag{3.21}$$

Put all together, we see that the radiation energy density of a volume with N photons scales as

$$\rho_R = \frac{N E_{\text{photon}}}{V} \propto a^{-4}, \tag{3.22}$$

since $V \propto a^3$. Comparing this equation to Eq. (3.13), we see that the equation-of-state parameter of radiation is

$$w = \frac{1}{3} \qquad \text{(radiation)}. \tag{3.23}$$

This result may be familiar; you may recall that the radiation field has equation of state $P = (1/3)\rho c^2$ (reinserting for the moment the speed of light c). Or else, you have surely heard of solar sails, a proposed method of powering spacecraft by deploying a large polymer sheet coated with metal which (along with the spacecraft to which it is attached) is pushed by solar radiation – by the Sun's photons. This radiation pressure is precisely the same effect as what we are talking about here; it is the reason why the equation of state w of radiation is nonzero.

Just like we did for matter, we can also define the "omega in radiation" parameter defined today:

$$\Omega_R \equiv \frac{\rho_R(t_0)}{\rho_{\text{crit}}(t_0)}. \tag{3.24}$$

3.2.3 Lambda-Dominated Universe

"Lambda" stands for the **cosmological constant** introduced by Einstein. The famous story is that Einstein added this term to his equations of general relativity in order to rescue a static universe in light of the fact that his equations predict the expanding or contracting one. After Hubble observed the expansion of the universe in 1929, Einstein was very sorry for his proposal, and called it the "**greatest blunder**" of his career[1] (Problem 3.10 investigates why). The cosmological-constant term is very much in vogue again due to the discovery about 25 years ago that

---

[1] Einstein's famous phrase was not found in any of his written correspondence. The words were uttered during Einstein's conversation with George Gamow, and were overheard by two other physicists who later recounted this.

the expansion of the universe is accelerating – this *might* be due to the small but nonzero cosmological constant. Much more on this in Chapter 12.

The cosmological-constant term Λ enters the Friedmann I equation as

$$\left(\frac{\dot{a}}{a}\right)^2 = \frac{8\pi G}{3}\rho(t) - \frac{\kappa}{R_0^2 a(t)^2} + \frac{\Lambda}{3}, \qquad (3.25)$$

corresponding to energy density

$$\rho_\Lambda = \frac{\Lambda}{8\pi G}. \qquad (3.26)$$

We see that the Λ term has (natural) units of $[\Lambda] = [\rho][G] = [M]^4/[M]^2 = [M]^2$.

Physically, the Lambda term represents the energy density of empty space – the vacuum. According to quantum field theory, vacuum is not quite empty, but rather filled with fields whose excitations we call particles. The Heisenberg uncertainty relation $\Delta E \Delta t \simeq \hbar$ implies that there are fluctuations of these fields in the vacuum; shorter-lasting fluctuations correspond to larger-energy excitations, and vice versa. Large-enough fluctuations correspond to particle–antiparticle pairs that are created which then, a moment later, annihilate. These fluctuations correspond to the energy of the vacuum.

By definition, the cosmological constant (or vacuum energy) has pressure equal to *negative* energy density, $P_\Lambda = -\rho_\Lambda$. That is,

$$w = -1 \qquad \text{(cosmological constant)}, \qquad (3.27)$$

and therefore,

$$\rho_\Lambda(t) = \text{const.} \qquad (3.28)$$

If you have a box of vacuum energy and increase its volume by a factor of two, its energy *density* will remain constant. Because the volume doubled, the total energy in the box will be two times bigger. This is how even a tiny amount of vacuum energy in the early universe starts to dominate at later times – it doesn't go away, scaling as $a^{-3(1+w)} = a^0$, while matter and radiation rapidly dilute, scaling as a^{-3} and a^{-4}, respectively.

You may have noticed that, either for the vacuum energy or for radiation, the total energy in the universe, $E \propto \rho V \propto \rho(a)a^3$, is not conserved. What is going on? It turns out that there is *no* energy conservation law in general relativity. Energy conservation is a non-relativistic concept, and holds in small areas of space. The universe as a whole does not conserve energy.

The other way to intuitively understand why the total energy in the universe is not conserved is to consider that, in general relativity, "pressure gravitates." Note the presence of pressure in the acceleration and continuity equations – we do not have that in classical mechanics. Nonzero pressure basically acts as a source of gravity in general relativity.

The takeaway message from all this is

> Energy densities of matter, radiation, and cosmological constant scale, respectively, as a^{-3}, a^{-4}, and a^0.

3.2.4 Age and Proper Distance

Certain quantities, while always computable numerically or using special mathematical functions, are particularly simple in the case of a spatially flat universe, with $\kappa = 0$. [We think we live in such a universe; see Eq. (2.55).]

In a flat universe, $\rho = \rho_{\mathrm{crit}}$ by definition, at all times; recall the basic message from Eq. (2.52) – once flat, always flat. We can then relate the Hubble parameter today to the critical density today,

$$H_0^2 = \frac{8\pi G}{3}\rho_{\mathrm{crit},0}. \tag{3.29}$$

Then the Friedmann I equation for constant-in-time equation of state, $w = \mathrm{const.}$, can be simplified to

$$\frac{H^2}{H_0^2} = \frac{\rho(z)}{\rho_0} = a^{-3(1+w)}. \tag{3.30}$$

It is easy to compute, for example, the age of the universe. Remembering that $H = d\ln a/dt$, we can solve the above equation to get $a \propto t^{2/[3(1+w)]}$, and generalize the result already obtained in Box 2.2 for the EdS case. We get

$$a(t) = \left(\frac{t}{t_0}\right)^{\frac{2}{3(1+w)}} \qquad \text{(flat, single component with } w = \mathrm{const.)}, \tag{3.31}$$

where t_0 is the age of the universe today; notice that this normalization assures $a(t_0) = 1$.

Integrating Eq. (3.30) with the help of Eq. (3.31). we get the age of the universe:

$$t_0 = \frac{2}{3(1+w)}H_0^{-1} \qquad \text{(flat, single component with } w = \mathrm{const.)}. \tag{3.32}$$

Problem 3.4 investigates how these equations evaluate when $w = -1$ in a single-component model.

For example, in a matter-dominated universe (visible or dark matter, doesn't matter!), the age is $t_0^{\mathrm{MD}} = 2/3H_0^{-1}$. This translates to

$$t_0^{\mathrm{MD}} \simeq 9.7\,\mathrm{Gyr} \qquad \text{(flat, matter dominated)}. \tag{3.33}$$

This is problematic, as t_0^{MD} is less than the age of some of the oldest objects in the universe, like globular clusters (of stars, in our galaxy), which are $\sim 12\,\mathrm{Gyr}$ old. This realization was one of the key reasons why a matter-dominated, flat universe was deemed in conflict with available data in the late 1980s and 1990s. The issue was definitively resolved with the discovery of dark energy, which predicts an older universe ($t_0 \simeq 14\,\mathrm{Gyr}$, which you will be asked to check in Problem 3.6), in agreement with modern measurements.

It is also easy to compute the proper distance (evaluated today, when it is equal to the comoving distance) in a spatially flat universe:

$$d_p = \int \frac{dt}{a(t)} = \int_{t_e}^{t_0} \frac{dt}{(t/t_0)^{2/(3(1+w))}} = t_0 \frac{3(1+w)}{1+3w} \left[1 - \left(\frac{t_e}{t_0} \right)^{(1+3w)/(3(1+w))} \right].$$

(3.34)

Now, recall that

$$1 + z = \frac{1}{a(t_e)} = \left(\frac{t_0}{t_e} \right)^{2/(3(1+w))}$$

(3.35)

and $t_0 = [2/3(1+w)]H_0^{-1}$, so that, finally

$$d_p(z) = \frac{2}{1+3w} H_0^{-1} \left[1 - \frac{1}{(1+z)^{(1+3w)/2}} \right] \quad \text{(flat; } w = \text{const.).} $$

(3.36)

As with all single-component results, this equation does not apply to our own (multi-component) universe, yet its closed form makes it very handy, especially in situations where we are making an approximate argument.

3.2.5 (Particle) Horizon Distance

The most distant object we can see, in theory, is one for which the light starts going toward us at the Big Bang (at $t = 0$) and reaches us today (at t_0). Proper distance for that case, in a given cosmological model, is fixed and is called the **particle horizon distance**.[2] Evaluated today (at t_0), this is

$$d_{\text{part hor}} = \int_0^{t_0} \frac{dt}{a(t)} \quad \text{(particle horizon distance today).} $$

(3.37)

In a flat universe with $a(t) = (t/t_0)^{2/(3(1+w))}$ and $w = \text{const.}$, this evaluates to

$$d_{\text{part hor}} = \frac{2}{1+3w} H_0^{-1}.$$

(3.38)

In other words, a sphere centered at us with radius d_{hor} represents all points in space that could have been in causal contact with us.

Note the subtle difference between the particle horizon distance, $d_{\text{part hor}}$, and the

---

[2] Also known as the particle horizon, or the horizon distance, or else simply the **horizon**. Cosmologists are not very good at agreeing on naming conventions.

Hubble distance, H_0^{-1}. These two quantities clearly differ; in a single-component case, their ratio is $2/(1 + 3w)$, as Eq. (3.38) shows. Yet cosmologists often mix the two and refer to a "horizon" when they mean the Hubble distance (*or* when they mean the particle horizon). Fortunately, in most situations – say, in the currently favored cosmological model with dark matter and dark energy – the particle horizon and the Hubble distance are comparable, and the order–unity difference between them is typically unimportant for arguments made by theoretical cosmologists. The particle horizon and the Hubble distance actually coincide for the purely radiation-dominated universe ($w = +1/3$), while the particle horizon is two times bigger for the purely matter-dominated universe ($w = 0$). However, for $w \leq -1/3$, the particle horizon distance is infinite – we are causally connected to all space, and we can, in principle, see every point in space. To summarize:

> Particle horizon encodes "how far out we can see."

Let us also introduce the notion of **event horizon**, which is the largest comoving distance from which light emitted *now* can ever reach the observer in the future. The definition of event horizon is therefore complementary to that of particle horizon:

$$d_{\text{event hor}} = \int_{t_0}^{\infty} \frac{dt}{a(t)} \qquad \text{(event horizon distance)}. \qquad (3.39)$$

Event horizon is certainly more important and more popular in the context of studies of individual black holes, while in cosmology particle horizon plays a more prominent role.

Particle horizon is often of interest in situations where we want to determine roughly the size of the causally connected region at any point in the history of the universe. Because $d_{\text{part hor}}(t) \simeq 1/H(t)$, this is just the inverse Hubble parameter at the given time. One specific example of the utility in considering the particle horizon is in considerations of the events during and after inflation, something that we will do in Chapter 8.

3.3 Multiple-Component Universes

Finally, we consider our own universe, which contains multiple components with mass/energy and pressure. A pie chart summarizing the components *today* is given in Fig. 3.1.

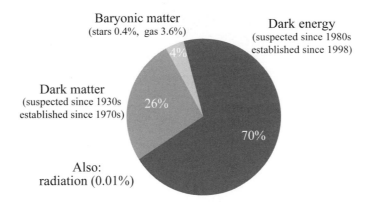

Fig. 3.1   Pie chart summarizing the components of the universe *today*. The percentages refer to fractions of the total energy density.

3.3.1 General Formulae

In a universe with multiple components, *the total equation of state is in general not constant* even if individual-component equations of state are constant. This is because, at any given epoch, more than one component may be significant, and the total equation of state, $w_{\text{TOT}} = \sum_i P_i / \sum_i \rho_i$, is changing with time even if $w_i \equiv P_i/\rho_i = \text{const.}$ individually. Note that, from here on, we follow this widely accepted convention:

> We reserve w to refer to the equation-of-state parameter of dark energy.

We do this because the equations of state of other components (matter, radiation) are fixed and known, and we do not need to parameterize them.

The non-negligible components today are dark energy (about 70 percent of energy density) and matter (about 30 percent). In the past, radiation was very important too; nowadays, the small amount of radiation is contributed by the CMB photons, and also by the cosmic background of *neutrinos*. In what follows, we assume that the universe consists of the following.

- Matter, both dark and luminous, described by time-dependent energy density $\rho_M(a) \propto a^{-3}$.
- Radiation (CMB photons and neutrinos, if/when they are relativistic): $\rho_R(a) \propto a^{-4}$.
- Curvature (i.e., the possibility that the universe is open or closed): $\rho_k(a) \propto a^{-2}$.
- Dark energy with the equation of state w: $\rho_{\text{DE}}(a) \propto a^{-3(1+w)}$.

Let us rewrite the Friedmann I equation in a more useful form:

$$H^2(a) = \frac{8\pi G}{3}\left(\rho_{M,0}a^{-3} + \rho_{R,0}a^{-4} + \rho_{\text{DE},0}a^{-3(1+w)} - \rho_{k,0}a^{-2}\right)$$

$$= \frac{8\pi G\rho_{\text{crit},0}}{3}\frac{\left(\rho_{M,0}a^{-3} + \rho_{R,0}a^{-4} + \rho_{\text{DE},0}a^{-3(1+w)} - \rho_{k,0}a^{-2}\right)}{\rho_{\text{crit},0}} \qquad (3.40)$$

$$= H_0^2\left(\Omega_M a^{-3} + \Omega_R a^{-4} + \Omega_{\text{DE}}a^{-3(1+w)} + \Omega_k a^{-2}\right),$$

where we made use of the density parameters Ω_i for each component, and adopted the standard convention for the sign of Ω_k so that

$$\Omega_k = 1 - \Omega_M - \Omega_R - \Omega_{\text{DE}} \equiv 1 - \Omega_{\text{TOT}}. \qquad (3.41)$$

Recall that $\Omega_{\text{TOT}} > 1$ for the closed case and < 1 for the open case.

Since $H = \dot{a}/a$, one can now easily relate the cosmic proper time t to the scale factor of the universe. First, the Friedmann I equation can be written as

$$\frac{da}{a\,dt} = H_0\left(\Omega_M a^{-3} + \Omega_R a^{-4} + \Omega_{\text{DE}}a^{-3(1+w)} + \Omega_k a^{-2}\right)^{1/2}, \qquad (3.42)$$

and solved to give the **age of the universe** in a general, multiple-component cosmological model:

$$t(a) = H_0^{-1}\int_0^a \frac{du}{\left[\Omega_M u^{-1} + \Omega_R u^{-2} + \Omega_{\text{DE}}u^{-(1+3w)} + \Omega_k\right]^{1/2}}. \qquad (3.43)$$

Setting $a = 1$ in this equation gives the age of the universe today.

3.3.2 The Best-Fit (and the Fiducial) Model

We finally consider the makeup of *our* universe, and specifically the **measured cosmological parameter values**. From Fig. 3.1 we already know that the model that best fits the data consists of about 70 percent dark energy, 26 percent dark matter, 4 percent baryonic matter (some, though not all of it, is luminous), and about 0.01 percent radiation.

The cosmological parameters and their current measurements are presented in Table 3.1. We also show the **fiducial values** of these parameters, which we adopt for all calculations in this book. The fiducial parameter values are very similar to the measured parameters, but are typically slightly rounded.

The measured parameter values, along with their 68 percent credible error bars, come from the analysis of Planck and other cosmological data. A curved ΛCDM (Lambda cold dark matter[3]) cosmological model has been assumed. The best-fit model corresponds to Planck constraints `base_omegak_plikHM_TTTEEE_lowl_lowE_BAO`, except for the measurement of the equation of state of dark energy w, where a flat wCDM

---

[3] "Cold" refers to the non-relativistic property of dark matter, as we will discuss in Chapter 11.

Table 3.1 Basic cosmological parameters, as well as a few commonly used derived parameters. The measured values are from Planck, while the fiducial values, adopted in plots and calculations in this book, are similar but slightly rounded. See text for details.

Parameter name	Symbol	Measured value	Fiducial value
Spatial curvature	Ω_k	0.001 ± 0.002	0
Matter density rel. to critical	$\Omega_{\rm M}$	0.310 ± 0.007	0.30
Baryon density	$\Omega_{\rm B} h^2$	0.0224 ± 0.0002	0.0224
Hubble constant	H_0	$(67.9 \pm 0.7)\,{\rm km/s/Mpc}$	$67\,{\rm km/s/Mpc}^{**}$
$P(k)$ amplitude at $k_{\rm piv} = 0.05$	A_s	$(2.10 \pm 0.03) \times 10^{-9}$	2.1×10^{-9}
Scalar spectral index	n_s	0.966 ± 0.005	0.966
Age of universe	t_0	$(13.76 \pm 0.08)\,{\rm Gyr}$	derived
Amplitude of mass fluctuations	σ_8	0.810 ± 0.007	derived
CMB temperature	T_0	$(2.7255 \pm 0.0006)\,{\rm K}$	$2.725\,{\rm K}$
Photon density	$\Omega_\gamma h^2$	derived from T_0	2.47×10^{-5}
Assumed-massless neutrino density	$\Omega_{\nu,\rm rel} h^2$	derived from T_0	1.68×10^{-5}
Equation of state of dark energy	w	-1.04 ± 0.06	-1

model is assumed with the w constraint in `base_w_plikHM_TTTEEE_lowl_lowE_BAO`; see Aghanim *et al.* (2020b) and `https://wiki.cosmos.esa.int/planck-legac y-archive/index.php/Cosmological_Parameters`. The asterisks next to the adopted fiducial value of the Hubble constant remind the reader that a different set of measurements in cosmology – from measuring a distance scale in the local universe – reports a higher value, $H_0 \simeq 73\,{\rm km/s/Mpc}$. The discrepancy between that and the value measured by Planck ($H_0 \simeq 67\,{\rm km/s/Mpc}$) is referred to as the "**Hubble tension**," which we already mentioned in Sec. 2.2. The fiducial values are noted as "derived" in cases where they can be computed from the values of other fiducial parameters in our adopted cosmological model. Finally, recall that $h = H_0/(100\,{\rm km/s/Mpc})$.

All calculations in this book adopt the fiducial values of the cosmological parameters, unless otherwise indicated.

3.3.3 Matter–Radiation Equality

Next we examine an important moment in the history of the universe: the epoch when the universe went from radiation domination to matter domination. As we will see in upcoming chapters, it is at this epoch when the matter density fluctuations finally got a chance to grow significantly.

Note that today, matter is much more significant than radiation: $\Omega_M \approx 0.3$, while $\Omega_\gamma h^2 \simeq 2.47 \times 10^{-5}$ (as will be demonstrated in Chapter 4 and is listed in Table 3.1). At earlier times, when $T \gg m_\nu \sim 0.1\,\mathrm{eV}$ (corresponding to redshifts $z \gg 500$), the neutrinos were relativistic, so they need to be added to the cosmic energy budget, as we will discuss in Chapter 5. Because the matter–radiation equality time will be in the era when the neutrinos were relativistic, let us add them to the radiation budget. Then $\Omega_R = (\Omega_\gamma h^2 + \Omega_{\nu,\mathrm{rel}} h^2)/h^2 = 4.15 \times 10^{-5}/h^2 \approx 8.5 \times 10^{-5}$, following the numbers from Table 3.1. Moreover, matter energy density dilutes slower than radiation; $\rho_M \propto a^{-3}$ while $\rho_R \propto a^{-4}$. Thus, early in the history of the universe radiation was actually more dominant than matter. The two components were equal at the redshift of equality z_{eq}, when

$$\rho_M(z_{\mathrm{eq}}) = \rho_R(z_{\mathrm{eq}})$$

$$\rho_{M,0}(1+z_{\mathrm{eq}})^3 = \rho_{R,0}(1+z_{\mathrm{eq}})^4$$

$$\rho_{\mathrm{crit},0}\Omega_M(1+z_{\mathrm{eq}})^3 = \rho_{\mathrm{crit},0}\Omega_R(1+z_{\mathrm{eq}})^4 \tag{3.44}$$

$$z_{\mathrm{eq}} = \frac{\Omega_M}{\Omega_R} - 1 \approx 3250.$$

Therefore, matter–radiation equality happened at around redshift 3500 (where the exact number depends on the value of the Hubble constant, whose measurements currently range from ~ 67 to $\sim 73\,\mathrm{km/s/Mpc}$). Feeding $a \simeq 1/(1+3500)$ into Eq. (3.43), we see that this corresponds to about $t = 50,000$ years after the Big Bang. Prior to this time (and since the end of inflation, in fact), the universe was radiation dominated. *After* this time and up until $z \simeq 0.5$, the universe was matter dominated. And since $z \simeq 0.5$ (or a few billion years ago, practically "yesterday" in cosmological terms), the universe has been dark-energy dominated.

3.3.4 Deceleration Parameter

We now consider a new, derived[4] parameter in cosmology, the **deceleration parameter** q, defined as

$$q_0 \equiv -\left.\frac{\ddot{a}a}{\dot{a}^2}\right|_{t=t_0} \equiv -\left.\frac{\ddot{a}}{aH^2}\right|_{t=t_0}. \tag{3.45}$$

Typically, when cosmologists talk about the deceleration parameter, they are referring to its value today, q_0. In general, the deceleration parameter is an arbitrary function of time or redshift, $q = q(z)$.

Let us relate the deceleration parameter to the various energy densities omega.

---

[4] We refer to a parameter as *derived* if it doesn't correspond to a set of fundamental parameters (roughly, the first six parameters listed in Table 3.1). A derived parameter can be computed given values of all of the fundamental parameters in a given cosmological model (say ΛCDM).

Dividing the acceleration equation (3.7) by the Friedmann I equation (2.46), and taking the negative sign, we get

$$-\frac{\ddot{a}a}{\dot{a}^2} = \frac{4\pi G}{3H^2} \sum_i (\rho_i + 3P_i), \tag{3.46}$$

where the sum goes over all components. Given that we are going to evaluate this expression today (so we will have $H = H_0$, etc.), the left-hand side is clearly just q_0, while the first term on the right-hand side is one half of the inverse critical density today (see Eq. (2.47)). Hence

$$q_0 = \frac{1}{2} \sum_i \frac{\rho_{i,0}(1 + 3w_i)}{\rho_{c,0}} = \frac{1}{2} \sum_i \Omega_i (1 + 3w_i). \tag{3.47}$$

The deceleration parameter has been very important historically, especially in the 1970s when all of cosmology was called "the quest for two numbers," q_0 and H_0, by astronomer Allan Sandage. Here is why: for a universe that is matter dominated,

$$q_0 = \frac{\Omega_M}{2} \qquad \text{(matter dominated)}. \tag{3.48}$$

In that scenario, q_0 effectively tells us about the matter density of the universe, while H_0 gives us (from the Friedmann I equation) the curvature. Therefore, q_0 and H_0 are indeed the only two parameters that are sufficient to describe the universe *in the matter-only universe.*

The deceleration parameter today is simple to evaluate even if we throw the cosmological-constant (Λ) term into the mix. Since $w = -1$ for the cosmological constant, we have

$$q_0 = \frac{\Omega_M}{2} - \Omega_\Lambda \qquad \text{(matter+Lambda)}. \tag{3.49}$$

However, when we allow for the possibility of dark energy with a time-variable equation of state, $w(t)$, the deceleration parameter *today* is no longer sufficient to describe the dynamics of the expanding universe (though one could always go to a time-dependent $q(t)$, which would still contain all information). This is one of the reasons why the deceleration parameter has largely been relegated to the dustbin of history, in favor of the density parameters Ω_i and the dark-energy equation of state w.

In summary, while the deceleration parameter is no longer a member of the standard set of cosmological parameters, it is important to at least be acquainted with it as it has had a strong "historical performance." For the best-fit ΛCDM cosmological model with $\Omega_M \approx 0.3$ and $\Omega_\Lambda \approx 0.7$, the deceleration parameter is negative ($q_0 \approx -0.55$), encoding the fact that the expansion of the universe is accelerating.

3.3.5 Distance–Redshift Relation at Low z

Let us Taylor-expand the scale factor around the present time:

$$a(t) = a(t_0) + \dot{a}|_{t=t_0} (t - t_0) + \frac{1}{2}\ddot{a}\Big|_{t=t_0} (t - t_0)^2 + \cdots \tag{3.50}$$

or

$$\frac{a(t)}{a(t_0)} = 1 + \frac{\dot{a}}{a}\Big|_{t=t_0} (t - t_0) + \frac{1}{2}\frac{\ddot{a}}{a}\Big|_{t=t_0} (t - t_0)^2 + \cdots$$

$$= 1 + H_0(t - t_0) - \frac{1}{2}q_0 H_0^2 (t - t_0)^2 + \cdots . \tag{3.51}$$

Remembering that $a(t_0) = 1$, we finally get

$$a(t) \approx 1 + H_0(t - t_0) - \frac{1}{2}q_0 H_0^2 (t - t_0)^2, \tag{3.52}$$

which holds for times in the recent past relative to today (can you quantify that?). Inverting this relation, we get

$$\frac{1}{a(t)} \approx 1 - H_0(t - t_0) + \frac{1 + q_0}{2} H_0^2 (t - t_0)^2. \tag{3.53}$$

Now we are ready to compute the approximate proper distance (note that we need to collect terms when integrating),

$$d_p(t) = \int_t^{t_0} \frac{dt'}{a(t')} \approx (t_0 - t) + \frac{H_0}{2}(t_0 - t)^2 \qquad \text{(for small distances).} \tag{3.54}$$

The first term on the right-hand side is what would be expected from a static universe. The second term accounts for the expansion, and says that the proper distance is actually larger since the universe has expanded since the emission time t_e.

The difficulty with the equation for the proper distance is that astronomers typically don't measure ages of objects – galaxies do not have a signature that says "Hello there, I am sending this light 12.53 billion years after the Big Bang." Rather, we typically measure redshifts of objects; therefore, we would like to get an (approximate, low redshift/distance) equation for $d_p(z)$. This is possible if we invert the relation in Eq. (3.53), shifting also from the scale factor to redshift via $1/a = 1 + z$, to get

$$t_0 - t_e \approx H_0^{-1}\left[z - \frac{1 + q_0}{2}z^2\right], \tag{3.55}$$

then one can show that

$$d_p(z) \approx H_0^{-1}z\left[1 - \frac{1 + q_0}{2}z\right] \qquad \text{(for small distances).} \tag{3.56}$$

The important takeaway from Eq. (3.56) is actually its lowest-order dependence on redshift. Because all distances (proper, comoving, angular-diameter, luminosity) agree at $z \ll 1$, this can be summarized as

> For $z \ll 1$, all distances in cosmology are a linear function of redshift: $d \simeq z/H_0 \simeq 4500z$ Mpc.

3.4 Observable Distances

In Chapter 2 we discussed the proper distance, as well as the coordinate distance (and time). There is, however, the question of which of these quantities (if either) can be measured using typical observations in cosmology. The answer is – neither. Rather, two other distances that are related to them – the angular-diameter and luminosity distance – are the ones that can be measured. In order to introduce them, however, we first cover the all-important stepping stone – the comoving distance.

3.4.1 Comoving Distance

Let us first obtain the grand final, user-friendly, exact expression for the comoving distance as a function of redshift.[5] We first recall the "omega in curvature" parameter, defined following Eq. (2.52):

$$\Omega_k \equiv 1 - \Omega_{\text{TOT}} = -\frac{\kappa}{R_0^2 H_0^2}. \tag{3.57}$$

Here Ω_k is a measure of openness or closedness of the universe. If the universe is closed, $\Omega_k < 0$, $\kappa = 1$ and $R_0 = H_0^{-1}/\sqrt{|\Omega_k|}$. If the universe is open, $\Omega_k > 0$, $\kappa = -1$, and $R_0 = H_0^{-1}/\sqrt{\Omega_k}$. Therefore, for either geometry, the curvature radius is given by

$$R_0 = \frac{H_0^{-1}}{\sqrt{|\Omega_k|}}. \tag{3.58}$$

The proper distance (which, when evaluated at the present time, is equal to the coordinate distance χ) is

$$d_p(t_0) \equiv \chi = \int \frac{dt}{a(t)} = \int \frac{da}{a^2 H} = -\int_z^0 \frac{dz'}{H(z')} = \int_0^z \frac{dz'}{H(z')} \tag{3.59}$$

$$= H_0^{-1} \int_0^z \frac{dz'}{\sqrt{\Omega_M(1+z')^3 + \Omega_{\text{DE}}(1+z')^{3(1+w)} + \Omega_R(1+z')^4 + \Omega_k(1+z')^2}}.$$

---

[5] We are effectively getting the exact expression for the proper distance too, but from here on we emphasize the comoving distance because it is directly related to the observable luminosity and angular-diameter distances, as we will see in a moment.

Recall next that the comoving distance r is related to the coordinate distance χ via

$$
r = \begin{cases}
R_0 \sin(\chi/R_0) & (\kappa = +1) \\
\chi & (\kappa = 0) \\
R_0 \sinh(\chi/R_0) & (\kappa = -1).
\end{cases} \tag{3.60}
$$

From the above two equations, we get a general form of the comoving distance as a function of redshift and cosmological parameters:

$$
r(z) = \frac{H_0^{-1}}{\sqrt{\Omega_k}} \sin\left[\sqrt{\Omega_k} \int_0^z \frac{dz'}{E(z')}\right], \qquad \text{where} \tag{3.61}
$$

$$
E(z') \equiv \sqrt{\Omega_M(1+z')^3 + \Omega_{\mathrm{DE}}(1+z')^{3(1+w)} + \Omega_R(1+z')^4 + \Omega_k(1+z')^2}.
$$

The sine in the $r(z)$ expression explicitly accounts for the possibility of a positively curved universe with $\Omega_k > 0$. For the negatively curved case, the square root of Ω_k is imaginary, but note that $\sin(ix) = i\sinh(x)$, and this i cancels that in the prefactor term (the square root of Ω_k), thus leading to a manifestly real answer. Finally, the flat case obtains by applying the limit $\lim_{x\to 0}\sin(x)/x = 1$. Nevertheless, when implementing the comoving distance in a computer code – an essential task for a budding cosmologist! – it is usually easier to enable options for three different expressions (with a sine, a sinh, or the argument without the curvature term), depending on the sign of Ω_k.

In short:

> Equation (3.61) – the formula for comoving distance as a function of redshift for arbitrary curvature – is an essential ingredient for many calculations in cosmology.

Moreover, recall that

$$
\Omega_k = 1 - \Omega_M - \Omega_{\mathrm{DE}} - \Omega_R, \tag{3.62}
$$

so that we can always calculate Ω_k given all of the other omegas. The comoving distance in representative cosmological models is shown in Fig. 3.2.

The master expression in Eq. (3.61) explains cosmologists' choice of their favorite distance units, $h^{-1}\mathrm{Mpc}$. Given a redshift of some object and fixed cosmological model (specifically the omegas), there is a direct relation between redshift z and distance r if the Hubble constant is also known. Because H_0 was poorly measured over most of the past century, the $r(z)$ relation is still exact if the distance is measured in units of H_0^{-1}, that is, in $h^{-1}\mathrm{Mpc}$. And, at $z \ll 1$, $r \simeq z/H_0 = 2997.9z\,(h^{-1}\mathrm{Mpc})$, so that in this limit, redshift immediately gives the distance regardless of either the cosmological model or the Hubble constant. We work out an example illustrating this in Box 3.1.

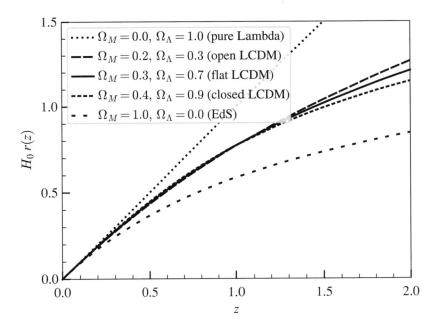

Fig. 3.2   Comoving distance as a function of redshift for several cosmological models. From top to bottom: flat Lambda-only model; negatively curved, flat, and positively curved ΛCDM models; and flat, matter-only (Einstein–de Sitter) model.

3.4.2 Comoving Volume Element

We often need to calculate the volume of some chunk of the universe – say, the volume bounded by a minimal and a maximal redshift and covering some solid angle that a given cosmological survey probes. From the metric in Eq. (2.19), we see that the physical volume element is

$$dV_{\text{phys}} = d\Omega \left(aS_\kappa(\chi)\right)^2 (ad\chi) = d\Omega\, a^3 r^2 d\chi, \tag{3.63}$$

where χ is the radial coordinate, $r \equiv S_\kappa(\chi)$ is the comoving distance, and Ω is the solid angle. We then recognize that $d\chi = dz/H(z)$, where z and dz are the corresponding redshift and thickness of the shell at coordinate distance χ and thickness $d\chi$. Thus, $dV_{\text{phys}} = d\Omega a^3 r^2 dz/H$. Turning our attention to the comoving volume element dV, which does not have the three powers of a and which is most commonly used, we have a remarkably simple expression:

$$\frac{dV}{d\Omega dz} = \frac{r^2(z)}{H(z)} \qquad \text{(comoving volume element)}. \tag{3.64}$$

| Box 3.1 | **Worked Example: Cosmology at Low Redshift** |

Calculations of distances and volumes at low redshift can usually be dramatically simplified, to the point that no calculator is needed. The key thing to remember is that, when $z \ll 1$, the comoving distance becomes

$$r(z) \simeq \frac{cz}{H_0} \equiv \frac{z}{H_0}, \tag{B1}$$

where we temporarily restored SI units in the first equality above. The same $z \ll 1$ expression holds for the angular and luminosity distance, $d_A(z)$ and $d_L(z)$, as they differ from $r(z)$ only at order z^2. Similarly, the recession velocity of an object at low z is (with d_p the proper distance)

$$v = H_0 d_p(z) \simeq H_0 r(z) \simeq cz \equiv z. \tag{B2}$$

Additionally, the universe was kind to us in making the inverse Hubble constant in our favorite units very nearly a round number,

$$H_0^{-1} = 2997.9 \, h^{-1} \mathrm{Mpc} \simeq 3000 \, h^{-1} \mathrm{Mpc}. \tag{B3}$$

Let us do a few simple $z \ll 1$ calculations. Consider a galaxy at $z = 0.1$. We would measure its recession velocity to be

$$v \simeq z = 0.1 = 30,000 \mathrm{km/s}. \tag{B4}$$

Given that a typical peculiar velocity of galaxies is $300\mathrm{km/s}$, a galaxy at $z = 0.1$ would contribute only about 1 percent to the expansion velocity, which can usually be neglected. Next, the distance to this galaxy is simply

$$r \simeq z H_0^{-1} \simeq 300 \, h^{-1} \mathrm{Mpc}. \tag{B5}$$

Finally, the volume element at $z \ll 1$ also takes a simple form

$$\frac{dV}{d\Omega dz} = \frac{r^2(z)}{H(z)} \simeq \frac{z^2}{H_0^3}. \tag{B6}$$

For our $z = 0.1$ galaxy, this evaluates to $0.27 \, (h^{-1}\mathrm{Gpc})^3$. When multiplied by the desired solid angle and redshift range, this can be used to evaluate the volume of a shell centered at this redshift, for example. For more practice with low-redshift calculations, see Problem 3.2.

Evaluating the volume element is therefore very easy, given the always indispensable expressions for r and H. Note also that $V \propto H_0^{-3}$ (since $r \propto H_0^{-1}$ and $H(z) \propto H_0$), as required on dimensional grounds. See Box 3.1 and Problem 3.2 for more exploration on this topic.

3.4.3 Luminosity Distance

In cosmology, we do not measure the proper distance. The reason is that we cannot tell the universe "stop expanding, I need to get my tape measure and measure the distance to that galaxy." Rather, we measure distances inferred from the luminosities of distant objects, or else from the angular sizes of objects (or other features on the sky – for example, cold and hot spots in the cosmic microwave background).

The **luminosity distance** d_L is inferred by looking at flux f coming from an object that has some luminosity L:

$$f = \frac{L}{4\pi d_L^2}. \tag{3.65}$$

Let us now explore in more detail what flux is received from the object. First, the area of the sphere centered at our location with radius going out to our object is

$$A = 4\pi S_\kappa(\chi)^2 \equiv 4\pi r^2, \tag{3.66}$$

where χ is the radial and r the comoving distance to the object. This area is what needs to be in the denominator of the right-hand side of Eq. (3.65). Moreover, photon wavelengths that we receive from the object are stretched by $1 + z$ because the universe has expanded by that factor since the emission; hence, each photon's energy $E = hc/\lambda$ is lower by that factor on observation. But the photons are also received at a *rate* that is lower by $1 + z$ than that at emission. In other words, if the proper interval between the emission of two photons is δt_e, the interval at their observation is extended to $\delta t_o = \delta t_e(1 + z)$ simply because there is more physical separation between two photons, and you have to wait longer for the next one to arrive. Putting these two effects together, the observed luminosity of incoming photons is lower by $(1 + z)^2$ than that at emission:

$$L_o \equiv \frac{\Delta E_o}{\Delta t_o} = \frac{\Delta E_e/(1 + z)}{\Delta t_e(1 + z)} = \frac{L}{(1 + z)^2}, \tag{3.67}$$

where $L \equiv L_e = \Delta E_e/\Delta t_e$ is the original luminosity – the one at emission.

Putting things together, the observed flux is

$$f = \frac{L_o}{A} = \frac{L}{4\pi r^2(1 + z)^2} \tag{3.68}$$

and, therefore, comparing to Eq. (3.65):

$$d_L(z) = (1 + z)r(z). \tag{3.69}$$

The luminosity distance as a function of redshift is shown in Fig. 3.3, and you will have a chance to plot it in Problem 3.5.

3.4.4 Angular-Diameter Distance

One way to measure distance is through luminosities and fluxes, as explained just above. The *other* way to measure distance is through inference from the angular extent $\Delta\theta$ of an object with proper length Δl on the sky. This defines the **angular-diameter distance**

$$d_A = \frac{\Delta l}{\Delta\theta}. \tag{3.70}$$

Let us compute the proper length of the object. From the (spatial part of the) metric,

$$ds^2 = a^2(t)\left[\frac{dr^2}{1-\kappa r^2} + r^2\left(d\theta^2 + \sin(\theta)^2 d\phi^2\right)\right], \tag{3.71}$$

we get, for the arc length

$$ds = a(t)\, r\, d\theta, \tag{3.72}$$

or

$$\Delta l = a(t) r\, \Delta\theta, \tag{3.73}$$

so that the angular-diameter distance is $\Delta l/\Delta\theta = ar = r/(1+z)$, or

$$d_A(z) = \frac{1}{1+z}\, r(z). \tag{3.74}$$

Note a fundamental relation in FLRW cosmology:

$$d_L(z) = (1+z)^2 d_A(z). \tag{3.75}$$

In principle, we could measure the two distances, $d_L(z)$ and $d_A(z)$, at any given redshift, and test whether this relation holds. Its breakdown would imply violation of the assumption of isotropic, homogeneous, FLRW expansion of the universe.

The two relevant distances, luminosity and angular-diameter distance, are shown in Fig. 3.3 for our fiducial ΛCDM model. We also show the comoving distance $r(z)$. The three distances simply differ by successive factors of $1+z$. Note that the angular-diameter distance does not indefinitely increase with redshift, but in fact peaks at $z \simeq 1.60$ (in the fiducial model that we assumed). This is, very loosely, often explained by the intuition that test objects in the universe – say, galaxies – were physically closer in the past.

We end this chapter with an important message about the nature of observations in astrophysical cosmology:

> Fundamentally, we observe only two things: angles on the sky, and fluxes (of photons from objects). The distance measures corresponding to these two observables are the angular-diameter and luminosity distance.

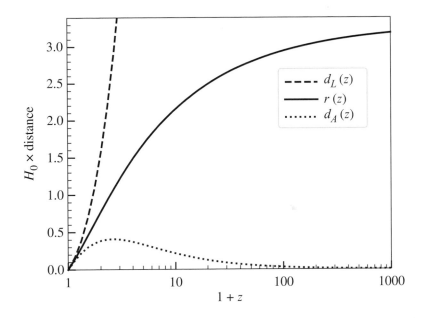

Fig. 3.3
Luminosity distance, comoving distance, and angular-diameter distance as a function of $1+z$, calculated in our fiducial cosmological model and shown out to $1+z = 1000$.

Bibliographical Notes

We again single out Ryden (2016) as the book that informed and influenced the presentation in this chapter. The material is also covered in a number of other textbooks. The general-relativistic angle is particularly clearly laid out in Carroll (2019).

Problems

3.1 Deriving the Friedmann II equation. Use the (flat-universe) version of the first Friedmann equation, Eq. (2.46), along with the continuity equation, Eq. (3.6), to derive the second Friedmann equation, Eq. (3.7). *Hint:* Start by taking the time derivative of Friedmann I.

3.2 Distance and volume calculations without a calculator. In Box 3.1 we explained how calculations at low redshift simplify dramatically. With that in mind, approximately evaluate without using any electronic gadgets:

(a) The comoving distance to $z = 0.1$, expressed in $h\,\mathrm{Mpc}^{-1}$.

(b) The comoving volume out to $z = 0.1$ (assuming the full solid angle of 4π steradians).

(c) The redshift at which a typical galaxy peculiar velocity, $v_{\mathrm{pec}} \simeq 300\,\mathrm{km/s}$, is 10 percent of its expansion velocity.

3.3 Correction to low-redshift distance. In Box 3.1 and Problem 3.2 we explored the low-redshift ($z \ll 1$) approximation for the distance, $H_0 r(z) \simeq z$. Find the next-order correction to this, that is, identify the constant C in the expression

$$H_0 r(z) \simeq z + C z^2 + O(z^3).$$

Assume a flat universe with dark matter and dark energy represented by the cosmological constant, so that $\Omega_M + \Omega_\Lambda = 1$.

3.4 Age and expansion for single component with $w = -1$. Investigate what happens in the case of a single-component universe with fluid with an equation of state $w = -1$ (see Sec. 3.2.4).

(a) Write an expression for the scale factor $a(t)$.

(b) What is the age of the universe in this scenario?

The result in part (a) will foreshadow what happens during inflation, discussed in Chapter 8.

3.5 [Computational] Distance–redshift relation in different cosmologies. In this problem and the next, you are asked to make plots of functions for different cosmological models. For each plot, you should produce a separate curve for each of the following FLRW cosmological models.

(a) Plot the luminosity distance–redshift relation in the following cosmologies – note all of them assume that the universe is flat.

- Einstein–de Sitter (so flat with matter only; $\Omega_M = 1$).
- Entirely cosmological constant-dominated universe (so flat with $\Omega_\Lambda = 1$).
- Currently favored ΛCDM model – the fiducial model from Table 3.1 (so flat with $\Omega_M = 0.3$ and $\Omega_\Lambda = 0.7$).
- Flat radiation-dominated universe ($\Omega_R = 1$; note, we definitely don't live in this one!).

On the x-axis you should have redshift z, going from 0 to 6. On the y-axis you should have the luminosity distance in units of H_0^{-1}, that is, $H_0 d_L(z)$ which is dimensionless. If you can, make the y-axis go from 0 to about 18.

(b) Make another distance–redshift plot as above, except with angular-diameter distances rather than luminosity distances.

3.6 [**Computational**] **Age in different cosmologies.** Get a better understanding of how the age of the universe depends on the cosmological model, as follows.

(a) Plot the age vs. scale factor relation, $t(a)$, in the following cosmologies (as in the previous problem, assume a flat universe throughout).

- Einstein–de Sitter (so flat with matter only; $\Omega_M = 1$).
- Our fiducial ΛCDM model, with $\Omega_M = 0.3$ and $\Omega_{\rm DE} = 0.7$.
- Fiducial model with $\Omega_M = 0.3$ and $\Omega_{\rm DE} = 0.7$, but with dark energy with the equation of state $w = -1.2$.
- Fiducial model with $\Omega_M = 0.3$ and $\Omega_{\rm DE} = 0.7$, but with dark energy with the equation of state $w = -0.8$.

On the y-axis, you should have age in units of giga-years (Gyr). The x-axis should have the scale factor a ranging from zero (Big Bang) to one (today).

(b) At what scale factor is the universe half of its present age in the fiducial cosmological model?

(c) What is the problem with the (present-day) age in the Einstein–de Sitter ($\Omega_M = 1$) model?

(d) Just for fun, could you precisely mimic the age in the Einstein–de Sitter case (i.e., age $t(a)$) with a dark-energy model with $\Omega_M = 1 - \Omega_{\rm DE} = 0.3$? If so, which other cosmological parameter would you need to change, and to which value? Explain (no need to plot anything).

(e) From your numerical results (or plots), what is the difference in the present-day age between the dark-energy universes with $w = -1.2$ and $w = -0.8$?

3.7 **Time-varying equation of state.** In 1998, cosmologists discovered that the universe is accelerating, and is therefore apparently filled with a new component dubbed dark energy. Dark energy has a strongly negative equation of state, which corresponds to $\ddot{a}/a > 0$ in the acceleration equation. A lot of effort is dedicated to measuring the equation of state from cosmological data. Since there is in principle no reason why w shouldn't change with time, cosmologists have decided that measuring the *temporal variation* of the equation of state is very important to look out for new physics.

(a) Writing $w(z) = w_0 + w'z$, we are describing the time evolution of w with two parameters, w_0 and w'. Derive the energy density of dark energy, $\rho_{\rm DE}(z)$, in terms of w_0, w', and z. [Recall that the density today, at $z = 0$, should be $\rho_{\rm DE,0}$, which should also feature in your equation.]

(b) Repeat the same exercise from part (a) for an alternative form: $w(a) = w_0 + w_a(1-a)$, where a is the scale factor (i.e., derive $\rho_{\rm DE}(a)$ or, if you prefer, make that $\rho_{\rm DE}(z)$).

(c) Which of the two forms, (a) or (b), can better describe a realistic physical component in our universe? Explain in a sentence. *Hint:* This is trivial; we are hoping to describe dark energy in the early universe as well, where $w(z)$ of DE presumably takes "reasonable" values comparable to those of other components we discussed in the book.

3.8 Particle horizon and event horizon. Calculate both the particle horizon and the event horizon in the following two spatially flat cosmological models:

- Einstein–de Sitter model ($\Omega_M = 1$);

- de Sitter model ($\Omega_\Lambda = 1$). [Be careful, as this model does not have a Big Bang in the past.]

3.9 Distant quasars. [Adapted from Guth (2013).] Quasars, or quasi-stellar objects (QSOs), are extremely luminous events that are caused by million- to billion-solar-mass black holes. These supermassive black holes swallow surrounding gas in a process that accelerates the gas particles and releases prodigious amounts of electromagnetic radiation. Because of their great luminosity ($L \sim 10^{40}$W $\simeq 10^{14} L_\odot$) and emission that falls across the electromagnetic spectrum, quasars can be observed out to cosmological distances. The first quasar ever identified, 3C 273, is at $z = 0.158$. In 1963, its discovery sent shock waves through the astronomy community due to the fact that "quasi-stellar objects" can be see out to such distances. A recent distance record-holder is quasar ULAS J1342+0928, at $z = 7.54$. It is simply amazing that a single black hole can produce radiation that can be seen tens of billions of light-years away!

In this problem, for simplicity, assume a purely matter-dominated (EdS) universe with

$$a \propto t^{2/3}.$$

(a) Let t_e be the time of emission from the quasar, while t_0 is the time today. Calculate the physical distance to the quasar, d_p, in terms of these two times.

(b) Find the redshift of the quasar z in terms of t_e and t_0.

(c) Rewrite the proper distance d_p in terms of the Hubble constant H_0 and the quasar's redshift z. [Note that this will basically recover the familiar EdS expression for distance vs. redshift.]

(d) Calculate the rate at which the proper distance to the quasar at $z = 7.54$ is changing with time, $d(d_p)/dt_0$. Is the result that you get concerning?

3.10 Greatest blunder. In 1917 Einstein realized that his equations predict an expanding universe, something that was thought to be in conflict with known properties of the universe at the time. He proposed adding a cosmological-constant term to his equations. This term was declared unnecessary once

Hubble discovered the expansion of the universe. To make things more painful for Einstein, it was also pointed out that the new term leads to an unstable equilibrium – that is, an unstable static solution, where even the smallest perturbation will lead to either an expanding or a contracting universe. You will now explore this latter fact, which is the principal reason for this "greatest blunder" (in Einstein's own words).

Assume a curved matter+Lambda cosmological model which is not necessarily expanding. The cosmological constant contributes the amount of $\Lambda/3$ on the right-hand side of the Friedmann I equation:

$$\left(\frac{\dot{a}}{a}\right)^2 = \frac{8\pi G}{3}\rho_M + \frac{\Lambda}{3} - \frac{k}{a^2},$$

where ρ_M is the energy density in matter and k is curvature.

(a) Remembering that the cosmological-constant term has $P_\Lambda = -\rho_\Lambda$, write down the second Friedmann equation for this cosmological model.

(b) Is it possible to find a value of Λ so that both $\dot{a} = 0$ and $\ddot{a} = 0$? If so, express Λ in this scenario in terms of the matter density ρ_M and any other parameters.

(c) Now perturb the scale factor

$$a(t) \simeq 1 + \delta a \qquad (\delta a \ll 1).$$

Therefore, $a - 1$ may well be constant on average, but some regions of the universe will have a slightly lower or higher a perturbed around this value. Because $\rho_M \propto a^{-3}$, this corresponds to the perturbation in matter energy density of

$$\rho_M(t) \simeq \rho_{M,0}(1 - 3\delta a)$$

(to lowest order in Taylor expansion). Now perturb the Friedmann II equation as well, and find the solution of the form

$$\frac{d^2\delta a}{dt^2} = \text{function of } \Lambda \text{ and } \delta a.$$

(d) Then solve this equation for $\delta a(t)$. If it helps, you may assume some reasonable initial conditions, such as $\delta a(t_0) = (\delta a)_0$, $d(\delta a)/dt_{t_0} = 0$. Does this lead to a situation where the universe evolves away from the static solution? Explain what happens for $\delta a > 0$ and for $\delta a < 0$.

PART II

THE EARLY UNIVERSE

4 Early-Universe Thermodynamics

We now turn to the evolution of particle species throughout the history of the universe. After the Big Bang, the key event in the early universe was cosmological inflation (Table 4.1), a topic that we will cover in detail in Chapter 8. Inflation is the ultimate "sweeper" since it dilutes the abundance of any particles present before inflation. It then generates the particles and fields of the Standard Model through the process called reheating. The processes that we cover in this chapter all happen *after* inflation.

Reheating after inflation therefore leaves behind particles of the Standard Model of particle physics: quarks and leptons; the Higgs boson; and the associated carriers of the fundamental forces – gluons (for the strong force), W and Z bosons and the neutrinos (weak force), and photons (electromagnetic force). The question that we will consider is what happens to the abundance of these species as the universe expands and cools.

In studying the thermodynamic processes in the early universe, we shall rely heavily on the basic results from statistical mechanics, particularly properties of integer and half-integer spin particles, bosons, and fermions. The cosmology-specific input will be furnished by the Hubble parameter H evaluated at the desired time; its competition with the rate of interactions determines the efficiency of those interactions for any particle process. To begin, however, we first have to consider the microscopic variables in the thermal bath, and count the number of states in the phase space.

4.1 Thermodynamics Basics

We now review aspects of statistical mechanics that will be building blocks for our understanding of thermodynamics in the early universe.

4.1.1 From Phase Space to Density

From statistical mechanics, you may recall that the density of states in phase space is $d^6N/(d^3p\,d^3x)$, where $\mathbf{x}$ and $\mathbf{p}$ are the position and momentum. Each state is assigned the Planck constant h cubed, that is, $1/h^3$. Then the phase-space density of states is

$$\frac{d^6 N_{\text{states}}}{d^3p\,d^3x} = \frac{g}{(2\pi)^3},\tag{4.1}$$

where we still work in natural units with $\hbar \equiv h/(2\pi) = 1$. Here the g is a multiplicity factor that counts the level of redundancy of the internal degrees of freedom (e.g., spin, color), assuming they are populated equally. For example, a particle with two spin states and three colors has $g = 2 \times 3 = 6$.

We next consider the **phase-space distribution function** $f(\mathbf{x}, \mathbf{p}, t)$, which defines how the states in our system are distributed. In a homogeneous universe, the distribution function does not depend on position $\mathbf{x}$ at all; in the isotropic universe, it depends only on the magnitude of momentum $\mathbf{p}$. Adopting these assumptions, and further keeping the time dependence implicit, we can write down the density of particles in phase space as

$$\frac{d^6 N}{d^3 p\, d^3 x} = \frac{g}{(2\pi)^3} \times f(p, t), \tag{4.2}$$

where the variable N refers to the number of particles. Then the number density of particles is (suppressing the time label from here on)

$$n = \frac{g}{(2\pi)^3} \int f(p)\, d^3 p. \tag{4.3}$$

The energy density ρ is equal to the number density weighted with energy E,

$$\rho = \frac{g}{(2\pi)^3} \int f(p) E(p)\, d^3 p, \tag{4.4}$$

where the relation between the energy and momentum of particles is

$$E(p) = \sqrt{p^2 + m^2}. \tag{4.5}$$

It is also useful to calculate the pressure of the system at hand. The pressure (denoted with a capital P to distinguish it from momentum p) is given by

$$P = \frac{g}{(2\pi)^3} \int f(p) \frac{p^2}{3E(p)}\, d^3 p, \tag{4.6}$$

where the $p^2/(3E)$ term can be derived from a simple consideration of particles in a box, as we do in Box (no pun intended) 4.1.

Table 4.1 Key events in the early history of the universe, along with respective times, redshifts, and temperatures. The correspondence between these quantities has been computed using our fiducial model from Table 3.1 and slightly rounded in most cases.

Event	Time	Redshift	Temp (eV)	Temp (K)
Big Bang	0	∞	∞	∞
Inflation	$10^{-35}\,$s ??	?	?	?
Baryogenesis	?	?	$\sim 100\,$GeV ??	?
QCD phase transition	10 µs	1.5×10^{12}	150 MeV	10^{12}
Dark-matter freezeout	?	?	$\sim 100\,$MeV ??	?
Neutrino decoupling	0.7 s	6×10^9	1 MeV	10^{10}
Electron–positron annihilation	2 s	3×10^9	0.5 MeV	6×10^9
Big Bang nucleosynthesis	2 min	5×10^8	0.1 MeV	10^9
Matter–radiation equality	50,000 yr	3500	0.8 eV	9000
Recombination	290,000 yr	1275	0.3 eV	3500
Photon decoupling	380,000 yr	1090	0.25 eV	3000
Dark energy dominates	10 Gyr	0.3	$3.1 \times 10^{-4}\,$eV	3.6
Present day	13.8 Gyr	0	$2.35 \times 10^{-4}\,$eV	2.725

4.1.2 Kinetic Equilibrium

In **kinetic equilibrium**, particles undergo frequent interactions and exchange energy and momentum efficiently. In that case, the distribution function for each particle species takes the famous form

$$f(p) = \frac{1}{e^{(E-\mu)/T} \pm 1}, \tag{4.7}$$

where the plus sign is for fermions, and minus for bosons, and μ is chemical potential that we shall discuss just below. This takes a pure-exponent form when temperature falls well below $E - \mu$. Moreover, note that each species of particles has its own

Box 4.1   **Pressure of Particles in Thermodynamic Equilibrium**

Consider particles close to a wall of area A. The density of particles per unit momentum-space element (per d^3p) is $dn = g/(2\pi)^3 \times f(p)$, and the differential number is that times the volume containing the particles that will collide with the wall in time dt: $dV = A\,|\mathbf{v} \cdot \hat{\mathbf{n}}|\,dt$, where $\mathbf{v}$ is the particles' velocity and $\hat{\mathbf{n}}$ is the normal unit vector in the wall direction. The number of particles that collide with the wall in time dt is then

$$dN = dndV = \frac{g}{(2\pi)^3} f(p) A |\mathbf{v} \cdot \hat{\mathbf{n}}|\, dt. \tag{B1}$$

These particles give their momentum to the wall, $dp = dN|\mathbf{p} \cdot \hat{\mathbf{n}}| = g/(2\pi)^3 f(p) A\, (|\mathbf{p} \cdot \hat{\mathbf{n}}|^2 / E)dt$, where we used $\mathbf{v} = \mathbf{p}/m \simeq \mathbf{p}/E$ for non-relativistic particles. The pressure is given by force divided by area:

$$P = \frac{1}{A}\frac{dp}{dt} = \frac{g}{(2\pi)^3} f(p) \frac{|\mathbf{p} \cdot \hat{\mathbf{n}}|^2}{E}. \tag{B2}$$

Now we have to average this over all directions of $\mathbf{p}$. This can be done formally (average over $\cos^2(\theta)\sin(\theta)$, which is $1/3$, or else using an elegant symmetry argument: imagine $\hat{\mathbf{n}}$ is, say, in the x-direction; then $|\mathbf{p} \cdot \hat{\mathbf{n}}|^2 = p_x^2 = p^2/3$. Integrating further over all momentum states gives the desired result:

$$P = \frac{g}{(2\pi)^3} \int f(p) \frac{p^2}{3E} d^3p. \tag{B3}$$

mass m_i, chemical potential μ_i, and perhaps even temperature T_i, and thus also its own distribution function $f_i(p)$, as well as the quantities n_i, ρ_i, and P_i.

We now deconstruct the concept of the chemical potential.

4.1.3 Chemical Potential

Given that **chemical potential** is an often-ignored quantity in standard courses on statistical mechanics, it is worth outlining its basic *raison d'être*. We start from the (Helmholtz) free energy of the system E; its variation is

$$dE = TdS - PdV + \mu dN, \tag{4.8}$$

where T and S are the temperature and entropy, P and V are the pressure and volume, and N is the number of particles in the system. The chemical potential is thus defined as

$$\mu = \left(\frac{\partial E}{\partial N}\right)_{S,V}. \tag{4.9}$$

That is, chemical potential is the energy that the system gains when one particle is added while the total entropy and volume of the system are held fixed.

In cosmology, it is useful to think of chemical potential as the quantity that enforces a specific conservation law – say, particle number conservation, or else charge conservation. At a given temperature T, we can calculate the chemical potential $\mu(T)$ of each species so that the species' number density, or charge, or another quantity that needs to be conserved in a reaction is indeed conserved. As we vary T, chemical potential adjusts so that the conservation law continues to hold. For example, consider the ionized plasma in the universe at $T \sim 10\,\mathrm{keV}$ (so well before recombination at $T \sim 1\,\mathrm{eV}$ when electrons and protons combine into hydrogen), which consists of electrons, protons, and photons. Because this period is at $T \ll m_e = 511\,\mathrm{keV} \ll m_p = 938\,\mathrm{MeV}$, both electrons and protons are non-relativistic. In Problem 4.4, you can work out what the net charge of the plasma would be with zero chemical potentials, and how the chemical potentials adjust so as to preserve charge conservation and thus a charge-neutral plasma.

Moreover, the chemical potential of photons is zero. This can be seen from the fact that the (electron–proton) Brehmstrahlung scattering

$$e^- + p \rightleftharpoons e^- + p + \gamma, \tag{4.10}$$

or double Compton scattering

$$e^- + \gamma \rightleftharpoons e^- + 2\gamma, \tag{4.11}$$

imply that the number of photons is not conserved, that is, adding a photon doesn't increase the energy of the system. Hence $\mu_\gamma = 0$.

An important special case is the relation between the chemical potential of a particle and that of its antiparticle. From annihilation reactions for any particle X and its antiparticle $\bar{X}$, such as

$$X + \bar{X} \longrightarrow 2\gamma, \tag{4.12}$$

the fact that $\mu_\gamma = 0$ implies that the chemical potential of particles is minus that of corresponding antiparticles:

$$\mu_X = -\mu_{\bar{X}}. \tag{4.13}$$

This result will simplify some equations in what follows. In fact, we will be able to ignore the chemical potentials of species in most of our discussions that follow.

4.1.4 Interaction Rate

Imagine a bath of particles X interacting with their antiparticles $\bar{X}$,

$$X + \bar{X} \rightleftharpoons 2\gamma. \tag{4.14}$$

The relation moving to the right is particle–antiparticle annihilation, while the reaction moving to the left is pair creation. Under what conditions will the annihilation reaction proceed?

It turns out that there are two timescales involved: one for particles to "find each other," and one for them to "escape from each other." The former timescale is given by the inverse of the annihilation rate Γ, where

$$\Gamma = n\sigma v, \tag{4.15}$$

with n the number density of particles, σ their cross-section, and v their velocity. Clearly, Γ has units of inverse time. Meanwhile, the timescale for escape is given by the Hubble parameter H, whose (natural) units are also inverse time. It then makes sense to guess that the X and $\bar{X}$ annihilations are efficient only when they occur rapidly relative to the local age of the universe, $t_{\text{age}} \simeq H^{-1}$; that is, when $t_{\text{rate}} \ll t_{\text{age}}$, or $\Gamma \gg H$. In other words,

> Particle–antiparticle annihilations can only occur when $\Gamma \gg H$.

This conjecture can be proven essentially by inspection of the Boltzmann equation for this annihilation process, as we will see in Chapter 6.

4.2 Thermal Equilibrium

We define **thermal equilibrium** as both chemical and kinetic equilibrium, as well as the $\Gamma \ll H$ condition for the relevant species. In thermal equilibrium, all species share the same temperature, so that $T_i = T$ and, moreover, all reactions are in chemical equilibrium. Thermal equilibrium makes the statistical mechanics of the system easy to analyze, since we can fall back on the familiar Bose and Fermi–Dirac distributions, to be discussed just below. Conversely, a lack of thermal equilibrium implies that, at least in principle, we have to go back to the Boltzmann equation to solve for the properties of particle species as a function of time. Finally, we can ignore the chemical potential of species in the early universe, which is a good approximation that also simplifies our calculations. You will prove this for the case of electrons in Problem 4.5.

Setting the chemical potentials to zero and adopting the kinetic-equilibrium distribution functions, the general formulae for the number density and energy density of any species with mass m, evaluated when the thermal bath has temperature T, are

$$
n = \frac{g}{(2\pi)^3} \int_0^\infty \frac{1}{\exp(E/T) \pm 1}\, d^3p = \frac{g}{2\pi^2} \int_0^\infty \frac{p^2}{\exp(\sqrt{p^2 + m^2}/T) \pm 1}\, dp
$$

$$
\rho = \frac{g}{(2\pi)^3} \int_0^\infty \frac{E}{\exp(E/T) \pm 1}\, d^3p = \frac{g}{2\pi^2} \int_0^\infty \frac{p^2\sqrt{p^2 + m^2}}{\exp(\sqrt{p^2 + m^2}/T) \pm 1}\, dp,
$$

(4.16)

where $E = \sqrt{p^2 + m^2}$ is the energy of particles and p is their momentum. One can immediately make additional progress by going to dimensionless variables, defining

$$
y \equiv \frac{p}{T}; \qquad x \equiv \frac{m}{T}
$$

(4.17)

and obtaining

$$n = \frac{gT^3}{2\pi^2} \int_0^\infty \frac{y^2}{\exp(\sqrt{y^2 + x^2}) \pm 1} \, dy$$

$$\rho = \frac{gT^4}{2\pi^2} \int_0^\infty \frac{y^2 \sqrt{y^2 + x^2}}{\exp(\sqrt{y^2 + x^2}) \pm 1} \, dy. \tag{4.18}$$

To convert these expressions to MKS units, we would multiply each power of T by $(\hbar c)/k_B$.

We next perform two important limits of the above equations: the relativistic $(T \gg m)$ and non-relativistic $(T \ll m)$ case.

4.2.1 Relativistic Limit

In Problem 4.2, you can show that Eqs. (4.18) can be drastically simplified in the relativistic limit, $T \gg m$. The result is that the number density is directly proportional to cube of the temperature:

$$n = \frac{\zeta(3)}{\pi^2} g T^3 \times \begin{cases} 1 & \text{(bosons)} \\ \dfrac{3}{4} & \text{(fermions)}, \end{cases} \tag{4.19}$$

where $\zeta(3) \simeq 1.202$ is the zeta function of argument three, while the energy density scales as the fourth power of T:

$$\rho = \frac{\pi^2}{30} g T^4 \times \begin{cases} 1 & \text{(bosons)} \\ \dfrac{7}{8} & \text{(fermions)}. \end{cases} \tag{4.20}$$

Note the interesting feature of these limits – that the answers for fermions are the same as those for bosons, save for the overall factors of 3/4 and 7/8 for the number and mass density, respectively.

One immediate result that follows from these equations is the number density of photons today. Adopting the temperature of the cosmic microwave background radiation, $T_0 = 2.725 \, \text{K}$, and $g_\gamma = 2$ degrees of freedom (a photon is a spin-1 particle with two polarization states), we have

$$n_{\gamma,0} = \frac{\zeta(3)}{\pi^2} g_\gamma T_0^3 \simeq 411 \, \text{cm}^{-3}$$

$$\rho_{\gamma,0} = \frac{\pi^2}{30} g_\gamma T_0^4 \simeq 36.28 \, \text{K}^4 \simeq 4.64 \times 10^{-31} \, \frac{\text{kg}}{\text{m}^3} \implies \Omega_\gamma h^2 = 2.47 \times 10^{-5}, \tag{4.21}$$

where in the second line we converted to energy density relative to critical today

$\Omega_\gamma = \rho_{\gamma,0}/\rho_{\text{crit},0} = 8\pi G/(3H_0^2) \times \rho_{\gamma,0}$. Here we followed convention by reporting the combination $\Omega_\gamma h^2$ which (unlike Ω_γ) is H_0-independent.

Finally, it can straightforwardly be shown that, in the relativistic limit,

$$P = \frac{\rho}{3},\tag{4.22}$$

or, restoring the speed of light, $P = \rho c^2/3$. This confirms what we had been assuming about the equation of state of radiation in Sec. 3.2.2, where we introduced the fact that the equation of state of radiation is $1/3$.

4.2.2 Non-relativistic Limit

When $m \gg T$, the number and energy density can also be dramatically simplified. In addition to working in this limit, we also restore the chemical potential μ of the species in question because (as we discuss below) it may be important in commonly considered situations. The result is (see Problem 4.3)

$$n = g\left(\frac{mT}{2\pi}\right)^{3/2} e^{-(m-\mu)/T}\tag{4.23}$$

$$\rho = mn.\tag{4.24}$$

Therefore, the number density is exponentially suppressed relative to the relativistic case, and there is no more dependence on the boson/fermion property of the particles. The exponential suppression is due to the annihilation of particles with antiparticles. [In the relativistic case the annihilation also occurs, but is counteracted by the particle–antiparticle pair production.] Moreover, the pressure of non-relativistic particles is negligible,

$$\frac{P}{\rho} = \frac{nT}{nm} = \frac{T}{m} \ll 1,\tag{4.25}$$

confirming our arguments in Sec. 3.2.1.

Consider for a moment the case of $\mu = 0$, which holds in thermal equilibrium, at $T \gg m$. Comparing Eqs. (4.19) and (4.20) to Eqs. (4.23) and (4.24), we see that both the number density and the energy density dramatically drop when the temperature falls below the mass; they scale as $\exp(-m/T)$. This is because particle–antiparticle annihilations are no longer balanced by particle pair production. This is a major lesson of early-universe thermodynamics:

> When the temperature of the universe falls below the particle mass and the particle is still in thermal equilibrium, particle–antiparticle (pair) creation stops and the particle's abundance and energy density rapidly drop.

The key condition in the statement above is that the particle be in thermal equilibrium. We will learn in Chapter 6 that a particle can fall out of equilibrium at late times, when the rate of the reaction that keeps it in equilibrium falls below the Hubble rate. When that happens, the particle density "freezes out" and stops falling exponentially. This is where the chemical potential comes in! The particle's chemical potential μ becomes nonzero, and is in fact a function of time (or temperature) that takes such values to enforce the required out-of-equilibrium density of particle species.

From the discussion above, it seems a good idea to seek a parameterization that keeps track of all species which non-negligibly contribute to the relativistic energy density. This is achieved with the effective number of relativistic species.

4.2.3 Number of Relativistic Degrees of Freedom

Motivated by Eq. (4.20) for the energy density of a single relativistic species, we can write a general formula for the energy density that includes all species/particles of relevance:

$$\rho_R \equiv \sum_i \rho_i = \frac{\pi^2}{30} g_* T^4, \tag{4.26}$$

where the parameter g_*, called the **effective number of relativistic degrees of freedom**, is a weighted sum of the multiplicity factors of all particles. Specifically, this factor is defined as

$$g_*(T) = \sum_{i\in\text{bosons}} g_i \left(\frac{T_i}{T}\right)^4 + \frac{7}{8} \sum_{i\in\text{fermions}} g_i \left(\frac{T_i}{T}\right)^4, \tag{4.27}$$

where the factor of 7/8 multiplies the fermionic contribution (see Eq. (4.20)). Here we are allowing the possibility that the species have a different temperature T_i from the photon temperature T, hence the power-law factors in this definition of g_*.

We show the evolution of g_* with time in Fig. 4.1. At $T > 100$ GeV, all of the particles are relativistic and contribute to g_*. The Standard Model particles' g-factors are listed in Appendix C. Specifically, quarks contribute 72 to this sum: six quarks times two spin states each (spin $\pm 1/2$) times two particle/antiparticle states times three color states each. Their fermion cousins, leptons, contribute $g = 12$ (three leptons times two particle/antiparticles, times two for spin 1/2). The neutrinos corresponding to those leptons contribute another $g = 6$. All fermions therefore contribute $g = 90$. On the bosons' side, gluons contribute $g = 16$, gauge bosons (W and Z) a total of $g = 9$, photons $g = 2$, and the Higgs boson $g = 1$. The total boson contribution is then $g = 28$. Therefore

$$g_*(T > 100\,\text{GeV}) = 28 + \frac{7}{8} \cdot 90 = 106.75. \tag{4.28}$$

The chain of events then unfolds as follows (keep in mind that the processes are not instantaneous so all temperature marks are approximate):

- At $T \sim 100\,\mathrm{GeV}$, electroweak phase transition takes place. Here, all particles except the photons, gluons, neutrinos, and the Higgs acquire their mass by the mechanism of spontaneous symmetry breaking. No change in g_* occurs.

- At around the same temperature, top quarks annihilate with antitops. The relativistic degrees of freedom go down to $g_* = 106.75 - (7/8) \cdot 12 = 96.25$.

- At $T \sim 50\,\mathrm{GeV}$, the Higgs boson H, as well as the gauge bosons $(Z, W^{\pm})$, annihilate; then $g_* = 96.25 - (1 + 6 + 3) = 86.25$.

- Shortly thereafter, at $T \sim 1\,\mathrm{GeV}$, the bottom and charm quarks and the tau leptons annihilate, leaving $g_* = 86.25 - (7/8) \cdot (2 \cdot 12 + 4) = 61.75$.

- At $T \simeq 150 - 200\,\mathrm{MeV}$, a very important event for both particle physics and cosmology takes place, the so-called QCD phase transition (QCD stands for quantum chromodynamics, the theory of strong interactions). Here, quarks essentially get bound into baryons (qqq) and mesons (qq) and no longer exist as free particles. Gluons follow where quarks go. The relativistic degrees of freedom take a big hit: $g_* = 61.75 - 16 - (7/8) \cdot (3 \cdot 12) = 14.25$. However, things are a bit more complicated than that: first, a few mesons that come to life in the QCD mass transition have masses low enough that they contribute *positively* to g_*; these are primarily the pi mesons or pions ($\pi^{\pm}$, $m = 140\,\mathrm{MeV}$ and $\pi^0, m = 135\,\mathrm{MeV}$) and, to a smaller extent, the next-lightest mesons, the kaons ($K^{\pm}$ and K^0, $m \sim 495\,\mathrm{MeV}$ for all three). Taking these mesons into account leads to a slower ramp-down observed at these temperatures in Fig. 4.1. Gradually, the pions and the kaons stop contributing, and at $T \sim 100\,\mathrm{MeV}$ the μ leptons also drop out, leaving $g_* = 14.25 - (7/8) \cdot 4 = 10.75$.

- At $T \sim 0.5\,\mathrm{MeV}$, electrons annihilate with positrons, removing a further $(7/8) \cdot 4 = 3.5$ to leave $g_* = 7.25$ degrees of freedom.

- However, just prior to electron–positron annihilation, neutrinos *decouple*, complicating slightly our sequence of events.

A bit further below, we will explain the concept and discuss the process of decoupling, and what it does to the relativistic degrees of freedom count g_*. We also need to discuss why there are two lines at the lowest temperatures shown in Fig. 4.1. In order to do all that, we first discuss entropy.

Before we do that, however, let us note an interesting fact: all particles heavier than the proton and the neutron decay over the evolution of the universe. They either decay into the lighter leptons and photons, or else (in the case of lighter quarks) make up the protons and neutrons around us today. Therefore, to the best of our knowledge, there are no massive relics left over from the early universe that are around us today. [A few of them may be produced by collisions at the Large Hadron Collider provided their mass is less than about 1 TeV.]

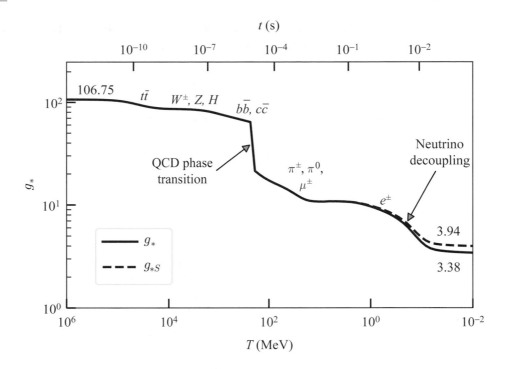

Fig. 4.1 Evolution of $g_*(T)$ with temperature, as particle species drop out due to becoming non-relativistic. The dashed line denotes $g_{*S}(T)$, the entropy-based number of effective degrees of freedom.

4.3 Entropy in the Universe

Entropy is a measure of disorder, and is defined as the logarithm of the number of states of a system. In the early-universe cosmology, entropy is a very useful concept. We start from the second law of thermodynamics,

$$TdS = dU + PdV, \tag{4.29}$$

where S is entropy. The second law states that the increase in entropy is due to the sum of the increase in internal energy of the system ($dU = d(\rho V)$) and work done on the system (PdV). Using basic mathematical operations, the entropy variation can be expressed as

$$dS = \frac{1}{T}\left[d[(\rho + P)V] - VdP\right]$$

$$= \frac{1}{T}d[(\rho + P)V] - \frac{V}{T^2}(\rho + P)dT \tag{4.30}$$

$$= d\left(\frac{\rho + P}{T}V\right),$$

where in the second line we used $\partial P/\partial T = (\rho + P)/T$ (which can be proven separately). In Problem 4.1, you are asked to prove that the right-hand side of Eq. (4.30) is constant, that is, $dS/dt = 0$.

It is useful to define the entropy *density*

$$s \equiv \frac{S}{V} = \frac{\rho + P}{T}. \tag{4.31}$$

Using Eq. (4.26) and the fact that $P = \rho/3$ for relativistic species, one can express the total entropy density as

$$s = \sum_i \frac{\rho_i + P_i}{T_i} = \frac{2\pi^2}{45} g_{*S} T^3, \tag{4.32}$$

where the parameter g_{*S}, defined as

$$g_{*S}(T) = \sum_{i \in \text{bosons}} g_i \left(\frac{T_i}{T}\right)^3 + \frac{7}{8} \sum_{i \in \text{fermions}} g_i \left(\frac{T_i}{T}\right)^3, \tag{4.33}$$

is subtly different from g_* defined in Eq. (4.27). In particular, for species in thermal equilibrium with the photons (and each other), $T_i = T$ and so $g_* = g_{*S}$, but if $T_i \neq T$ for any species, then the two gs are different. In our universe, $g_* = g_{*S}$ up until about 1 s after the Big Bang ($T \simeq 1\,\text{MeV}$), when the neutrinos decouple; see Fig. 4.1. We will discuss neutrino decoupling in Sec. 5.3.

From the facts that entropy S is conserved and entropy density is $s = S/V$, we see that $s \propto a^{-3}$. It is useful to define the **number-density-to-entropy ratio**

$$Y_i \equiv \frac{n_i}{s}. \tag{4.34}$$

Given that the entropy density scales as a^{-3}, Y_i is proportional to $n_i a^3$, so this is a good quantity to define as the number of some species in a given (comoving) volume. In general, if particles are neither produced nor destroyed, then $n_i \propto a^{-3}$ and Y_i is constant.

Given $S = sV = \text{const.}$ and using Eq. (4.32), we obtain a very general relation between temperature and scale factor:

$$g_{*S}(T)\,T^3 a^3 = \text{const.} \quad \Longrightarrow \quad T \propto g_{*S}^{-1/3} a^{-1}. \tag{4.35}$$

Therefore, the temperature in general scales as inverse scale factor, *except* around the times when a particle species becomes decoupled from the thermal bath. At such times, g_{*S} (rather abruptly) decreases, and the temperature decreases less rapidly over a period of time. We give an example a bit further below, and illustrate it in Fig. 5.3 in Chapter 5. But we first emphasize the takeaway from our hard work here:

The quantity $g_{*S}(T)\,T^3 a^3$ is conserved throughout the expansion history of the universe.

Finally, it is useful to simply have a relation between the Hubble parameter and temperature in the radiation-dominated epoch (in the early universe). From the Friedmann equation and $T \propto g_{*S}^{-1/3} a^{-1}$, we get

$$H \simeq \left(\frac{8\pi \rho_R}{3 m_{\rm Pl}^2} \right)^{1/2} = 1.66 g_*^{1/2} \frac{T^2}{m_{\rm Pl}} \quad \text{(in the radiation-dominated era)}, \quad (4.36)$$

while the temperature scales with time according to

$$\frac{t}{1\,{\rm s}} \simeq 2.42\, g_*^{-1/2} \left(\frac{T}{\rm MeV} \right)^{-2}, \quad (4.37)$$

obtained from the fact that $a \propto t^{1/2}$ in the radiation-dominated era, so that $H = 1/(2t)$. Equations (4.36) and (4.37) are both quite useful in converting from temperature/energy to cosmic time and vice versa.

Bibliographical Notes

The granddaddy of all presentations of thermal history of the universe is Kolb and Turner (1994), which covers it in depth and at a fairly technical level. Both it and the superb lecture notes by Baumann (2013; his Chapter 3) inspired the presentation in this chapter. Inspired and clear discussion of the subject is also to be found in Dodelson and Schmidt (2020).

Problems

4.1 Fill in the blanks. Evaluate the temporal variation of total entropy in volume V which is given in Eq. (4.30),

$$\frac{dS}{dt} = \frac{d}{dt} \left[\frac{\rho + P}{T} V \right].$$

You may find this equality helpful:

$$\frac{\partial P}{\partial T} = \frac{\rho + P}{T},$$

as well as the continuity equation. Remember also that the physical volume V scales as the cube of the scale factor a.

4.2 **Equilibrium distributions in relativistic limit.** Starting from the equilibrium expressions for the number and energy density (with zero chemical potentials),

$$n = \frac{g}{(2\pi)^3} \int f(p)\, d^3 p$$

$$\rho = \frac{g}{(2\pi)^3} \int f(p) E\, d^3 p,$$

where $f(p)$ is the Bose–Einstein or the Fermi–Dirac distribution function and $E = \sqrt{p^2 + m^2}$, integrate to get the expressions for n and ρ in terms of T alone for both the boson and the fermion cases *in the relativistic limit.* You may find the following integrals useful:

$$\int_0^\infty \frac{y^n}{e^y - 1}\, dy = \zeta(n+1)\Gamma(n+1)$$

$$\int_0^\infty y^n e^{-y^2}\, dy = \frac{1}{2}\Gamma\left(\frac{n+1}{2}\right),$$

where ζ is the Riemann zeta function and Γ is the Gamma function. You may also find it useful that

$$\frac{1}{z+1} = \frac{1}{z-1} - \frac{2}{z^2 - 1}.$$

4.3 **Equilibrium distributions in non-relativistic limit.** This is the same as the previous question, but for non-relativistic particles (and nonzero chemical potential).

Starting from the equilibrium expressions for the number and energy density

$$n = \frac{g}{(2\pi)^3} \int f(p)\, d^3 p$$

$$\rho = \frac{g}{(2\pi)^3} \int f(p) E\, d^3 p,$$

where $f(p)$ is the Bose–Einstein or the Fermi–Dirac distribution function and $E = \sqrt{p^2 + m^2}$, integrate to get the expressions for n and ρ in terms of T alone for both the boson and the fermion case *in the non-relativistic limit.* You may find the following relation useful:

$$\int y^2 e^{-ay^2}\, dy = -\frac{d}{da} \int e^{-ay^2}\, dy.$$

Do not ignore the chemical potential; thus, you want to start with

$$f(p) = \frac{1}{e^{(E-\mu)/T} \pm 1}.$$

4.4 Chemical potential of the electron–proton plasma. Consider an electron–proton–photon plasma at a temperature of $T \simeq 10\,\text{keV}$. At this temperature both the protons and the electrons are non-relativistic, since $m_p = 938\,\text{MeV}$ and $m_e = 511\,\text{keV}$. Find the relation between the chemical potentials of electrons and protons, $\mu_e(t)$ and $\mu_p(t)$, so that the plasma remains *neutral*. It is sufficient to express your answer as a function of temperature T.

4.5 Chemical potential of electrons and positrons. Get a better understanding of the chemical potential, as follows.

(a) Calculate the difference between the number densities of electrons and positrons

$$n_e - \bar{n}_e$$

in the relativistic limit ($m_e \ll T$). Express the result in terms of the electron chemical potential μ_e and T. *Hints:* (1) Recall that the electrons have chemical potential μ_e, while the positrons have $-\mu_e$. (2) You may find it useful that

$$\int_0^\infty \frac{y}{e^y + 1}\, dy = \frac{\pi^2}{12}.$$

(3) If you are having a hard time evaluating integrals exactly, feel free to start with the approximation $\mu_e/T \ll 1$. (4) The result is a sum of two power laws in μ_e/T.

(b) The electrical neutrality of the universe implies that the number of protons n_p (or nearly equivalently, the number of baryons n_B) is equal to $n_e - n_{\bar{e}}$. Use this to estimate μ_e/T. Which assumption in this chapter does your result justify? *Hint:* A significant shortcut is to express $n_B = \eta n_\gamma$, where η is the baryon-to-photon ratio and you can approximately take $\eta \simeq 10^{-9}$.

4.6 Graviton decoupling. Gravitons are massless particles which are the mediators of the gravitational force. They have not been detected yet, and they probably never will as they have mindbogglingly small interaction cross-sections (due basically to the fact that gravity is so weak compared to all fundamental forces). Here we will nevertheless show that one can readily work out a few basic properties of the graviton field.

   The graviton annihilation rate is given by a Feynman diagram with two gravitons going in, annihilating at a vertex, and two going out. The amplitude goes as Newton's gravitational constant G, and the cross-section therefore as G^2.

(a) If we write

$$\Gamma \simeq G^2 T^n,$$

based on purely dimensional arguments find the exponent n. *Hint:* Remember that $G = 1/m_{\text{Pl}}^2$ in natural units.

(b) At what temperature/mass scale, roughly, do the gravitons decouple? Remember that the Hubble parameter is given by Eq. (4.36).

(c) After gravitons decouple, other particles that decouple transfer their energy to photons (but not gravitons, since they are decoupled). With that knowledge, calculate the temperature of the relic graviton particles today. Assume just the Standard Model particles, that is, the $g_*(T)$ (and $g_{*S}(T)$) plot given in Fig. 4.1.

(d) Let us say that there exist supersymmetric particles at $E \simeq 1000\,\text{TeV}$ so that the total multiplicity factor of all particles, g_{*S}, is eight times bigger than in the Standard Model. What do you expect for the graviton temperature today?

(e) Calculate the relic number density of gravitons today, in units of cm^{-3}. Note that gravitons, like photons, are spin-1 bosons with two polarization states so that $g = 2$.

5 Neutrinos in Cosmology

In Chapter 4 we talked about the thermal history of the universe. This sets the stage to talk about neutrino decoupling and the cosmic background of neutrinos, two important case studies of the processes we covered in that chapter. Before we do that, however, we step back and properly introduce neutrinos. We cover the historical early evidence for these particles, then discuss neutrino oscillations and how cosmological observations can help learn more about neutrinos. With that useful background, we finally discuss neutrino decoupling and the cosmic neutrino background.

5.1 Neutrinos: A Brief History

Physicists knew about **beta decay** since the beginning of the twentieth century: an atom X emits an electron, leaving another atom Y with one extra proton:

$$^A_Z X \longrightarrow \, ^A_{Z+1} Y + e^-, \tag{5.1}$$

where Z is the atomic number (the number of protons) and A is the mass number (the number of protons plus neutrons). If atom X is at rest initially, then atom Y and the electron will move in opposite directions in the frame where X is at rest. Conservation of energy and momentum for this process imply that the energy of the outgoing electron is fixed, and given by (Problem 5.2)

$$E_e = \left(\frac{m_X^2 - m_Y^2 + m_e^2}{2m_X} \right). \tag{5.2}$$

In 1911, Lise Meitner and Otto Hahn performed experiments and observed a *continuous* spectrum of the outgoing electrons, finding that the energy E_e in Eq. (5.2) is actually only the *maximum* energy of the electrons, and that their actual observed energy varies between zero and that value; see Fig. 5.1. These experimental results were controversial for a while, but were definitively confirmed in the 1920s. Was energy conservation violated (as Niels Bohr suggested at the time)?!

No. In the famous 1930 letter to a conference that he couldn't attend, Wolfgang Pauli proposed the existence of a neutral, extremely feebly interacting new particle that today we call a neutrino (confusingly, Pauli at first called them neutrons; they were renamed neutrinos, or "little neutrons," by Fermi shortly thereafter, and the name neutron came to refer to massive particles that reside in the nucleus).

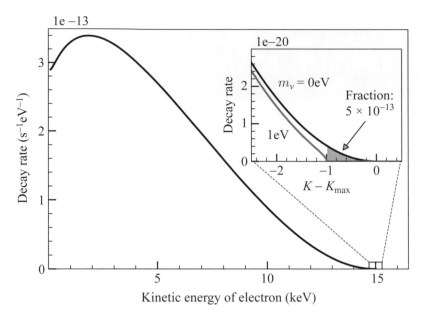

Fig. 5.1 Electron spectrum (i.e., distribution of events vs. their energy) in the decay of tritium, $^3\mathrm{H} \to {}^3\mathrm{He} + e^- + \bar{\nu}_e$ ($\bar{\nu}_e$ refers to an antineutrino of the electron flavor). The continuous distribution up to some fixed energy is due to the fact that the neutrino picks up the remaining energy. The inset shows the dependence of the tail of the distribution on the neutrino rest mass; higher neutrino mass means more energy taken up by the neutrino, and thus the maximum electron energy is reduced.

Neutrinos were eventually discovered in the 1950s by Cowan and Reines who looked at the inverse beta decay

$$\bar{\nu} + p^+ \longrightarrow n + e^+, \tag{5.3}$$

where ν generally refers to the neutrino, and the overbar indicates that an antineutrino is involved. The positron e^+ thus generated subsequently annihilates with an electron, creating two photons that can be detected. Cowan and Reines set up a 200-liter water tank near the Savannah River nuclear reactor in South Carolina. Even with the neutrino flux of 10^{13} neutrinos/cm^2/s which react with the protons in the water, Cowan and Reines were only expecting – and detected – 2–3 positrons/hour. This established the existence of neutrinos – the particles that Pauli postulated.

Electron neutrinos, together with electrons, form the so-called first (of the three) **generation** of leptons, light fundamental particles that also include electrons, muons, and tau particles. [Generations are also referred to as **flavors**, namely the electron, muon, and tau flavor. We discuss these just below.]

The first experimental evidence for the *second* family of neutrinos came in 1962. Lederman, Schwartz, and Steinberger observed the following reaction at Brookhaven

National Laboratory:

$$\bar{\nu} + p^+ \to \mu^+ + n, \tag{5.4}$$

but did not observe the reaction

$$\bar{\nu} + p^+ \to e^+ + n. \tag{5.5}$$

If there existed only one kind of neutrino, the second reaction would have been as common as the first one. Therefore, by 1962, two generations of neutrinos were discovered: the electron neutrino ν_e and the muon neutrino ν_μ, which respectively enter the above two reactions. [Muons themselves had already been discovered in 1936 by Anderson and Neddermeyer.]

Discovery of the third generation of neutrinos had to wait for several more decades. Its existence was certainly strongly suspected when, in the mid-1970s, a surprisingly massive tau lepton ($m_\tau = 1.77\,\text{GeV}$) was discovered by Martin Perl and collaborators. The corresponding tau neutrino was not discovered until the year 2000, by the DONUT experiment at Fermilab.

How do we know there are no more than three generations of leptons and, thus, three "flavors" (e, μ, and τ) of neutrinos? The principal evidence comes from the measurements of the Z-boson decays at the Large Electron–Positron collider (LEP at CERN, the machine which had to be decommissioned in 2000 to leave room for the Large Hadron Collider). The four experiments at LEP (see Fig. 5.2(a)) measured the Z cross-section as a function of beam energy – this is often called the "line width" or "line shape" of the particle. The line width Γ has units of energy[1] and is related to the lifetime of the particle by the Heisenberg uncertainty relation, $\Gamma\tau \simeq \hbar$, so the larger the width is, the shorter the lifetime is ("particle exists for a shorter period of time"). The Z boson can decay into various lepton–antilepton and quark–antiquark pairs, including the neutrinos of each generation G:

$$Z \to \nu_G + \bar{\nu}_G. \tag{5.6}$$

A large number of different light neutrino generations will consequently reduce Z's lifetime τ, thereby increasing its decay width Γ. The LEP experiments (in 1989) were thus able to determine that only three neutrinos exist with a mass *smaller than half* the Z mass.

This still leaves the possibility that neutrinos heavier than $m_Z/2 \simeq 45\,\text{GeV}$ exist. However, it would be very bizarre if neutrinos that heavy are around, given that the known neutrino species are more than a billion times lighter. There are also other, separate reasons why the existence of the fourth generation is very unlikely – though not impossible.

In any case, the Z-boson measurements provide the best indication to date for three **weakly interacting** neutrino flavors. There is a further possibility that there exists a fourth generation of neutrinos that do *not* participate in weak interactions,

---

[1] The shape of the line – the probability dP of finding a particle at energy E in the interval dE – typically has the Breit–Wigner profile, $\frac{dP}{dE} = \frac{\Gamma}{(E-E_0)^2 + \Gamma^2}$.

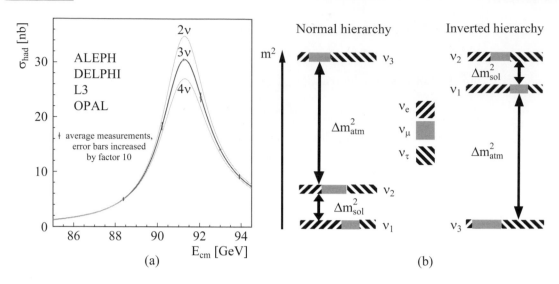

Fig. 5.2 (a) Spectrum of Z particles measured by four experiments at LEP in the late 1980s. The error bars are blown up by a factor of 10 for better visibility. The three curves show theory predictions for two, three, and four neutrino generations. A large number of different light neutrino generations will consequently reduce the Z lifetime, thereby increasing its decay width. The data are in excellent agreement with three neutrino generations. Reprinted from Schael *et al.* (2006), with permission from Elsevier.
(b) Two possible neutrino hierarchies; both explain the solar and atmospheric neutrino data where $\Delta m^2_{\text{solar}} \simeq 8 \times 10^{-5}\,\text{eV}^2$ and $\Delta m^2_{\text{atm}} \simeq 2 \times 10^{-3}\,\text{eV}^2$. The striped/shaded divisions of each mass eigenstate show the flavor-eigenstate (e, μ, or τ) contributions.

thereby evading the LEP Z-boson limit discussed above. These neutrinos are referred to as **sterile neutrinos** and are the subject of much contemporary research.

Finally, it is useful to connect how the LEP evidence for three neutrino flavors conceptually relates to the measurement of the number of relativistic species N_{eff} obtained from the CMB power spectrum measurements; see Eq. (5.26) near the end of this chapter. The quantity N_{eff} measures the abundance of *all* relativistic species at $z \sim 1000$, not just neutrinos. Therefore, the current bound ($N_{\text{eff}} \simeq 3 \pm 0.3$; see Eq. (5.28)) points to no evidence for additional sterile neutrinos or other relativistic particles.

5.2 Neutrino Oscillations

It turns out that a single neutrino can oscillate from one flavor to another as it travels through space. This is a purely quantum-mechanical effect, and showcases the awesome ability of quantum mechanics to produce new non-intuitive physical phenomena.

5.2.1 Theory Behind Neutrino Oscillations

The existence of oscillations depends on whether the quantum-mechanical wavefunction of the neutrino flavor state is an eigenstate of the Hamiltonian of the system or not. If the flavor state were an eigenstate of the Hamiltonian, then neutrinos would not oscillate. But since it is not, then the **mass eigenstates** are linear combinations of the **flavor eigenstates**, and the oscillations take place. This was first pointed out by physicist Bruno Pontecorvo in 1968, and we now explain it in some detail.

For simplicity, let us assume two neutrino flavors e and μ (the formalism easily generalizes to three). We will use the quantum-mechanical bra-ket notation, and label the flavor eigenstates as $|\nu_e\rangle$ and $|\nu_\mu\rangle$, and let the mass eigenstates be $|\nu_1\rangle$ and $|\nu_2\rangle$. Each set of eigenstates forms a complete basis, and the two sets are therefore mutually related by a linear transformation. Then we can write

$$|\nu_1\rangle = \cos\theta|\nu_\mu\rangle - \sin\theta|\nu_e\rangle$$
$$|\nu_2\rangle = \sin\theta|\nu_\mu\rangle + \cos\theta|\nu_e\rangle, \tag{5.7}$$

where θ is some constant. Because $|\nu_1(t)\rangle$ and $|\nu_2(t)\rangle$ are eigenstates of the Hamiltonian with some corresponding energies E_1 and E_2, they evolve simply as

$$|\nu_1(t)\rangle = |\nu_1(0)\rangle\, e^{iE_1 t}$$
$$|\nu_2(t)\rangle = |\nu_2(0)\rangle\, e^{iE_2 t}. \tag{5.8}$$

Let us assume that the initial state is a pure $|\nu_e\rangle$. Going from the flavor basis to the mass basis via the projection operator and using this initial condition, we get

$$|\nu_e\rangle = \sum_{i=1}^{2} |\nu_i\rangle\langle\nu_i|\nu_e\rangle = -\sin\theta\,|\nu_1\rangle + \cos\theta\,|\nu_2\rangle$$
$$|\nu_\mu\rangle = \sum_{i=1}^{2} |\nu_i\rangle\langle\nu_i|\nu_\mu\rangle = \cos\theta\,|\nu_1\rangle + \sin\theta\,|\nu_2\rangle, \tag{5.9}$$

where all bra-ket states in the equations above are evaluated at $t = 0$. Because we know how the mass eigenstates evolve – as in Eqs. (5.8) – we know $|\nu_e(t)\rangle$ and $|\nu_\mu(t)\rangle$ as well. So, for example,

$$|\nu_\mu(t)\rangle = \cos\theta\, e^{iE_1 t}\, |\nu_1(0)\rangle + \sin\theta\, e^{iE_2 t}\, |\nu_2(0)\rangle. \tag{5.10}$$

In Problem 5.3 you will be able to complete this calculation and show that the probability of the electron neutrino oscillating to a muon neutrino at time t, which is given by the square of the amplitude of the overlap of $|\nu_e(0)\rangle$ and $|\nu_\mu(t)\rangle$, is equal to

$$P_{\nu_e \to \nu_\mu} \equiv |\langle \nu_\mu(t) | \nu_e(0) \rangle|^2 = \left[\sin(2\theta) \sin\left(\frac{E_2 - E_1}{2} t \right) \right]^2$$

$$= \left[\sin(2\theta) \sin\left(1.27 \frac{\Delta m^2 L}{E} \right) \right]^2 ,$$

(5.11)

where $\Delta m^2 \equiv m_1^2 - m_2^2$ is the square of the difference of masses of the neutrinos corresponding to the two mass eigenstates (measured in eV^2), L is the length over which the neutrino has traveled at that time (measured in km), and E is the energy of the neutrino (measured in GeV). The key new result here is that the neutrino-flavor oscillation probability depends on the difference of *squares* of the neutrino masses.

The relation between the flavor and mass eigenstates can be described succintly with a **mixing matrix**. For the two-generation case that we are studying here, the mixing matrix, $U_{2\,\text{flav}}$, has one free parameter, the "mixing angle" θ:

$$\begin{pmatrix} \nu_e \\ \nu_\mu \end{pmatrix} = \begin{pmatrix} \cos\theta & \sin\theta \\ -\sin\theta & \cos\theta \end{pmatrix} \begin{pmatrix} \nu_1 \\ \nu_2 \end{pmatrix} \equiv U_{2\,\text{flav}} \begin{pmatrix} \nu_1 \\ \nu_2 \end{pmatrix} .$$

(5.12)

In our world we observe three generations of neutrinos, and the mixing matrix generalizes to

$$\begin{pmatrix} \nu_e \\ \nu_\mu \\ \nu_\tau \end{pmatrix} = \begin{pmatrix} U_{e1} & U_{e2} & U_{e3} \\ U_{\mu1} & U_{\mu2} & U_{\mu3} \\ U_{\tau1} & U_{\tau2} & U_{\tau3} \end{pmatrix} \begin{pmatrix} \nu_1 \\ \nu_2 \\ \nu_3 \end{pmatrix} \equiv U_{3\,\text{flav}} \begin{pmatrix} \nu_1 \\ \nu_2 \\ \nu_3 \end{pmatrix} ,$$

(5.13)

and it turns out that the matrix $U_{3\,\text{flav}}$ has *four* free parameters – three angles and a phase.[2] The phase is particularly important, as it describes the level of charge-parity (CP) violation in the neutrino sector. Accurate experimental measurement of these parameters and, on the other hand, their prediction from theory are among the important topics of research in particle physics, but we will not pursue them further here.

5.2.2 Experimental Confirmation of Oscillations

In the 1960s, physicist Ray Davis ran an experiment using a chlorine tank in Homestake Mine in South Dakota. Davis was trying to count neutrinos from the Sun by counting argon atoms produced in the reaction

$$\nu_e + {}^{37}\text{Cl} \longrightarrow {}^{37}\text{Ar} + e^- .$$

(5.14)

---

[2] This mixing matrix between neutrino flavors should not be confused with (but is conceptually similar to) the mixing matrix between *quark* flavors – the so-called Kobayashi–Maskawa matrix – that we do not cover in this book. Moreover, the matrix $U_{3\,\text{flav}}$ has two additional parameters (six total) if the neutrinos are **Majorana particles** – that is, if they are their own antiparticles.

Davis found about a third of the neutrinos that the so-called standard solar model (and theorist John Bahcall) predicted. For a long time, this was considered to be an anomaly possibly driven by systematic uncertainties associated with solar modeling. The missing neutrino flux was later explained via neutrino oscillations, where electron neutrinos convert to other flavors. Because the Davis experiment was only sensitive to electron neutrinos, it observed their lower flux than that expected without oscillations.

Later experiments, including at Sudbury Neutrino Observatory (measuring neutrino interactions with heavy water, in Sudbury Mine, Canada) confirmed the general picture of solar neutrino oscillations. These experiments were also sensitive to other neutrino flavors, and thus measured the elements of the matrix $U_{3\,\text{flav}}$ in Eq. (5.13). For solar neutrino oscillations, the characteristic distance is that to the Sun ($L \sim 150$ million km), and the characteristic neutrino energy is $E \sim 10\,\text{MeV}$.

Separately from the solar neutrino story, we also observe atmospheric neutrinos – those created primarily by the decay of pions and muons when they enter Earth's atmosphere. These decays also produce neutrinos, and these neutrinos oscillate to other flavors. For atmospheric oscillations, the characteristic distance is the diameter of Earth ($L \sim 10,000$ km), while the characteristic neutrino energy is $E \sim 1\,\text{GeV} - 1\,\text{TeV}$. In 1998, the Super-Kamiokande experiment, which was sensitive to neutrino interactions with water in Kamioka Mine in Japan, observed the disappearance of about 50 percent of the expected μ-neutrinos coming from the atmosphere.

The solar and atmospheric oscillation experiments are both sensitive to the square of the mass differences between two neutrino mass eigenstates according to Eq. (5.11). These two kinds of experiments, however, point to two *different* values of Δm^2:

$$\Delta m^2_{\text{solar}} \simeq (7.53 \pm 0.18) \times 10^{-5}\,\text{eV}^2$$
$$\Delta m^2_{\text{atm}} \simeq (2.45 \pm 0.05) \times 10^{-3}\,\text{eV}^2. \tag{5.15}$$

Note that two square mass differences can be nicely explained by three neutrino generations – no more, no less! If a third measurement of Δm^2 came along with a value different from these two, then either one of the Δm^2 measurements would be wrong, or else there would exist a fourth neutrino species (which must be "sterile" – that is, not subject to weak interactions). Such a measurement did in fact take place, by the Liquid Scintillator Neutrino Detector (LSND) experiment in the 1990s, and has recently been strengthened by results from the MiniBooNE experiment, indicating oscillations with $\Delta m^2 \sim 1\,\text{eV}^2$ – a value different from either of the two in Eq. (5.15). Therefore, the hunt for all neutrino species continues.

Even apart from the possibility of the existence of sterile neutrinos, there is the question of what the individual mass eigenstates of the three weakly interacting generations are, given the oscillation measurements in Eq. (5.15). Figure 5.2(b) shows the two possible neutrino hierarchies that both explain the solar and atmospheric neutrino data. You will have a chance to work out these neutrino hierarchies in Problem 5.4.

The solar and atmospheric experiments' discovery of neutrino oscillations proved that neutrinos have mass. But neutrino-oscillation experiments have no sensitivity to *absolute* masses of individual neutrino mass eigenstates, as a simple inspection of Eq. (5.15) reveals (see also Problem 5.4). It *is* possible to measure the absolute neutrino masses to high precision by a very careful mapping of the endpoint of the beta-decay spectrum shown in the inset of Fig. 5.1; this is currently being carried out by the KATRIN experiment. But, at least so far, it has proven challenging to infer absolute neutrino masses from particle-physics experiments.

This is where cosmological measurements come in. As we will see later in Chapters 9 and 11, galaxy clustering is sensitive to the *sum* of the neutrino masses, $m_1 + m_2 + m_3$, as larger masses suppress clustering at small angular scales. Thus, a combination of neutrino-oscillation (particle) experiments and cosmological measurements of the distribution of galaxies in the universe provides a unique "inner space–outer space" combination that will allow us to pin down the individual masses of the three neutrino particles.

5.3 Decoupling of Neutrinos

Neutrino interactions in the very early universe – before about a second after the Big Bang, or at temperatures greater than about 1 MeV or higher than 10^{10} K – are determined by typical weak-interaction reactions such as

$$e^- + \nu_e \rightleftharpoons e^- + \nu_e$$
$$\nu_e + \bar{\nu}_e \rightleftharpoons e^- + e^+. \tag{5.16}$$

The cross-section for such weak reactions mediated by the W-boson is $\sigma \sim G_F^2 T^2$, where $G_F \sim \alpha/M_W^2 \simeq 1.1 \times 10^{-5}\,\mathrm{GeV}^{-2}$ is the Fermi constant, $\alpha \simeq 1/137$ is the fine-structure constant, and M_W is the W-boson mass,[3] and the temperature dependence can be derived from dimensional analysis.

The interaction rate Γ then scales as

$$\Gamma = n_\nu \sigma v \simeq T^3\, G_F^2 T^2 = G_F^2 T^5, \tag{5.17}$$

since the number density is, in the relativistic regime, $n_\nu \simeq T^3$ (recall Eq. (4.19), and note that the prefactor is a constant of order unity), and for relativistic particles $v \simeq 1$.

The interaction rate Γ has units of 1/time. It has to be compared with the

---

[3] The quadratic dependence of the cross-section on the fine-structure constant can be obtained from the fact that the amplitudes of the relevant tree-level Feynman diagrams have two vertices, each contributing a $\sqrt{\alpha}$ to the cross-section. Squaring the amplitude to get the cross-section gives the α^2 dependence, and the single W-boson propagator brings the factor of $1/M_W^2$ to the amplitude.

only other global quantity with the same units in the expanding universe, which is the Hubble rate $H = \dot{a}/a$. The Hubble parameter scales with temperature as in Eq. (4.36), $H \simeq T^2/m_{\mathrm{Pl}}$. Therefore

$$\frac{\Gamma}{H} \simeq \frac{G_F^2 T^5}{T^2/m_{\mathrm{Pl}}} \simeq \left(\frac{T}{1\,\mathrm{MeV}}\right)^3. \tag{5.18}$$

When this ratio drops below unity, the expansion rate is so high that the particles "can't find each other" to interact any more, and the interactions stop. For neutrinos, Fig. 5.3 shows that this happens at $T \simeq 1\,\mathrm{MeV}$, or about $1\,\mathrm{s}$ after the Big Bang.

After decoupling, at $t > t_{\mathrm{dec}}$, the number density of neutrinos scales as $n_\nu \propto a^{-3}$. Their (relativistic) momentum scales inversely with scale factor[4]

$$p_\nu = p_{\nu,\mathrm{dec}} \frac{a_{\mathrm{dec}}}{a}, \tag{5.19}$$

where $p_{\nu,\mathrm{dec}}$ is the momentum at decoupling, p_ν is the momentum at any time $t > t_{\mathrm{dec}}$, and $a \equiv a(t)$ as usual. Then it follows that the neutrinos' distribution function after decoupling is a snapshot of the thermal spectrum at decoupling:

$$f_\nu(p, t > t_{\mathrm{dec}}) = \frac{1}{e^{p_{\nu,\mathrm{dec}}/T_{\nu,\mathrm{dec}}} + 1} = \frac{1}{e^{(p_\nu a/a_{\mathrm{dec}})/T_{\nu,\mathrm{dec}}} + 1}$$
$$\equiv \frac{1}{e^{p_\nu/T_\nu(a)} + 1}, \tag{5.20}$$

where we have defined the effective neutrino temperature

$$T_\nu(a) \equiv T_{\nu,\mathrm{dec}} \frac{a_{\mathrm{dec}}}{a}. \tag{5.21}$$

Therefore, massless neutrinos retain their Fermi–Dirac distribution, with temperature that decreases with the inverse of the scale factor, $T_\nu(a) \propto 1/a$.

You may have noticed from Eq. (5.21) that the neutrino effective temperature falls off as $1/a$, while Eq. (4.35) indicates that the photon temperature does not quite decrease with the same law, scaling instead as $T_\gamma \propto 1/(ag_{*S}^{1/3})$, where g_{*S} is also a function of time. Therefore, the neutrino and photon temperatures evolve differently around the time of decoupling. We now study this in more detail.

5.3.1 Electron–Positron Annihilation

The electron–positron annihilation and its inverse process, the e^+e^- creation,

$$e^- + e^+ \rightleftharpoons 2\gamma, \tag{5.22}$$

stop at temperature $T \simeq m_e \simeq 0.5\,\mathrm{MeV}$. At that point, the energy density and entropy of the electrons and positrons are definitively transferred to the photon

---

[4] It is an excellent approximation to assume that neutrinos are relativistic in this context even if they have mass $m_\nu \simeq 0.1\,\mathrm{eV}$ as indicated by current experiments. You can explicitly check this.

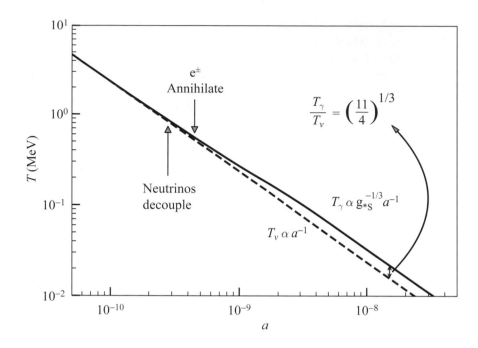

Fig. 5.3   Temperature of photons (solid) and neutrinos (dashed) as a function of scale factor around the time of electron–positron annihilation.

bath, thus heating the photons relative to the other relativistic particles present at the time – the neutrinos.

Let us account for the effective number of degrees of freedom before and after e^+e^- decoupling. At that point in time, other heavier species (such as the Higgs or the baryons) have long ago become non-relativistic. In addition to electrons/positrons and photons, there are also the (now decoupled) neutrinos, but we ignore them since entropy is separately conserved for thermal (coupled) and decoupled species. Therefore we can just consider the electron–positron–photon bath, for which

$$g_{*S}^{\rm th} = \begin{cases} 2 + \dfrac{7}{8} \times 2 \times 2 = \dfrac{11}{2} & (T \gtrsim m_e) \\[2mm] 2 & (T \lesssim m_e), \end{cases} \tag{5.23}$$

where, in the latter case, only two polarizations of photons contribute, while in the former case the factors of two are also accounting for the electrons and positrons, each with two spin states.

Because the entropy $S = g_{*S}^{\rm th}(aT_\gamma)^3$ is conserved, a step down in $g_{*S}^{\rm th}$ implies a step up in the photon temperature, so that $T_\gamma = (11/4)^{1/3}T_\nu$; see Fig. 5.3. Most often, we think of the photon temperature as something fixed (the famous $T_{\gamma,0} = 2.725\,{\rm K}$

today, for example), and we like to express the neutrino temperature in terms of that. Then

$$T_\nu = \left(\frac{4}{11}\right)^{1/3} T_\gamma, \tag{5.24}$$

which is valid at all times after the e^+e^- decoupling.

Therefore the e^+e^- annihilation, which occurs slightly after the neutrinos are decoupled, pumps energy to the photons, leading to a temporary slower-than-a^{-1} cooling of the photon temperature. The neutrino temperature, on the other hand, continues to scale as a^{-1} since the neutrinos are decoupled. The result is that the neutrino background temperature today is lower than that of the photons:

$$T_{\nu,0} = \left(\frac{4}{11}\right)^{1/3} 2.725\,\mathrm{K} = 1.96\,\mathrm{K} \qquad \text{(neutrino background temp. today)}. \tag{5.25}$$

If a species were to decouple earlier in the history of the universe, later particle annihilations would impart a relatively larger fraction of energy to photons, and the present-day temperature of that species, relative to the photons, would be even lower than that of the neutrinos. I asked you to work out such an example in Problem 4.6 at the end of Chapter 4.

5.3.2 Effective Number of Neutrino Species

Including the decoupled neutrinos to account for all relativistic species, the present-day effective number of degrees of freedom is

$$g_*^{\text{today}} = 2 + \frac{7}{8} \times 2N_{\text{eff}} \left(\frac{4}{11}\right)^{4/3} \simeq 3.38$$

$$g_{*S}^{\text{today}} = 2 + \frac{7}{8} \times 2N_{\text{eff}} \left(\frac{4}{11}\right) \simeq 3.94. \tag{5.26}$$

Here, N_{eff} is the **effective number of neutrino species**. It is a useful quantity, and one of the more interesting early-universe quantities to measure with the CMB experiments. N_{eff} is not actually precisely equal to 3 (for the three neutrino flavors) due to subtle quantum effects: the neutrino spectrum after decoupling deviates slightly from the Fermi–Dirac distribution, because the energy dependence of the weak interaction causes neutrinos in the high-energy tail to interact more strongly. Careful calculation shows that

$$N_{\text{eff}} \simeq 3.046 \qquad \text{(theory expectation in the Standard Model)}. \tag{5.27}$$

The effective number of neutrino species is often promoted to an **effective number of *relativistic* species**, and treated as a free parameter in cosmological data analyses. The motivation is to search for evidence of additional relativistic species due to new physics (either due to decays of new particles into photons, or new relativistic particles in their own right). These hypothetical additional relativistic species go under the umbrella name **dark radiation**.

The current constraint from Planck and other cosmological data is

$$N_{\text{eff}} = 2.99^{+0.34}_{-0.33} \quad \text{(Planck plus other data, 95 percent credible interval)}, \tag{5.28}$$

and is therefore, as of this writing, consistent with the expectation from the Standard Model of particle physics and cosmology, with no evidence for the presence of dark radiation.

5.4 Neutrino Density and Cosmic Background

"Empty space" is filled with not only photons left over from the early universe, but also with primordial neutrinos. Their number density today can be easily related to that of photons with the aid of Eqs. (4.19), (4.21), and (5.24):

$$n_{\nu,0} = \frac{3}{4} N_{\text{eff}} \times \left(\frac{4}{11}\right) n_{\gamma,0} \simeq 336 \, \text{cm}^{-3}, \tag{5.29}$$

that is, around 112 neutrinos (56 νs and 56 $\bar{\nu}$s) of each flavor per cubic centimeter.

Even though we have not yet *directly* detected the background of cosmic neutrinos, we are essentially certain that it exists, and that the universe is indeed awash in neutrinos with number density given in Eq. (5.29). Our confidence comes not only because the standard thermodynamics of the early universe predicts the neutrino background, but also because subtle signatures of the neutrino background (having to do with the properties of their general-relativistic stress energy) have been detected in the CMB data. Nevertheless, there is much interest in also directly detecting the cosmic neutrino background, and we discuss this in Box 5.1.

If the neutrinos were massless, they would be relativistic today, and their energy density would be (following similarly Eqs. (4.20), (4.21), and (5.24))

$$\rho_{\nu,0} = \frac{7}{8} N_{\text{eff}} \times \left(\frac{4}{11}\right)^{4/3} \rho_{\gamma,0} \simeq 3.16 \times 10^{-31} \frac{\text{kg}}{\text{m}^3}, \tag{5.30}$$

or, in more friendly dimensionless units

$$\Omega_\nu h^2 = 1.68 \times 10^{-5} \quad \text{(assuming } m_\nu = 0\text{)}, \tag{5.31}$$

where, following the usual notation convention for the omegas, Ω_ν refers to the density relative to critical today.

| Box 5.1 | Direct Detection of the Cosmic Neutrino Background? |

We now further discuss the background of cosmic neutrinos. First, recall that cosmic neutrinos decouple at temperature $T \simeq 1\,\text{MeV}$ and that, at the present time, they have a temperature of $T_{\nu,0} \simeq 1.96\,\text{K}$ and fill all space with number density of $112\,\text{cm}^{-3}$ per neutrino flavor.

The cosmic neutrino background (CνB) has been discovered indirectly in the WMAP data (around 2005) via its effects on the cosmic *microwave* background power spectrum. As alluded to in Chapter 4, the main effect of the CνB is on the radiation density of the universe, and hence the precise time of matter–radiation equality (which, recall, in the standard cosmological model happens at redshift $z \simeq 3500$).

Here, we outline a very interesting particle-physics method to directly detect the CνB, first proposed by Steven Weinberg more than 50 years ago. He considered the neutrino capture on tritium (hydrogen atom with two neutrons and a proton, ^{3}H), which creates helium-3:

$$\nu_e + {}^3\text{H} \longrightarrow {}^3\text{He} + e^-, \tag{B1}$$

which is really based on the beta decay, $\nu_e + n \to p + e^-$.

The kinematics of this equation can be worked out as an exercise, and the results are graphically illustrated in the figure below, adapted from Long *et al.* (2014). It shows that the endpoint of the electron spectrum in the "normal" beta decay is followed by a separate bump whose peak is separated by $2m_\nu$ (relative to the end of the electron spectrum). This $2m_\nu$ shift is a tell-tale signature of the CνB, and its location can therefore be used to measure the mass m_ν.

Compared to other beta-decaying nuclei, tritium is an attractive target because of its availability, high neutrino capture cross-section, and long lifetime (12 years). For a 100-gram target, the expected capture rate is approximately 10 events per year. However, energy resolution on the order of the

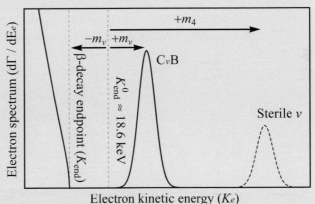

desired error in the neutrino mass is required which, at $\sim 0.1\,\text{eV}$, presents a challenge here. Note also that solar and atmospheric (so, non-cosmic) neutrinos are not a contaminant here, since they have a huge kinetic energy relative to the CνB, and hence would appear far to the right in the figure.

There is an ongoing experiment called PTOLEMY (Princeton Tritium Observatory for Light, Early-Universe, Massive-Neutrino Yield) which is based on precisely this idea. The constraints from a PTOLEMY-type experiment are expected to be comparable to those from the effect of the CνB on cosmology, but the two are highly complementary as they exploit different temporal and spatial scales.

Evidence from neutrino oscillation experiments, however, points to neutrinos having a small (sub-eV) mass of order $(0.05-0.10)\,\text{eV}$. Given that the temperature today is only $T_0 \simeq 0.24\,\text{meV}$ (*mili*-eV), the neutrinos are non-relativistic today. Therefore, the expression in Eq. (5.31) does not apply. In Problem 5.5 you will have a chance to derive a simple expression for the energy density of non-relativistic neutrinos in terms of their mass. The result is

$$\Omega_\nu h^2 = \frac{\sum_i m_{\nu,i}}{94\,\text{eV}} \qquad (\text{massive and non-rel., } m_\nu > T_0 \simeq 0.24\,\text{meV}), \qquad (5.32)$$

where the sum runs over the three neutrino mass eigenstates. Given that $h \simeq 0.7$, we conclude that:

> Neutrinos whose sum of the masses is of order $50\,\text{eV}$ would close the universe.

The fact that we knew that Ω (total, today) is not much bigger than unity provided an important early constraint on the absolute masses of the neutrinos, and was in fact one of the pioneering connections between cosmology and particle physics. This bound is called the Cowsik–McClelland bound, although it was first put forth by Gershtein and Zeldovich in 1966.

Today, a much stronger constraint on the sum of the masses is obtained from the cosmic microwave background anisotropies combined with the clustering of galaxies and their gravitational lensing. Specifically, there is a sharp upper limit

$$\sum_i m_{\nu,i} < 0.12\,\text{eV} \qquad (\text{Planck plus other data, 95 percent credible interval}).$$

$$(5.33)$$

This constraint uses key signatures of massive neutrinos: that they suppress galaxy clustering (this is further discussed in Chapter 11, particularly around Eq. (11.22) and in Fig. 11.4), and that they affect the epoch of matter–radiation equality and thus the sound horizon (discussed in Chapter 13). The constraint in Eq. (5.33) is very strong, yet it falls short of *detecting* the sum of the masses which, we know from oscillation experiments, must be nonzero and in the ballpark of $(0.05-0.10)\,\text{eV}$.

Leaving details of cosmological measurements of neutrino masses to these upcoming chapters, for the moment we conclude that:

> Neutrinos provide a small but non-negligible contribution to the energy-density budget of the universe. The sum of their masses is nonzero, currently constrained to be well below 1 eV, and expected to be precisely measured in upcoming measurements of the CMB and large-scale structure.

Bibliographical Notes

Neutrinos are covered in various textbooks on particle physics and, in less detail, in some cosmology textbooks. Short, concise summaries are given in the neutrino chapter by the Particle Data Group (Zyla *et al.*, 2020), and the very clear and informative presentation in the WMAP 5-year paper (Komatsu *et al.*, 2009). A very comprehensive textbook on neutrinos in cosmology is Lesgourgues *et al.* (2013).

Problems

5.1 Neutrinos through your thumb. Roughly estimate the number of relic neutrinos going through your thumb every second. Assume that the neutrinos are moving with the speed of light.

5.2 Beta decay ignoring the neutrino. Assume that atom X emits an electron, leaving another atom Y with one extra proton:

$$_Z^A X \to \, _{Z+1}^{A}Y + e^-.$$

Let us be ignorant about the existence of the outgoing neutrino (as physicists were 100 years ago). Using conservation of energy and momentum, derive the outgoing electron's total energy in Eq. (5.2).

5.3 Neutrino oscillations with two flavors. Complete the calculation from Sec. 5.2:

(a) Starting from a pure electron neutrino state, first derive the amplitude of the overlap of the muon state at time t and the neutrino state at time zero:

$$\langle \nu_\mu(t)|\nu_e(0)\rangle = \sin\theta\cos\theta \left(-e^{iE_1 t} + e^{iE_2 t}\right).$$

Then show that the probability of conversion of the electron into a muon neutrino is

$$P_{\nu_e \to \nu_\mu} = \left[\sin(2\theta)\sin\left(\frac{E_2 - E_1}{2}t\right)\right]^2$$

and further, remembering that the neutrinos at these energies are highly relativistic so that $m_1 c^2, m_2 c^2 \ll pc \simeq E$ (note $p_1 = p_2$ even though masses aren't necessarily close), that this is equal to

$$P_{\nu_e \to \nu_\mu} = \left[\sin(2\theta)\sin\left(1.27\frac{\Delta m^2 L}{E}\right)\right]^2,$$

where Δm^2 is the square of the difference of masses of muon and electron neutrino (measured in eV^2), L is the length over which the neutrino has traveled at that time (measured in km), and E is the energy of the neutrino (measured in GeV).

(b) One of the goals of neutrino experiments is to measure the masses of neutrinos more precisely (note, using the neutrino oscillations we only get Δm^2 between the mass eigenstates, so there is still degeneracy in m_i).

Say that we would like to look more closely at "atmospheric" neutrino oscillations where we already know that $\Delta m^2_{\text{atm}} \simeq 3 \times 10^{-3}\,\text{eV}^2$. If a beam of neutrinos is generated at Fermilab (Illinois) and the neutrinos are being detected at the Soudan Mine in northern Minnesota about 700 km away, roughly what energy should the neutrinos have in order for atmospheric oscillation (and hence Δm^2) to be *most easily measurable*? [Think about what the condition for that is.]

(c) What is the relativistic factor γ for the heavier neutrino from part (b), assuming that the two flavors are not nearly degenerate in mass (so that $m^2_{\text{heavy}} \approx \Delta m^2_{\text{atm}}$)?

5.4 **Breaking the degeneracy between neutrino masses.** The neutrino-oscillation experiments have measured two distinct mass-squared-difference values:

$$\Delta m^2_{\text{solar}} = (7.50 \pm 0.20) \times 10^{-5}\,\text{eV}^2$$

$$\Delta m^2_{\text{atm}} = (2.32 \pm 0.10) \times 10^{-3}\,\text{eV}^2,$$

from the solar and atmospheric experiments, respectively. In the following, make the following assumptions:

- There are three species of neutrinos.
- Ignore the errors in Δm^2 above (i.e., assume them to be zero).
- Assume that the lightest neutrino species has zero mass, $m_3 = 0$ (this is not necessarily the case in reality, but simplifies the analysis here).

(a) How many mathematically distinct possibilities are there for (m_1, m_2, m_3) and what are they?

(b) Among the triplets found in part (a), how many *qualitatively different* classes are there, and what are the *names* given to these mass "hierarchies"?

(c) Imagine now a *third* measurement of mass squared, call it Δm^2_{new}, comes along tomorrow, and that it disagrees with both values (i.e., lies well outside of their error bars) from Eqs. (5.15). Report what value(s), if any, Δm^2_{new} can take and yet *not* be in conflict with the three-neutrino-species scenario.

(d) Cosmological measurements are sensitive to the sum of the neutrino masses, $m_{\text{tot}} = m_1 + m_2 + m_3$. How accurate (i.e., how low in m_{tot}, given that there are only upper limits so far) do these measurements have to reach in order to decide between the possibilities in part (b)?

5.5 **Omega in massive neutrinos.** Assuming that the neutrinos are non-relativistic, which is true today given the results from neutrino oscillation experiments, derive Eq. (5.32):

$$\Omega_\nu h^2 = \frac{\sum_i m_{\nu,i}}{94\,\text{eV}},$$

where, recall, $\Omega_\nu \equiv \rho_{\text{nu},0}/\rho_{\text{crit},0}$. [Note that, because the sum in the above equation runs over the neutrino species, in your calculation of $\rho_{\nu,0}$ you will want to use the neutrino density *per species*, discussed in the chapter.]

6 The Boltzmann Equation and Baryogenesis

We now discuss two topics that are only loosely related, but both fall in the early-universe era: the Boltzmann equation and baryogenesis. The Boltzmann equation is a very general result from statistical mechanics that enables cosmologists to calculate the evolution and interaction of species (baryonic and dark matter, neutrinos, radiation, etc.) in an expanding universe. Baryogenesis, on the other hand, is a general framework for how baryons were created. As we will discuss in this chapter, there was a very small excess of baryons to their antiparticles – antibaryons – in the early universe, something that must be a key building block of a successful baryogenesis theory. There are many ways to make baryogenesis part of the Standard Model of particle physics or, alternatively, beyond-the-Standard-Model (BSM) theories. At the same time, however, there is not a "standard baryogenesis" model.

6.1 The Boltzmann Equation

The **Boltzmann equation** is an indispensable tool in cosmology.[1] It allows tracking the evolution of various species in the expanding universe, even as they interact with each other or annihilate with other particles or themselves. The equation is powerful and applies in very general situations. We first introduce it in all generality, then specialize in a simplified form that holds in many situations of interest to us.

The Boltzmann equation ultimately traces its lineage to the **Liouville theorem** of classical statistical mechanics, which (in one formulation) reads:

> In the absence of collisions, the distribution function $f(\mathbf{x}, \mathbf{p}, t)$ is unchanged along the particle trajectory.

Here

$$f(\mathbf{x}, \mathbf{p}, t) \equiv \frac{d^6 N}{d^3 p\, d^3 x} \tag{6.1}$$

is our old friend from the beginning of Chapter 4, and refers to the phase-space density of particles. The boxed statement above effectively says that f behaves like

---

[1] In the words of cosmologist Scott Dodelson who was the author's graduate-school teacher: "Whenever I have questions about life, whenever I feel lonely or sad I... go and solve the Boltzmann equation!"

an incompressible fluid. The Liouville theorem is remarkably general, as it only requires the assumptions of (1) conservation of particle number (in the absence of annihilations/creations) and (2) Hamilton equations of motion. In particular, it applies both in and out of equilibrium, and for both massive and massless particles.

The Boltzmann equation is a special case of the hierarchy of equations (the so-called Bogoliubov–Born–Green–Kirkwood–Yvon hierarchy, or BBGKY) that are obtained as a corollary of the Liouville theorem. The general statement of the Boltzmann equation is

$$\frac{df}{dt} \equiv \frac{\partial f}{\partial t} + \frac{dx_i}{dt}\frac{\partial f}{\partial x_i} + \frac{dp_i}{dt}\frac{\partial f}{\partial p_i} = \mathbf{C}[f], \qquad (6.2)$$

where summation over index i is assumed. Here $\mathbf{C}$ is known as the *collision term*, and takes into account particle creation and annihilation; the version with $\mathbf{C} = 0$ is called the *collisionless* Boltzmann equation. At its heart, the Boltzmann equation is a statement about the conservation of particle number[2] (f integrated over $d^3x d^3p$), stating that it is conserved unless there is particle creation or annihilation ($\mathbf{C} \neq 0$).

6.1.1 The Boltzmann Equation for Massive Particles

In anticipation of upcoming applications, notably particle "freezeout" to be discussed just below, we now specialize in the specific form of the Boltzmann equation for homogeneously distributed species in an expanding universe. We will leave out the derivation of this result, which includes a tedious-but-straightforward series of steps (integrating the Boltzmann equation with $g/(2\pi)^3 \int d^3p$ in order to go from f to the number density n, then integrating by parts; see Kolb and Turner, 1994). The result is the following equation for the number density of some species, n:

$$\dot{n} + 3Hn = \mathbf{C}[\{n_j\}], \qquad (6.3)$$

where $\{n_j\}$ is referring to number densities of other species or, in a more compact form,

$$\frac{1}{a^3}\frac{d(na^3)}{dt} = \mathbf{C}[\{n_j\}]. \qquad (6.4)$$

Let us now further specialize in a process involving four particles:

$$1 + 2 \longleftrightarrow 3 + 4. \qquad (6.5)$$

Let us say we wish to track the abundance of species 1. Its rate of change is given by the difference between the rates for producing and destroying the species. The

---

[2] Actually the Boltzmann equation is more general, as integrating it over momentum leads to the conservation-of-number equation as a derived result. Nevertheless, we find it most intuitive to think about the Boltzmann equation in terms of number conservation.

Boltzmann equation simply formalizes this statement:

$$\frac{1}{a^3}\frac{d(n_1 a^3)}{dt} = -\alpha n_1 n_2 + \beta n_3 n_4,$$ (6.6)

where α and β are some coefficients. The first term, $-\alpha n_1 n_2$, describes the destruction of particles 1, while the second term, $+\beta n_3 n_4$, describes their creation. Note also that in fact

$$\alpha \equiv \langle \sigma v \rangle$$ (6.7)

is the **thermally averaged cross-section**. Moreover, the second coefficient β can be related to α by noting that the collision term has to vanish in (chemical) equilibrium, denoted "eq":

$$\beta = \left(\frac{n_1 n_2}{n_3 n_4}\right)_{eq}\alpha.$$ (6.8)

With that, the Boltzmann equation for the process in Eq. (6.5) becomes

$$\frac{1}{a^3}\frac{d(n_1 a^3)}{dt} = -\langle \sigma v \rangle \left[n_1 n_2 - \left(\frac{n_1 n_2}{n_3 n_4}\right)_{eq} n_3 n_4\right].$$ (6.9)

One more simplification can be applied to Eq. (6.9). Recall our old friend, the number-density-to-entropy ratio $Y_i = n_i/s$ from Eq. (4.34). Remembering also that $s \propto T^3 \propto a^{-3}$, the left-hand side of Eq. (6.9) becomes $n_1 d\ln Y_1/dt$. Also recall that $d/dt = H d/d\ln a$. If we also ignore any change in the number of relativistic degrees of freedom g_*, Eq. (6.9) can be rewritten as

$$\frac{d\ln Y_1}{d\ln a} = -\frac{\Gamma_1}{H}\left[1 - \left(\frac{Y_1 Y_2}{Y_3 Y_4}\right)_{eq}\frac{Y_3 Y_4}{Y_1 Y_2}\right],$$ (6.10)

where we defined $\Gamma_1 = n_2\langle \sigma v \rangle$, the rate of annihilation of particles 1 (with particles 2).

Equation (6.10) supports our earlier conjecture that $\Gamma \gtrsim H$ indicates particles in thermal equilibrium and vice versa. Consider, for example, the $\Gamma/H \gg 1$ case, and suppose that the density of particles 1 is, at some point in time, lower than the equilibrium value, $Y_1 < (Y_1)_{eq}$. Then the term in square brackets is negative, and the right-hand side is therefore positive. Thus, $d\ln Y_1/d\ln a > 0$ and the number (per entropy) of particles 1 increases until it reaches equilibrium. An opposite conclusion is reached if $Y_1 > (Y_1)_{eq}$ at some point in time; then the number density of particles 1 decreases to the equilibrium value.

6.1.2 Freezeout of Thermal Relics

We now derive and illustrate a classic result regarding evolution of species in an expanding universe. For a particle species initially in a thermal bath, its density falls off exponentially as the universe cools. However, when the annihilation rate of the particle, Γ, falls below the expansion rate, H, the particles "cannot find each other" any more, and the species freezes out – its abundance (more specifically, its density-to-entropy ratio Y) asymptotes to a constant, and remains such to the present day. The latter effect is called **freezeout**, and now we discuss it in some detail.

We will illustrate the freezeout on its probably most important example in cosmology – that of a dark-matter particle, discussed further in Chapter 11. Let us consider the annihilation of dark-matter particles X with their antiparticles $\bar{X}$, producing some Standard-Model particles p and their antiparticles $\bar{p}$:

$$X + \bar{X} \longrightarrow p + \bar{p}. \tag{6.11}$$

For such particle–antiparticle annihilation, the Boltzmann equation as stated in Eq. (6.9) becomes

$$\frac{1}{a^3} \frac{d(na^3)}{dt} = -\langle \sigma v \rangle \left[n^2 - n_{\text{eq}}^2 \right], \tag{6.12}$$

where n is the number density of particles X, n_{eq} is their equilibrium number density, and $\langle \sigma v \rangle$ is the thermally averaged annihilation cross-section. We next define

$$Y \equiv \frac{n}{s}, \qquad x \equiv \frac{m}{T}, \qquad \text{and} \qquad \lambda \equiv \frac{2\pi^2}{45} g_{*S} \frac{m^3 \langle \sigma v \rangle}{H(T = m)}, \tag{6.13}$$

where m is the mass of particles X, g_{*S} is the number of relativistic degrees of freedom introduced in Chapter 4, and the Hubble parameter is evaluated when $T = m$. This equation can be rewritten as (Problem 6.1)

$$\frac{dY}{dx} = -\frac{\lambda}{x^2} \left[Y^2 - Y_{\text{eq}}^2 \right]. \tag{6.14}$$

Equation (6.14) cannot be solved analytically, but we can further simplify it using the fact that, after freezeout, $Y \gg Y_{\text{eq}}$, so that Y_{eq} can be ignored. Then

$$\frac{dY}{dx} = -\frac{\lambda}{x^2} Y^2. \tag{6.15}$$

Integrating this from freezeout x_f to inifinite future x_∞ gives

$$\frac{1}{Y_\infty} - \frac{1}{Y_f} = \frac{\lambda}{x_f}. \tag{6.16}$$

Clearly $Y_f \gtrsim Y_\infty$. Making the assumption $Y_f \gg Y_\infty$ so as to get a simple result in this rough approximation, we get

$$Y_\infty \simeq \frac{x_f}{\lambda}, \tag{6.17}$$

where $x_f \simeq 10 - 50$, depending on the freezeout temperature. Y_∞ is the final, present-day abundance of dark-matter particles; it is

- constant (doesn't depend on x), and
- is smaller for stronger interaction rate (higher λ) and vice versa.

Realizing also that Y is roughly the comoving number density of particles, $Y \equiv n/s \simeq na^3$, we conclude:

> When the annihilation rate of a particle in thermal equilibrium falls below
> the Hubble rate, $\Gamma/H \ll 1$, the particle "freezes out" – its comoving
> number density asymptotes to a constant.

Figure 6.1 illustrates freezeout of a thermal relic for a few values of the parameter λ. Note that, for an $m = 1-100$ GeV particle (e.g., a dark-matter particle), the freezeout temperature is $T = m/x_f \simeq 10\,\mathrm{MeV} - 10\,\mathrm{GeV}$.

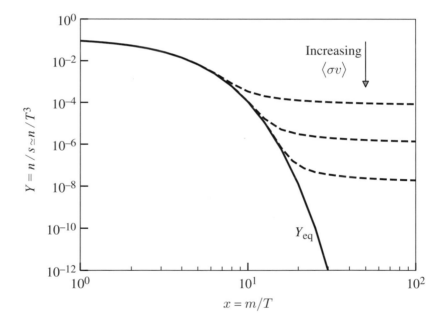

Fig. 6.1 Freezeout of particle species. The y-axis is the comoving number-density-to-entropy ratio $Y \equiv n/s$, which is proportional to the comoving number density of particles (na^3). The x-axis is the mass of the particle divided by the temperature of the universe, $x = m/T$; basically, time flows to the right. The freezeout abundance depends on the annihilation cross-section of the particle multiplied by its velocity; the more efficient annihilation is (larger $\langle \sigma v \rangle$), the lower the final abundance. The three dashed curves correspond respectively to cases $\lambda = 10^5, 10^7, 10^9$; see text for details.

We will apply many of the tools and results derived in this section to the case of dark matter in Chapter 11. Now we shift gear and talk about the mysterious ratio of baryons to photons in the universe.

6.2 Baryogenesis

In Chapter 7, we will see both the CMB and the light element abundances determine very accurately the baryon-to-photon ratio. The ratio is defined as

$$\eta \equiv \frac{n_B}{n_\gamma} \qquad \text{(baryon-to-photon ratio)}. \tag{6.18}$$

Note that η is constant throughout the evolution of the universe as both n_B and n_γ scale as a^{-3}. The measured value of η (by the cosmic microwave background) is both precise and surprising:

$$\eta = (6.12 \pm 0.04) \times 10^{-10}. \tag{6.19}$$

Is this what we would naturally expect starting with a hot, dense early universe? Why is the photon number so large relative to the number of baryons? Moreover, why are there so many more baryons than *antibaryons*?

Consider an example: initially, for each 1,000,000,000 quarks, let there be 999,999,997 antiquarks. At the time of QCD phase transition (energy $\sim 150\,\text{MeV}$, meaning some $10^{-5}\,\text{s}$ after the Big Bang; see Table 4.1), all the antiquarks annihilate with the quarks, leaving roughly 2,000,000,000 photons, and three remaining quarks. The three quarks combine into one baryon, and we have a baryon-to-photon ratio on the order of one in a billion. From these numbers, we see that the fine-tuning in terms of near-but-not-perfect symmetry between matter and antimatter is interesting, to say the least!

Clearly, we need a mechanism to generate a *slight* initial excess of baryons B over antibaryons $\bar{B}$:

$$\frac{n_B - n_{\bar{B}}}{n_B} \simeq 10^{-9}. \tag{6.20}$$

Such a mechanism goes under the name of "baryogenesis." With the initial condition in Eq. (6.20) in place, particle-physics processes will ensure that most of the antibaryons annihilate with baryons, leaving a small number of baryons relative to the number of photons – and essentially no antibaryons. In the present-day universe, protons and neutrons constitute most of the baryons, and we encounter their antiparticle partners mostly as rare events in high-energy cosmic-ray showers or at particle accelerators.

6.2.1 World Without Baryon Asymmetry

The best way to motivate baryogenesis is to consider the consequences of an early-universe scenario that did *not* include this process.

The state of things without baryogenesis can be calculated using the Boltzmann equation. For particle–antiparticle annihilation, the Boltzmann equation reads as Eq. (6.14):

$$\frac{dY}{dx} = -\frac{\lambda}{x^2}\left[Y^2 - Y_{\text{eq}}^2\right], \tag{6.21}$$

where, recall, $\lambda \simeq m^3\langle\sigma v\rangle/H(T=m)$, while the final abundance is $Y_\infty = x_f/\lambda$. Specializing now in nucleons, the thermally averaged cross-section is $\langle\sigma v\rangle \simeq 100\,\text{GeV}^{-2}$, the freezeout temperature is $T_f \simeq 20\,\text{MeV}$, and the nucleon mass is $m \simeq 1\,\text{GeV}$. Plugging in these numbers (see Problem 6.2), we get

$$\frac{n_B}{n_\gamma} \simeq 10^{-20} \qquad \text{(with no initial B asymmetry).} \tag{6.22}$$

This is about 10 orders of magnitude smaller than the observed baryon-to-photon ratio![3] This is why we require a violation of baryon–antibaryon symmetry in the initial conditions.

6.2.2 Sakharov Conditions

The general principles of baryogenesis have been understood for 50 years, yet there is no standard model of this process, as the physics at its epoch takes place at extremely high energies that are difficult to probe experimentally. The temperature at which the baryogenesis occurs is not accurately known; it is presumed to be somewhere in the range

$$T_{\text{baryogenesis}} \sim (1-100)\,\text{GeV}, \tag{6.23}$$

with a big model dependence (it could be as low as $\sim 5\,\text{MeV}$). This corresponds to roughly a nanosecond after the Big Bang.

In 1967, famous Soviet physicist (and human rights activist) Andrei Sakharov outlined three crucial conditions for successful baryogenesis, now called the

Sakharov conditions:

1. Baryon number violation.
2. C *and* CP violation.
3. Interactions out of thermal equilibrium.

See Box 6.1 for a brief review of C, P, and T symmetries.

---

[3] To make things even worse and by the same argument, the abundance of *antibaryons* is also roughly equal to this, $n_{\bar{B}}/n_\gamma \simeq 10^{-20}$, implying that any excess of baryons to antibaryons would be presumably smaller than even this tiny number.

Box 6.1	C, P, T, and the Baryon Number

In particle physics, the charge conjugation operator C changes all of a particle's quantum numbers; the parity-reversal operator P changes the sign of coordinates, while the time-reversal operator T changes the direction of time. In other words,

$$\text{particle} \xrightarrow{\text{C}} \text{antiparticle}$$
$$(\mathbf{r}, t) \xrightarrow{\text{P}} (-\mathbf{r}, t)$$
$$(\mathbf{r}, t) \xrightarrow{\text{T}} (\mathbf{r}, -t).$$

Specifically, it is useful to see how the momentum $\mathbf{p}$ and spin $\mathbf{s}$ of a particle change under the three operators:

$$\mathbf{p} \xrightarrow{\text{C}} \mathbf{p} \qquad \mathbf{s} \xrightarrow{\text{C}} \mathbf{s}$$
$$\mathbf{p} \xrightarrow{\text{P}} -\mathbf{p} \qquad \mathbf{s} \xrightarrow{\text{P}} \mathbf{s}$$
$$\mathbf{p} \xrightarrow{\text{T}} -\mathbf{p} \qquad \mathbf{s} \xrightarrow{\text{T}} -\mathbf{s}.$$

The relations for the momentum are easy to understand once we recall that $\mathbf{p} = m(d\mathbf{r}/dt)$. For spin, a useful analogy would be that of angular velocity in classical mechanics, $\boldsymbol{\omega} = \mathbf{r} \times \mathbf{v}$; by considering how $\mathbf{r}$ and $\mathbf{v}$ transform under C, P, and T, we can find how $\boldsymbol{\omega}$ (and hence $\mathbf{s}$) transforms.

The baryon number, B, is the number of baryons minus that of antibaryons. Every quark has $B = 1/3$ and every antiquark has $B = -1/3$. Therefore, baryons (particles made up of three quarks) have $B = 1$, antibaryons have $B = -1$, and mesons (particles made up of a quark and an antiquark) have $B = 0$.

We now review the three Sakharov criteria in more detail, and discuss their modern understanding.

1. **Baryon number violation.** The first one is obvious: the baryon number B has to be violated in order for the baryon–antibaryon asymmetry to be generated, as a universe with a perfect conservation of B would not support it. Baryon number violation has not been observed in experiments, even though it is generally allowed in the Standard Model of particle physics, and in its various extensions.

2. **C and CP violation.** Even if the baryon number is violated, a perfect conservation of charge conjugation symmetry, C, and simultaneous charge and parity symmetries, CP, leads to an equal number of particles and antiparticles. This is because for any reaction $X \rightarrow p_1 + p_2$, the antiparticle reaction $\bar{X} \rightarrow \bar{p}_1 + \bar{p}_2$ would proceed at the same rate, leading to an equal number of particles and antiparticles. Interestingly, C and CP violations have been observed in historic experimental discoveries (1957 and 1964, respectively), and are a key building block of the Standard Model of particle physics. [The observed CP violation is slight, which has implications for difficulties of the Standard Model to generate sufficient baryogenesis; there are, however, ways to more easily arrange for the desired result in beyond-the-Standard-Model theories.]

3. **Out-of-equilibrium interactions.** Even if both conditions above were satisfied, particle interactions in thermal equilibrium would not lead to an overabundance of particles over antiparticles. This is because, in thermal equilibrium, the inverse process would destroy exactly as many units of baryon number as the forward process creates.

In more technical language, in thermal equilibrium, the phase-space distribution $f(p)$ is, at zero chemical potential, the same for particles and antiparticles: the two have the same mass and, since μ is negligible at this early time, $f(p)$ is the same (e.g., Bose–Einstein or Fermi–Dirac). Recalling that $n = \int f(p)d^3p$, we see that the number density of particles and antiparticles is the same in thermal equilibrium.

Fortunately, nature has provided us with a very simple way out of thermal equilibrium: the expanding universe! Recall that, when $\Gamma \lesssim H$, particle interactions freeze out. If at that time the temperature is smaller than the mass of the decaying particles, $T \lesssim m$, then the inverse decay will be blocked by the Boltzmann factor $e^{-m/T}$, and we have a departure from thermal equilibrium which allows for the generation of baryon number.

Box 6.2 gives an example that takes us through the logic of Sakharov conditions.

Box 6.2   **Sakharov Conditions: An Example**

Suppose that a heavy particle X decays into two particles p_1 and p_2 with branching ratios r and $1 - r$. Let the baryon numbers of the two daughter particles be B_1 and B_2. Note that $\bar{X}$ (the antiparticle of X) will decay into $\bar{p}_1$ and $\bar{p}_2$ with branching ratios $\bar{r}$ and $1 - \bar{r}$; note also that the baryon number of an antiparticle is of opposite sign to that of its corresponding particle. Then the change in the total baryon number per reaction is

$$\Delta B = rB_1 + (1 - r)B_2 - \bar{r}B_1 - (1 - \bar{r})B_2 = (r - \bar{r})(B_1 - B_2). \tag{B1}$$

This equation shows that we need:

- $B_1 \neq B_2$, so unequal baryon number of the daughter particles;
- $r \neq \bar{r}$, so different branching ratio for particles and antiparticles – this is what C and CP violation are required for.

Let us discuss this second bullet point a bit more. The CPT theorem from quantum field theory says that, in addition to masses of particles and their antiparticles being identical, the *total* branching ratios (and decay rates) are equal. However, the individual branching ratios, r and $\bar{r}$ above, do not have to be equal if C and CP are violated.

> **Box 6.2**   **Sakharov Conditions: An Example (continued)**
>
> Why both C and CP violation? Consider[a] a reaction R with rate $\Gamma[R]$ that changes baryon number by $\Delta B = +1$. Consider three additional reactions that are related to the original one by conjugation under C, P, and CP. Call them c(R), p(R), and cp(R). Note that p(R) also has $\Delta B = +1$ (since the baryon number charge operator commutes with the parity operator), but that c(R) and cp(R) have $\Delta B = -1$. If C, P, and CP are good symmetries then $\Gamma[c(R)] = \Gamma[p(R)] = \Gamma[cp(R)] = \Gamma[R]$. Since all the reactions occur at the same rate, the baryon number of the system just fluctuates around zero, and there is no baryogenesis. In order to bias the reactions toward generating $\Delta B > 0$, we need $\Gamma[R], \Gamma[p(R)] > \Gamma[c(R)], \Gamma[cp(R)]$. Then the interactions must violate both C and CP. To put it another way, suppose that we only violate CP and not C. Then we would have $\Gamma[R] = \Gamma[c(R)] > \Gamma[cp(R)] = \Gamma[p(R)]$, which does not lead to an excess of baryon number.
>
> [a] The following argument is due to Andrew Long.

We end the topic of baryogenesis with two parting thoughts. It is impressive that particle physicists elucidated the kinds of processes that must have been at play in order to create a slight overabundance of baryons to antibaryons in the early universe. And it is frustrating that baryogenesis is so challenging to probe experimentally or observationally.

Bibliographical Notes

The two topics covered in this chapter, the Boltzmann equation and baryogenesis, tend to be covered in the literature either not at all or at a very technical level. Accessible, first-principles introductions to the Boltzmann equation are in textbooks by Kardar (2007) and Thorne and Blandford (2017). At a more advanced level, it is treated in Kolb and Turner (1994) and Dodelson and Schmidt (2020), and the lecture notes by Baumann (2013). Baryogenesis, along with a variety of other interesting points of contact between particle physics and cosmology, is discussed thoroughly in Kolb and Turner (1994), Mukhanov (2005), and Weinberg (2008).

Problems

6.1 **Fill in the steps: the Boltzmann equation.** Fill in the steps leading from Eq. (6.12) to Eq. (6.14). Assume a radiation-dominated universe, and assume g_{*S} is constant. *Hint:* First express the temporal evolution of both the Hubble parameter H and the entropy density s in terms of variable x. Then express dt in terms of dx.

6.2 **Baryon number in a universe without initial B asymmetry.** Let us investigate the baryon-to-photon number in a universe without baryogenesis (so not ours!). Starting from the version of the Boltzmann equation in Eq. (6.14)

$$\frac{dY}{dx} = -\frac{\lambda}{x^2} \left[Y^2 - Y_{\mathrm{eq}}^2 \right],$$

where $Y \equiv n/T^3$ (and n is the number density of nucleons), and $x \equiv m/T$, where m is the nucleon mass. Further,

$$\lambda \equiv \frac{m^3 \langle \sigma v \rangle}{H(T = m)}$$

combines mass, cross-section, and Hubble parameter evaluated when $T = m$. Assuming that Y_{eq} is constant, the solution at large x (late times) is

$$Y_\infty = \frac{x_f}{\lambda},$$

where x_f is the value of x at freezeout.

(a) For protons, the thermally averaged cross-section is $\langle \sigma v \rangle \simeq 100\,\mathrm{GeV}^{-2}$, the freezeout temperature is $T_f \simeq 20\,\mathrm{MeV}$, and the proton mass is $m \simeq 1\,\mathrm{GeV}$. **Use these values to estimate the final proton-to-photon number density** n_p/n_γ. You may ignore factors of order unity, but show your work clearly, including canceling out of units, etc. You should need no calculator at all! You may find the formula for the Hubble parameter in the early universe (in terms of T), as well as the photon density formula, both covered earlier in this book, helpful. The Planck mass is $m_{\mathrm{Pl}} \simeq 10^{19}\,\mathrm{GeV}$ (note I am already ignoring a prefactor of $O(1)$).

(b) Does the ratio n_p/n_γ from part (a) vary with cosmic time?

6.3 **[Computational] Freezeout of particle species.** Reproduce Fig. 6.1. You will need to solve Eq. (6.14). Note that you will first have to calculate, and then tabulate/spline, the equilibrium abundance Y_{eq}. *Hint:* The integration of Eq. (6.14) may become more difficult for higher values of λ, but it works fine if you switch to integrating $d\ln Y/d\ln x$ – that is, use the logs of Y and x as variables in your integration.

6.4 **[Computational] Freeze-in of particle species.** An interesting alternative to the usual freezeout production of relic particles, notably dark matter, is the freeze-*in* mechanism (reviewed in Bernal *et al.*, 2017). Here, unlike in the standard freezeout scenario, the dark-matter particle was never in thermal equilibrium. Rather, the dark-matter particle χ is produced from the decay of some heavier, Standard-Model particle σ, so that $\sigma \to \chi\chi$ and $\delta n_\chi = -2\delta n_\sigma$. The interaction is characterized by a very small rate $\Gamma \equiv \Gamma_{\sigma \to \chi\chi}$ which, unlike in the freezeout scenario, does not depend on the number density (i.e., it is some constant, possibly related to the mass of particles σ). For small enough Γ, the number density of dark-matter particles χ increases, then flattens off when the number density of σ particles becomes Boltzmann-suppressed ($n_\sigma \propto e^{-m_\sigma/T} \ll 1$). You will now demonstrate and briefly investigate the feasibility of this scenario.

(a) For the freeze-in scenario, the Boltzmann equation reads

$$\frac{1}{a^3}\frac{d(na^3)}{dt} = 2\Gamma h(t)n_{\sigma,\mathrm{eq}}(t),$$

where you can take $h(x) \simeq x/(x+2)$ in terms of our usual time variable $x \equiv m_\sigma/T$ below. Note that the density entering on the left-hand side is that of decay-product particles χ ($n \equiv n_\chi$), while the equilibrium density on the right-hand side is that of the decaying particles σ. And, as mentioned above, Γ is time independent.

Starting from this equation and using a derivation similar to that in the freezeout scenario that goes from Eq. (6.12) to Eq. (6.14), demonstrate that the equivalent of the latter equation in the freeze-in picture is

$$\frac{dY}{dx} = \lambda_1 xh(x)Y_{\mathrm{eq}}(x),$$

where λ_1 is some constant. *Hint:* Because I am not asking that you derive λ_1 in terms of Γ and the Hubble parameter at $m_\sigma = T$, all you need to do is track the x-dependence.

(b) Plot $Y(x)$ for $\lambda_1 = 10^{-6}, 10^{-8}, 10^{-10}$. Take a smallish initial value, say $Y(x_0) = 10^{-20}$. For the curves to look nice, start integrating at $x_0 = 0.01$, but plot only from $x = 0.1$ to $x = 100$. As in Problem 6.3, it may be easier to integrate $d\ln Y/d\ln x$. Also plot $Y_{\mathrm{eq}}(x)$, thus again mimicking Fig. 6.1.

(c) Comparing the figure you generated in part (b) and Fig. 6.1: what is the key difference between the freezeout and freeze-in scenarios as the annihilation/decay rate Γ increases? Explain this qualitatively.

7

Big Bang Nucleosynthesis

In this chapter we study the primordial process in which nuclei of atoms formed – the **Big Bang nucleosynthesis**, or BBN for short. We will see that BBN results principally in the synthesis of hydrogen and helium, along with a few more of the lightest elements whose abundances are already much smaller. We will describe the basic thermodynamical and nuclear-physics conditions, and explain how they determine the primordial origin of nuclei in the universe. Finally, we will illustrate how the lightest-element abundances are in spectacularly good agreement with observations, making BBN one of the pillars of the hot Big Bang cosmological model.

7.1 The Case for Primordial Nucleosynthesis

The question of when the elements of the periodic table formed is both basic and profound, and has been on scientists' mind for a long time. The correct answer only arrived around the mid-twentieth century. *Most* of the familiar elements – those up to iron, the most stable element – formed in stars, in nuclear reactions that take place for up to millions of years. Stars burn hydrogen to helium (which the Sun is doing busily at the moment); sufficiently massive stars then burn helium to carbon and oxygen, and so on, all the way up to iron. Elements beyond iron have a slightly different origin. Some of these heavy elements form in the so-called s-process ("slow" process) which takes place in stars, in a sequence of beta decays and neutron captures, over a period of thousands of years. Other heavy elements form in the so-called r-process ("rapid" process), where protons and neutrons get jammed in nuclei during supernova explosions or neutron–star mergers in mere fractions of a second.

Is it possible that *all* elements heavier than hydrogen formed in stars, and not in any other way? The answer is negative, as correctly surmised by George Gamow, Fred Hoyle, and their colleagues around 1950. The issue is that there is simply too much helium in the universe. By mass, hydrogen makes up about three-quarters of the total baryons, helium about one-quarter, and all other elements much less (we will return to the exact proportions shortly). Stars could not produce enough helium even assuming optimistically that every star busily produces this element over the whole age of the universe (see Problem 7.2). Therefore, helium must have a primordial origin.

7.1.1 Energy and Time Scales

An important parameter that determines the stability of nuclei is the **binding energy per nucleon**, B/A, where B is the total binding energy and A is the mass number (see Box 7.1 to review the basics of nuclear physics). The quantity B/A increases as we go up in atomic number, and reaches its peak of about $B/A \sim 8.8\,\mathrm{MeV}$ for $^{56}\mathrm{Fe}$ and $^{62}\mathrm{Ni}$; see Fig. 7.1. So we would expect that elements are fused all the way up to iron and nickel during BBN. Big bang nucleosynthesis, however, is shockingly inefficient and incomplete – as we will soon see, nuclear reactions **freeze out** soon after the formation of hydrogen, helium, and some deuterium and helium-3 and traces of a few more elements.

Box 7.1   **Basics of Nuclear Physics**

An atom consists of a nucleus and electrons that are found in orbits around it. In the nucleus, there are Z protons and N neutrons, which are jointly called **nucleons**. So:

- $Z = $ **atomic number**, or proton number (number of protons in nucleus);
- $N = $ **neutron number** (number of neutrons in nucleus);
- $A = N + Z = $ **mass number** (or nucleon number).

In the standard notation, an element E is fully specified via its atomic and mass number as

$$\frac{A}{Z}\mathrm{E}. \tag{B1}$$

An **isotope** is any of the different types of atoms of the *same* chemical element (same atomic number Z), each having a *different* mass number (equivalently, different number of neutrons N). The notation in Eq. (B1) helps us distinguish isotopes simply as elements with the same Z but different A. Some of the best-known isotopes are those of hydrogen: deuterium ($\mathrm{D} \equiv {}^{2}_{1}\mathrm{H}$) and tritium ($\mathrm{T} \equiv {}^{3}_{1}\mathrm{H}$).

    Nuclear force has the characteristic U-shape: at large separation, nuclei attract regardless of their charge while at small separation, they repel. The typical distance separating the two regimes is around a femtometer, also known as the **fermi** ($1\,\mathrm{fm} = 10^{-15}\,\mathrm{m}$). At about that distance, the force between two nucleons is zero.

    There are three well-known nuclear processes, historically (before their true origin was necessarily known) named after the first three letters of the alphabet:

- **Alpha emission**, a process whereby the nucleus emits an α-particle (helium nucleus, or two protons and two neutrons). Thus ${}^{A}_{Z}\mathcal{P} \rightarrow {}^{A-4}_{Z-2}\mathcal{D} + {}^{4}_{2}\mathrm{He}$, where $\mathcal{P}$ and $\mathcal{D}$ are the "parent" and "daughter" nuclei.
- **Beta decay**, referring to the two reactions $n \rightarrow p + e^{-} + \bar{\nu}_e$ and $p \rightarrow n + e^{+} + \nu_e$. In the first instance an electron is emitted, and in the second a positron.
- **Gamma emission**, where the type of nucleus is not altered, but rather a very energetic photon, of energy $\sim 1\,\mathrm{MeV}$ (in the gamma-ray part of the electromagnetic spectrum) is emitted: ${}^{A}_{Z}\mathcal{P} \rightarrow {}^{A}_{Z}\mathcal{P} + \gamma$. Leptons possibly come out of the reaction as well, but not nucleons.

Box 7.1	Basics of Nuclear Physics (continued)

The total mass of an atom is not quite equal to the sum of proton and neutron masses (ignoring electrons, which are ~ 2000 times lighter). The difference between the sum of the proton and neutron masses and the total mass is the **binding energy** B, or

$$m(N, Z) = Nm_n + Zm_p - B. \tag{B2}$$

Binding energy is the energy required to "disassemble" the nucleus into its constituent nucleons by fighting against the attractive strong nuclear force. It is instructive to look at the binding energy per nucleon, B/A; see Fig. 7.1. At low A, B/A rises with nucleon number because each nucleon has more neighbors to interact with and the force between the nucleons is attractive. At high mass number ($A \gtrsim 60$), B/A decreases because the Coulomb repulsive electrostatic force between the protons starts to dominate, and also because the strong force has short range and the newly added nucleons only feel the force of their neighbors and not all nucleons in the atom. Finally, at $A \gtrsim 209$ (i.e., nuclei larger than about six nucleons in diameter) the nuclei are unstable, and spontaneously decay into lighter products.

A **magic number** is the number of nucleons (either protons or neutrons) such that they are arranged into complete shells within the atomic nucleus. The seven known magic numbers are $2, 8, 20, 28, 50, 82, 126$. Atomic nuclei consisting of such a magic number of nucleons have a higher average binding energy per nucleon than one would expect based upon simple predictions, and are hence more stable against nuclear decay. Nuclei which have *both* neutron number and proton (atomic) number equal to one of the magic numbers are called **doubly magic**, and are especially stable against decay. Examples of doubly magic isotopes include some of the most abundant and stable elements in the universe: helium-4 (^{4}He), oxygen-16 (^{16}O), calcium-40 (^{40}Ca), calcium-48 (^{48}Ca), nickel-48 (^{48}Ni), and lead-208 (^{208}Pb); see Fig. 7.1.

Let us quickly estimate the time at which BBN happens. Because $B/A \sim 1\,\text{MeV}$, we can guess – as George Gamow did in the 1940s – that BBN happens when the temperature is $T \sim 1\,\text{MeV}$, that is, roughly $T \sim 10^{10}\,\text{K}$. It turns out that the characteristic temperature is closer to $0.1\,\text{MeV}$, so let us adopt that value. At this temperature the universe was radiation dominated, so that $a \propto t^{1/2}$. We make use of the time–temperature relation in the radiation-dominated era in Eq. (4.37), and the fact that the BBN epoch comes shortly after electron–positron annihilation and neutrino decoupling (see Chapter 4), so that the effective number of relativistic degrees of freedom is $g_* \simeq 3.38$. Then

$$t \simeq 132\,\text{s} \left(\frac{0.1\,\text{MeV}}{T} \right)^2. \tag{7.1}$$

Hence, BBN takes place a couple of minutes after the Big Bang.

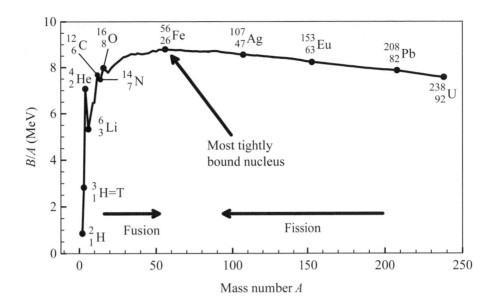

Binding energy per nucleon, B/A, of stable nuclei plotted as a function of mass number A. Fusion reactions merge lighter nuclei into heavier ones, while fission breaks up heavier nuclei into lighter ones.

7.1.2 Setting the Stage

Let us introduce the main players in the Big Bang nucleosynthesis story. First, we have the neutron and the proton. They are close in mass:

$$m_n = 939.565 \, \text{MeV}$$

$$m_p = 938.272 \, \text{MeV}, \tag{7.2}$$

with the mass difference

$$Q \equiv m_n - m_p = 1.29 \, \text{MeV}. \tag{7.3}$$

The proton is absolutely stable for all intents and purposes, as the current lower bound on its lifetime is $\sim 10^{33}$ yr. The neutron is not stable, and its decay time is about 15 minutes:

$$\tau_n \simeq 890 \, \text{s} \quad \text{(decay lifetime of neutron)}. \tag{7.4}$$

Neutrons decay via the (weak-interaction) beta decay

$$n \longrightarrow p + e^- + \bar{\nu}_e. \tag{7.5}$$

If neutrons were allowed to decay indefinitely, their remaining fraction after time t would be e^{-t/τ_n}, and would end up being negligible a few hours after the Big Bang.

The reason that neutrons do not disappear altogether is that they become bound in nuclei during BBN, and thus survive to the present-day era.

Beside protons (hydrogen nuclei, or $^1_1$H) and neutrons, we also have

- $Z = 1$: **Deuterium** ($^2_1$H; stable)
- $Z = 1$: **Tritium** ($^3_1$H; $t_{1/2} = 12$ yr)
- $Z = 2$: **Helium-4** ($^4_2$He; stable)
- $Z = 3$: **Lithium** ($^7_3$Li; stable)
- $Z = 4$: **Beryllium-7** ($^7_4$Be, $t_{1/2} = 52$ days).

Nuclei of these elements, along with neutrons, are the main players during BBN. At the absolute lowest level of approximation, it is really all about hydrogen and helium-4. In fact, as we will show in more detail in the remainder of this chapter:

> BBN results in about three-quarters of the total baryonic mass in hydrogen and one-quarter in helium-4. Several other elements (deuterium, tritium, helium-3, lithium-7) are generated in much smaller quantities, while the still heavier elements are not synthesized during this epoch.

Let us also take a moment to wax philosophical about the existence of stable deuterium. A proton and a neutron can be considered sister particles, as they differ by only 0.1 percent in mass.[1] We know however that p and n have positive binding energy, and can be combined to form a deuterium nucleus:[2]

$$p + n \rightleftharpoons D + 2.22\,\text{MeV}, \tag{7.6}$$

and that the binding energy of deuterium is

$$B_\text{D} = 2.22\,\text{MeV}. \tag{7.7}$$

The other two combinations of p and n are not stable however. The nn state is closest, being unbound (i.e., unstable) by $B_{nn} \simeq -50$ keV, or just 0.005 percent of the mass of either p or n! [The pp state is also unstable, but by a larger margin because of an additional Coulomb repulsion between the protons.] One could (and some cosmologists have), for fun, investigate what details of the early-universe physics would be different with the stable dineutron (the nn state).

---

[1] In fact, Heisenberg proposed in 1932 that p and n are related by the so-called isospin symmetry – basically, that they are different states of the same particle, thus potentially explaining their similar properties. In this picture, isospin I is a quantum number, somewhat similar to spin (hence the name), with a triplet of states for $I = 1$ and singlet for $I = 0$ all corresponding to different states of pp, pn, and nn. Isospin symmetry is useful in understanding many properties of nuclei, since the strong force is very similar for protons and neutrons, which can thus be treated as related to each other by "rotations" of isospin space. The isospin concept was also broadened to quarks after their discovery in the 1960s.

[2] The deuterium nucleus is technically called *deuteron*, but we will refer to it as deuterium to follow standard nomenclature in the field.

7.2 Leaving Thermal Equilibrium

The relative amounts of the two most abundant elements synthesized during BBN, hydrogen and helium, can be derived via a rather straightforward analysis featuring out-of-equilibrium dynamics, specifically the freezeout of neutrons. We now discuss this in some detail.

7.2.1 Thermal Equilibrium

At times safely earlier than BBN, at about 0.1 seconds after the Big Bang ($T \gg 10\,\mathrm{MeV}$), photons had much more energy than the rest energy of the electron ($0.511\,\mathrm{MeV}$). These photons were able to create electron–positron pairs via the pair production

$$\gamma + \gamma \rightleftharpoons e^- + e^+. \tag{7.8}$$

Further, neutrons and protons were in equilibrium with each other via the reactions

$$n + \nu_e \rightleftharpoons p + e^-$$
$$n + e^+ \rightleftharpoons p + \bar{\nu}_e. \tag{7.9}$$

Since neutrons and protons are in equilibrium at this time, we can calculate their abundances from the Maxwell–Boltzmann distribution

$$n_p = g_p \left(\frac{m_p T}{2\pi} \right)^{3/2} \exp\left(-\frac{m_p - \mu_p}{T} \right)$$
$$n_n = g_n \left(\frac{m_n T}{2\pi} \right)^{3/2} \exp\left(-\frac{m_n - \mu_n}{T} \right), \tag{7.10}$$

where $g_p = g_n = 2$ are the multiplicity factors for protons and neutrons, and $\mu_p(T)$ and $\mu_n(T)$ are the respective chemical potentials which ensure that $n_p, n_n \propto a^{-3} \propto T^3$. Then

$$\frac{n_n}{n_p} = \left(\frac{m_n}{m_p} \right)^{3/2} \exp\left(\frac{\mu_n - \mu_p}{T} \right) \exp\left(-\frac{m_n - m_p}{T} \right) \simeq \exp\left(-\frac{Q}{T} \right), \tag{7.11}$$

where $Q = 1.29\,\mathrm{MeV}$ is the difference between the neutron and proton masses.

7.2.2 Neutron Freezeout

If the neutron–proton fraction followed the equilibrium distributions, then the neutron fraction would just dive exponentially to zero. However, what happens here is the process described in Chapter 6: the nuclear reaction rate Γ becomes smaller than the expansion rate H, and particle species **decouple** – that is, their density **freezes out**. We now review decoupling/freezeout as applied to the case of neutrons during BBN.

Neutrinos play a key role here. At high temperature, neutrinos are kept in equilibrium with the reactions

$$\nu + \bar{\nu}_e \rightleftharpoons e^- + e^+ \tag{7.12}$$

but, more importantly, they also keep the protons and neutrons in equilibrium via the reactions listed in Eq. (7.9). Consider first the neutrino weak-interaction cross-section. It is – well, weak – and proportional to temperature squared. In particular,

$$\sigma_{\text{w}} \simeq 10^{-47}\,\text{m}^2 \left(\frac{T}{1\,\text{MeV}}\right)^2. \tag{7.13}$$

To illustrate the feebleness of σ_{w}, consider that it would evaluate to $10^{-60}\,\text{m}^2$ at recombination ($T \simeq 1\,\text{eV}$), thus being *much* smaller than the Thomson cross-section that is relevant during that time, $\sigma_{\text{T}} = 6.65 \times 10^{-29}\,\text{m}^2$.

During the BBN era the universe is radiation dominated, so that

$$T \propto a^{-1} \propto t^{-1/2}, \tag{7.14}$$

and the neutrino scattering rate Γ goes as

$$\Gamma = n_\nu c \sigma_{\text{w}} \propto a^{-3} T^2 \propto T^5 \propto t^{-5/2}. \tag{7.15}$$

Meanwhile, the expansion rate $H = \dot{a}/a$ is proportional to t^{-1}. Hence, the neutrino scattering rate falls off faster than the expansion rate and, when the two are about equal, the neutrino weak interactions decouple and the particles cannot "find each other" any more. We have already worked out, around Eq. (5.18), that neutrino decoupling happens at roughly $1\,\text{MeV}$.

Because the neutrinos' cross-section for coupling to nucleons is somewhat higher than that for coupling to leptons (electrons), and also because the nucleons are less numerous than any of the leptons and thus have a correspondingly larger number of collisions per particle, the decoupling of nucleons happens at a slightly later time than the neutrino decoupling. Quantitatively, the neutron **decoupling temperature** evaluates to

$$T_{\text{dec,n}} \simeq 0.8\,\text{MeV}, \tag{7.16}$$

so that

$$\frac{n_n}{n_p} \to \exp\left(-\frac{Q}{T_{\text{dec,n}}}\right) = \exp\left(-\frac{1.29\,\text{MeV}}{0.8\,\text{MeV}}\right) \simeq 0.2. \tag{7.17}$$

Thus, at $t > t_{\text{dec,n}}$ there is about one neutron for every five protons left.[3]

The small number of neutrons left explains why BBN is so inefficient. Given a chance, a neutron and a proton will much rather bond together to form deuterium,

$$p + n \rightleftharpoons \text{D} + \gamma, \tag{7.18}$$

_____
[3] A slightly more accurate estimate gives $n_n/n_p \simeq 1/7$.

than will two protons fuse,

$$p + p \longrightarrow D + e^+ + \nu_e, \qquad (7.19)$$

because the first reaction is fast, while the second is slow and thus has a much smaller cross-section. For the same reason, the neutron fusion reaction

$$n + n \longrightarrow D + e^- + \bar{\nu}_e \qquad (7.20)$$

is also suppressed.

This enables us to estimate the **mass fraction of helium** fused. This parameter is defined as[4]

$$Y_p \equiv \frac{\rho(^4\text{He})}{\rho_B}, \qquad (7.21)$$

where $\rho(^4\text{He})$ and ρ_B are the helium and total baryon energy density, respectively. Let us first estimate the *maximum* amount of helium that possibly could have been created. Let

$$X_n \equiv \frac{n_n}{n_{\text{nucleons}}} \simeq \frac{n_n}{n_n + n_p} \simeq \frac{1}{6} \qquad (7.22)$$

be the fraction of neutrons to all nucleons, and consider the case of 10 protons and two neutrons. The maximum helium fraction by mass is obtained when the two neutrons are combined with two of the 10 protons, yielding one helium nucleus (of mass number four), so that

$$Y_{p,\text{max}} = \frac{4}{12} = \frac{1}{3}. \qquad (7.23)$$

[See Problem 7.1 for $Y_{p,\text{max}}$ in terms of a general X_n.] This upper bound is consistent with the actual observed value of the helium mass fraction, $Y_p \simeq 0.24$. There are several corrections that will lower the actual Y_p from its theoretical maximum in Eq. (7.23) to the value today – notably, as we will see below, a fraction of the neutrons decays after neutron decoupling but before the assembly of nuclei is complete.

We will return to estimate Y_p more accurately further below. But first, we review the theoretical underpinnings that control production of the most abundant elements created during BBN.

7.3 Elemental Abundances: Theory

We now outline the reactions that lead to production of the lightest elements beyond hydrogen, and summarize our results in Figs. 7.2 and 7.3.

---

[4] The unusual notation Y_p for the primordial helium fraction is historical; "p" stands for "primordial," while X, Y, and Z traditionally refer to abundances of H, He, and "metals," respectively.

7.3.1 Deuterium Production

Let us first discuss the next significant nucleus produced during BBN – deuterium. The principal reaction that synthesizes deuterium is

$$p + n \rightleftharpoons D + \gamma, \tag{7.24}$$

and deuterium's binding energy is

$$B_D \equiv (m_n + m_p - m_D) = 2.22 \, \text{MeV}. \tag{7.25}$$

To track the equilibrium abundance of deuterium, we again resort to the Maxwell–Boltzmann distribution:

$$
\begin{aligned}
\frac{n_D}{n_p n_n} &= \frac{g_D}{g_p g_n} \left(\frac{m_D}{m_p m_n} \right)^{3/2} \left(\frac{T}{2\pi} \right)^{-3/2} \exp\left(\frac{B_D}{T} \right) \\
&\simeq 6 \left(\frac{m_n T}{\pi} \right)^{-3/2} \exp\left(\frac{B_D}{T} \right),
\end{aligned}
\tag{7.26}
$$

using the fact that $g_D = 3$ (as deuterium is spin one), $g_p = g_n = 2$, and $m_n \approx m_p \approx m_D/2$.

We now evaluate the deuterium abundance by first making use of the baryon-to-photon ratio η introduced in Chapter 6. Remembering that about five out of six baryons were unbound protons, we have

$$n_p \approx 0.8 n_B = 0.8\eta \, n_\gamma = 0.8\eta \left[0.243 T^3 \right] = 0.194 \, T^3, \tag{7.27}$$

where we adopted $n_\gamma \simeq 0.243 T^3$, which follows from Eq. (4.19) with $g_\gamma = 2$. Then Eq. (7.26) evaluates to

$$\frac{n_D}{n_n} \approx 6.6\eta \left(\frac{T}{m_n} \right)^{3/2} \exp\left(\frac{B_D}{T} \right). \tag{7.28}$$

The ratio n_D/n_n starts off $\ll 1$ before BBN, and becomes $\gg 1$ when most of the remaining free neutrons end up in helium. One can solve Eq. (7.28) to obtain the temperature for a given deuterium-to-hydrogen ratio. For example, we can define the moment of "deuterium synthesis" to be when $n_D/n_n = 1$. At this level of approximation, we get

$$
\begin{aligned}
T_{\text{BBN}} &\sim 0.07 \, \text{MeV} \\
t_{\text{BBN}} &\sim 300 \, \text{s}.
\end{aligned}
\tag{7.29}
$$

Notice the very late time (or low temperature) of the D formation – $T \simeq 0.07 \, \text{MeV} \ll B_D \simeq 2.3 \, \text{MeV}$. This is because of the small baryon-to-photon number, $\eta \simeq 10^{-9}$, or, equivalently, because of the large entropy in the universe (recall that n_B/s is a similarly small number). Indeed, η is the only free parameter in the standard BBN, and its smallness has various implications, late D formation being one of them.

We will adopt the characteristic time of BBN to be $t \sim 200 \, \text{s}$, which can be obtained from a more complete analysis than our rough estimate of deuterium synthesis above. At that time, a fraction of neutrons has decayed into protons (since, recall, $\tau_n \simeq 890\,\mathrm{s}$). So X_n goes from $\simeq 1/6$ to

$$X_n(t \sim 200\,\mathrm{s}) \equiv \frac{n_n}{n_p + n_n} \simeq \frac{\exp(-200/890)}{6} \simeq 0.13, \qquad (7.30)$$

so that the maximum helium mass fraction is in fact

$$Y_{p,\mathrm{max}} = 2X_n \approx 0.27, \qquad (7.31)$$

which is lower than the previously found maximum value in Eq. (7.23) of 0.33, and closer to the measured value of $\simeq 0.24$.

7.3.2 Creation of Helium-4

Once deuterium forms, there are other possible reactions that use deuterium, such as the creation of helium-3:

$$\mathrm{D} + \mathrm{D} \rightleftharpoons n + {}^3\mathrm{He}$$

$$\mathrm{T} + p \rightleftharpoons n + {}^3\mathrm{He}, \qquad (7.32)$$

or the creation of tritium $(\mathrm{T} = {}^3\mathrm{H})$:

$$^3\mathrm{He} + n \rightleftharpoons p + \mathrm{T}$$

$$\mathrm{D} + \mathrm{D} \rightleftharpoons p + \mathrm{T}. \qquad (7.33)$$

Tritium is unstable but, with a half-life of about 12 years, it is effectively stable during the BBN era.

Then there is the all-important synthesis of helium-4, the second most abundant element in the universe. Soon after they are formed, D, ${}^3\mathrm{He}$, and T are converted to ${}^4\mathrm{He}$ via

$$\mathrm{T} + \mathrm{D} \rightleftharpoons n + {}^4\mathrm{He}$$

$$^3\mathrm{He} + \mathrm{D} \rightleftharpoons p + {}^4\mathrm{He} \qquad (7.34)$$

$$\mathrm{T} + p \rightleftharpoons \gamma + {}^4\mathrm{He}.$$

All of these reactions involve strong nuclear force, so that they proceed with large cross-sections and efficiently, resulting in a lot of ${}^4\mathrm{He}$.

7.3.3 Abundance of Lithium and Beryllium

Lithium and beryllium are created in trace quantities (mass fractions of $< 10^{-10}$) via reactions such as

$$^4\mathrm{He} + \mathrm{T} \longrightarrow \gamma + {}^7\mathrm{Li}$$

$$^4\mathrm{He} + {}^3\mathrm{He} \longrightarrow \gamma + {}^7\mathrm{Be}, \qquad (7.35)$$

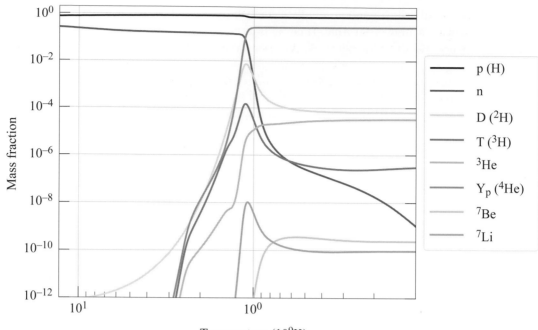

Fig. 7.2 Elemental abundances (that is, mass fractions ρ/ρ_B) produced using a computer code that takes into account many nuclear reactions occurring during the BBN era. The abundances are shown as a function of decreasing temperature (or increasing time). Numerical calculation and figure courtesy of Aidan Meador-Woodruff. A black and white version of this figure will appear in some formats. For the color version, please refer to the plate section.

but are also destroyed by reactions like

$$^7\text{Li} + p \longrightarrow {}^4\text{He} + {}^4\text{He}$$

$$^7\text{Be} + n \longrightarrow p + {}^7\text{Li}. \tag{7.36}$$

7.3.4 The Light-Element Bottleneck, and Elements Beyond ^{4}He

Big Bang nucleosynthesis produces no significant amount of elements heavier than beryllium due to the "light-element bottleneck," which is due to a few inter-related reasons:

- It takes a long time to obtain abundances of deuterium, tritium, and helium-3 that are "big" (order 10^{-4}); this happens at $T \sim 0.1\,\text{MeV}$, by which time essentially all free neutrons are locked up into ^{4}He.
- There do not exist stable elements with mass numbers $A = 5$ and $A = 8$ (i.e., those with five and eight nucleons, respectively). In stars, the bottleneck is bypassed by triple collisions of helium-4 nuclei, producing carbon (the **triple-alpha process**). However, this three-body process happens rarely in the BBN

environment with low baryon density ($n_B/s \simeq 10^{-9}$), so that the three-body collisions are unlikely.

- BBN takes place at a very low baryon-to-photon ratio ($\eta \sim 10^{-9}$), which favors[5] a high binding energy per nucleon, B/A. This helps the creation of ^{4}He, which has the highest B/A relative to its neighbors in the periodic table. There do exist elements with a higher B/A (see Fig. 7.1), but they are not reached during BBN for the reasons mentioned in the bullet points above.

The mass fractions of lithium and beryllium produced are of order $\sim 10^{-9}$. Even these at-first-sight tiny amounts are actually quite impactful, since they can be measured and compared to observations. Elements beyond beryllium are produced in negligible quantities, however.

7.3.5 BBN Predictions for the Elemental Abundances

The abundances of elements produced during BBN are shown in Fig. 7.2 as a function of (decreasing) temperature, that is, increasing time. To calculate the BBN abundances, we need to write a computer code and take into account a number of reactions, with cross-sections and reaction rates as measured in nuclear-physics experiments. These cross-sections and rates have non-negligible uncertainties in almost all cases. The first such codes were written in the late 1960s, ushering in an era of precision predictions for elemental abundances.

Keeping in mind the results shown in Fig. 7.2, we now continue our discussion of primordial abundances.

7.4 Elemental Abundances: Measurements

Measured abundances of lightest elements in the universe are in excellent agreement with the predictions of BBN, as we now describe.

7.4.1 ^{4}He Abundance Measurements

Helium-4 is the most straightforward element to test BBN observationally, since its abundance is essentially fixed by the number of free neutrons at the onset of nucleosynthesis and *that* number is determined by the decoupling of the weak-interaction $n \leftrightarrow p$ rates. Unfortunately, ^{4}He does not strongly depend on the abundance of baryons (see Fig. 7.3); a factor of 10 change in η leads only to a 15 percent change in Y_p. Hence even a precise measurement of ^{4}He cannot be used to infer η very accurately.

Helium abundance is measured by observing this element's abundance in HII (ionized hydrogen) nebulae around blue, star-forming galaxies. Because helium is produced in massive stars along with other heavy elements, its abundance needs

---

[5] Proving this statement is a bit more involved so we do not do it here, but see Kolb and Turner (1994; their Section 4.3).

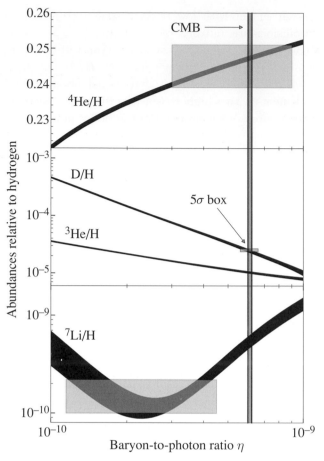

Fig. 7.3 Elemental abundances of ^{4}He, D/H, ^{3}He/H, and ^{7}Li/H, as a function of the baryon-to-photon ratio η. The curves show predictions from BBN theory, with line thicknesses representing 95 percent credible intervals due to the uncertainties in nuclear rates. The boxes show the light-element abundances measured directly from quasars and stars, with the vertical extent showing the measurement errors at 2σ (95 percent interval) and the horizontal extent showing the corresponding range in η for which the theory curve is consistent with the measurements. Note that the D/H measurement is exceptionally precise, and to make it more easily visible we show its 5σ measurement range (and the corresponding η range). The vertical line is the measurement of the baryon-to-photon ratio inferred from the Planck CMB data, again at 95 percent credible interval. The plot was made using the `alterbbn v2.2` code (Arbey *et al.*, 2020).

to be corrected for each star's metallicity in order to get the primordial amount. Current state-of-the-art measurements yield

$$Y_p = 0.245 \pm 0.003 \qquad \text{(measurements from stars).} \qquad (7.37)$$

[4]He can also be measured in the cosmic microwave background (CMB) temperature fluctuations, as it affects recombination, the process that we will study in detail in Chapter 13. The current constraint from Planck is

$$Y_p = 0.243 \pm 0.023 \qquad \text{(measurement from CMB; 95 percent C.L.).} \qquad (7.38)$$

Such precise agreement in the measurement of Y_p between stars and CMB is impressive. It tells us that the application of physics to the CMB era at $T \simeq 380,000$ years after the Big Bang agrees with the inferences applied to BBN, a process that took place $t \sim 3$ minutes after the Big Bang.

7.4.2 Deuterium as a Baryometer

Figure 7.3 reveals that the deuterium abundance is a particularly sensitive measure of the baryon-to-photon ratio η, and hence, because we know the photon density very accurately, the total density of baryons in the universe. This was first pointed out by Reeves *et al.* (1973).

In fact, the sensitivity of the deuterium abundance to η is easy to understand. After all the neutrons are locked up in nuclei and statistical equilibrium ends, there follows ∼15 minutes of deuterium burning; these are the D+D depletion reactions such as those in Eqs. (7.32) and (7.33). A higher baryon density implies it is easier for the nuclei to find each other, the deuterium burning lasts longer, and thus there is less deuterium left. And while the [4]He and [3]He are also sensitive to η for the same reason, deuterium is special because its density varies the most as a function of this all-important cosmological parameter.

But how can we measure the abundance of deuterium in the universe? The answer is provided by quasars, extremely luminous extragalactic sources thought to be powered by accreting black holes. As the light from distant quasars comes toward us, it passes through intergalactic clouds of gas. These clouds mostly contain hydrogen and helium, but they also contain a bit of deuterium – usually quoted as abundance of deuterium atoms to those of hydrogen, D/H.

When light from a quasar intercepts a neutral hydrogen cloud, it excites some hydrogen atoms in the cloud from the ground state, $n = 1$, to the first excited state, $n = 2$. This produces the absorption feature corresponding to the Lyman-α frequency, which in the rest frame is

$$\lambda_H = 1215.7\text{Å}. \qquad (7.39)$$

However, those same clouds also contain deuterium atoms, which are *also* kicked to the first excited state, leading to the Lyman-α absorption feature at a slightly shifted frequency

$$\lambda_D = 1215.4\text{Å}. \qquad (7.40)$$

When we inspect a quasar spectrum, we see a dominant Lyman-α absorption line[6] redshifted to $\lambda_H(z) = \lambda_H(1 + z)$, and then a much weaker deuterium absorption

---

[6] Ideally, *many* transitions of the Lyman ($n \to 1$) series are observed in order to constrain modeling of the absorbing system. This builds confidence that we are seeing the deuterium

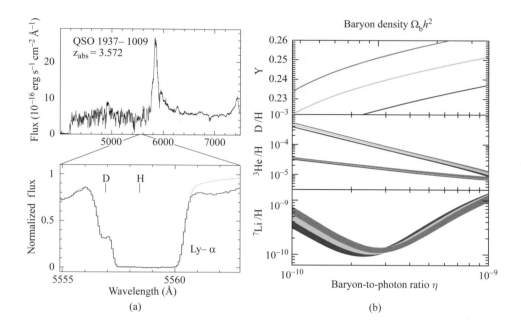

Fig. 7.4   (a) The signature that deuterium leaves in the spectrum of a distant quasar. Note that the deuterium line is very slightly shifted relative to the dominant hydrogen line, and is correspondingly hard to measure (notice the hugely stretched wavelength scale). Figure courtesy of Ken Nollett; see also Burles *et al.* (2000). (b) Similar to Fig. 7.3, except shown for three values of number of neutrino species: $N_\nu = 2$ (blue), 3 (green), and 4 (red). The line thicknesses are equal to the current observational uncertainties. Reprinted with permission from Cyburt *et al.* (2016), copyright (2016) by American Physical Society. A black and white version of this figure will appear in some formats. For the color version, please refer to the plate section.

line which is additionally shifted slightly blueward, at $\lambda_D(z) = \lambda_D(1 + z)$; see Fig. 7.4(a). The depth of the deuterium feature in the spectrum is proportional to the abundance of this element in the cloud. This measurement is challenging and requires access to the world's biggest telescopes with high-resolution spectrographs. Such measurements yield the modern value (Cooke *et al.*, 2018)

$$\left(\frac{D}{H}\right)_{\text{primordial}} = (2.53 \pm 0.03) \times 10^{-5} \qquad \text{(measurements from QSO spectra)}.$$

$$(7.41)$$

feature and not a smaller cloud of pure hydrogen separated from the main cloud by $\sim 40\,\text{km/s}$ in redshift due to either the physical (peculiar) velocity or radial separation in space.

This very precise (to about 1 percent!) astrophysical measurement can be used in conjunction with the BBN theory prediction to infer the baryon-to-photon ratio:[7]

$$\eta = (6.10 \pm 0.10) \times 10^{-10} \quad \text{(from BBN).} \tag{7.42}$$

One can further turn the measurement of η to that of the baryon density n_B, and hence to the parameter $\Omega_B h^2$:

$$\Omega_B h^2 = \frac{\rho_{B,0}}{\rho_{\text{crit},0}/h^2} = \frac{m_p(\eta\, n_{\gamma,0})}{\rho_{\text{crit},0}/h^2} \simeq 3.662 \times 10^7\, \eta, \tag{7.43}$$

where we used $m_p = 938.272\,\text{MeV}$, $n_{\gamma,0} = 4.11 \times 10^8\,\text{m}^{-3}$, and $\rho_{\text{crit},0} = 1.053\,h^2 \times 10^4\,\text{MeV}\,\text{m}^{-3}$. For the central value of η from Eq. (7.42), this gives $\Omega_B h^2 = 0.0223$.

In fact, one can go even further and compare the BBN determination of the baryon density to that from the CMB. As we will see in Chapter 13, the baryon density affects the morphology of the peaks in the CMB angular power spectrum, specifically the ratio of the heights of odd to even peaks. This allows a very precise determination of the physical baryon density. The agreement between BBN and CMB is superb – see the small rectangle in Fig. 7.3 whose vertical extent is the D/H measurement, the horizontal extent is the inference of η from BBN, and the long vertical band is the independent CMB measurement of η (converted from $\Omega_B h^2$, which the CMB actually constrains, as shown in Eq. (7.43)). The measurements are

$$\Omega_B h^2 = 0.0224 \pm 0.0002 \quad \text{(CMB)}$$
$$\Omega_B h^2 = 0.0223 \pm 0.0004 \quad \text{(BBN),} \tag{7.44}$$

where the BBN value is obtained by converting the range of η from Eq. (7.42) using Eq. (7.43).

This amazing agreement is another triumph of modern cosmology – in fact:

> The agreement between the baryon density obtained using the deuterium abundance measurements combined with BBN theory and, independently, using the CMB, provides a major, non-trivial validation of the Big Bang cosmological model.

Incidentally, the agreement in these two inferences also tells us that the assumption that deuterium is largely made during BBN (and not in stars' evolution) is correct.

7.4.3 Lithium-7 and Helium-3 Abundance Measurements

Lithium-7 is by far the least abundant primordial element whose abundance can actually be measured and compared to BBN theory. ^{7}Li is observed in the star spectra. Even more so than deuterium, ^{7}Li is subject to late-time production in a variety of astrophysical processes, but also destruction in stars, especially those of

---

[7] The error bar on η is slightly larger than that on the D/H measurement, as the BBN theory contributes error to the former. This is because of the uncertainty in some of the nuclear reaction rates.

higher metallicity. The modern value is subject to significantly more uncertainty than those of ^{4}He and D:

$$\left(\frac{^7\text{Li}}{\text{H}}\right)_{\text{primordial}} = (1.6\pm0.3)\times10^{-10} \qquad \text{(direct measurements from stars). (7.45)}$$

Note that the measured ^{7}Li abundance is apparently somewhat discrepant with BBN theory for the baryon-to-photon ratio η measured by the CMB. This so-called **"lithium problem"** may be due to unforeseen systematic errors in lithium measurements or perhaps new physics in the BBN era, and continues to be investigated by researchers in the field.

Observations of present-day abundances of helium-3 are also available, but extrapolating to the primordial value is subject to even more astrophysical systematic errors than in the case of lithium, so ^{3}He is presently not used as a probe of BBN.

7.4.4 BBN as the Probe of Beyond-the-Standard-Model Physics

BBN is a very effective probe of the beyond-the-Standard-Model (affectionately often called BSM) physics. This is because the element abundances are sensitive to the presence and properties of purported new particles or forces in the early universe.

Consider a typical case of some new particle that is relativistic around the time of BBN. This particle therefore contributes some amount g_*^{new} to the effective number of relativistic species g_*. At[8] $T \gtrsim 1\,\text{MeV}$,

$$g_*^{\text{BBN}} \simeq 5.5 + \frac{7}{4}N_\nu + g_*^{\text{new}} \simeq 10.75 + g_*^{\text{new}} \qquad (7.46)$$

for $N_\nu = 3$ (recall that 7/8 becomes 7/4 when neutrino–antineutrino multiplicity is included). Then using Eq. (4.26):

$$H^{\text{BBN}} \propto (\rho_R^{\text{BBN}})^{1/2} \propto (g_*^{\text{BBN}})^{1/2}. \qquad (7.47)$$

Recall that the decoupling/freezeout of neutrons is determined by the condition $\Gamma_{\text{weak}} \simeq H$, where Γ_{weak} is initially larger but is decreasing faster than H. Therefore, when g_* is larger, H^{BBN} is larger, and the decoupling happens earlier. This leaves more neutrons around and therefore a larger helium abundance (larger Y_p). This is how the helium abundance measurements (and those of other light elements) can be used to test for the presence of new relativistic species during BBN.

One can also test the *Standard* Model of particle physics, even without going to BSM land. For example, one can test the number of neutrino species (which, as we learned in Eq. (5.27), is $N_\nu = 3.046$, which includes a small, subtle correction that makes it non-integer). This is shown in Fig. 7.4(b). Notice that the line thicknesses reflect the nuclear-physics uncertainties. Therefore, BBN strongly favors $N_\nu \simeq 3$. Such a constraint on N_ν was first proposed in 1977, using BBN

---

[8] The value of g_* is rapidly decreasing around this time so it is hard to pin down (see Fig. 4.1). The point is that an additional particle species will increase g_* and thus H – at any given time – relative to the standard value.

theory and helium abundance measurements at the time to put an upper bound on the number of *additional* light species, $\Delta N_\nu < 5$ (Steigman *et al.*, 1977). While not impressive-sounding today, this pioneering constraint had been obtained well before particle-physics experiments, or any other method, informed us about the number of neutrino species.

Bibliographical Notes

A very short and accessible introduction to BBN is in Liddle (2015). In more detail, the topic is clearly and concisely reviewed, with emphasis on connections with data, by Cyburt *et al.* (2016), while Ryden (2016) discusses it at the level comparable to ours in this chapter. The standard reference for BBN physics is Kolb and Turner (1994), which contains a more thorough and technical treatment.

Problems

7.1 **Maximum mass fraction of helium.** Find the maximum helium fraction, $Y_{p,\mathrm{max}}$, as a function of an arbitrary neutron fraction $X_n = n_n/(n_n + n_p)$, using the same arguments as those leading to Eq. (7.23). Recall that Y_p is defined as the ratio of the mass in helium to that in all elements (essentially, helium plus hydrogen).

7.2 **Can stars create enough helium?** A very basic argument in favor of *primordial* generation of helium, made by Gamow, Hoyle, and others in the mid-twentieth century, is that fusion in stars simply cannot create the amount of ^{4}He equal to about a quarter of all baryonic mass (recall, $Y_p \simeq 0.243$).

Reconstruct this argument by considering stars in our galaxy. Assume that the baryonic matter M_B in the Milky Way consists purely of hydrogen when the latter was formed. Assume further that all of helium (and nothing but helium) forms from nuclear fusion reactions in stars. Adopt the overall mass-to-luminosity ratio in the Milky Way to be $M/L \simeq 100 M_\odot/L_\odot$ (we will talk more about M/L in Chapter 11). Finally, assume that the process of creating helium is uniform over the total age of the Milky Way, which you can conservatively take to be 10 Gyr.

What is the mass fraction in helium in our galaxy that would be generated by nuclear fusion? Is this fraction large enough to explain the observed helium abundance of $Y_p \simeq 0.243$? *Hint:* Don't forget that the baryonic and total mass are related by $M_B/M = \Omega_B/\Omega_M$.

7.3 **The what-if universe.** Imagine a parallel universe where everything is the same, down to the present author's love of basketball, except for some key differences in the early universe.

(a) Suppose that the neutron–proton mass difference is bigger. For example, take

$$m_n - m_p = 2Q$$

where $Q = 1.29\,\text{MeV}$ is the value of $m_n - m_p$ in our universe.

- Describe the BBN predicton of the helium (^4He) abundance. How would it change, and what would the new value be? Back up your conclusions with a simple analytic estimate.

- In terms of the measurement of the helium abundance quoted in Eq. (7.37), how far off is the abundance in this parallel-universe scenario – that is, "how many sigmas" is it away?

(b) Suppose the weak interactions were stronger than they actually are, so that the thermal equilibrium distribution between neutrons and protons is maintained until $T \simeq 0.25\,\text{MeV}$. Would the predicted helium abundance be larger or smaller than in the Standard Model? Explain, and be quantitative.

(c) Suppose an extra relativistic species (e.g., an extra neutrino) is added to the particle content of the universe. Would the predicted helium abundance go up or down? Explain.

7.4 **[Computational] BBN with the Boltzmann equation.** Perform a little more detailed calculation of the neutron abundance X_n, following arguments summarized here (adapted from Baumann, 2013). First, use the Boltzmann equation

$$\frac{1}{a^3}\frac{d(a^3 n_n)}{dt} = -\Gamma_n \left[n_n - \left(\frac{n_n}{n_p}\right)_{\text{eq}} n_p \right]$$

where $\Gamma_n = n_\ell \langle \sigma v \rangle$, and n_ℓ is the lepton number density. Defining

$$X_n \equiv \frac{n_n}{n_n + n_p}$$

as well as its equlibrium value discussed in the text, we can rewrite the above equation as

$$\frac{dX_n}{dt} = -\Gamma_n \left[X_n - (1 - X_n)e^{-Q/T} \right]$$

where Q is the difference in the neutron and proton mass. Defining a new variable

$$x \equiv \frac{Q}{T}$$

and making use of the chain rule

$$\frac{dX_n}{dt} = \frac{dx}{dt}\frac{dX_n}{dx} = -\frac{x}{T}\frac{dT}{dt}\frac{dX_n}{dx} = xH\frac{dX_n}{dx},$$

where we used $T \propto 1/a$ so that $\dot{T} = -H/a \propto -HT$, and where during BBN

$$H = \sqrt{\frac{8\pi}{3m_{\rm Pl}{}^2}\rho} = \frac{\pi}{3}\sqrt{\frac{g_*}{10}}\frac{Q^2}{m_{\rm Pl}}\frac{1}{x^2} \simeq (1.13\,{\rm s}^{-1})\frac{1}{x^2} \equiv H_1\frac{1}{x^2}$$

so that $H_1 \equiv 1.13\,{\rm s}^{-1}$. Then the above differential equation simplifies to

$$\frac{dX_n}{dx} = \frac{\Gamma_n}{H_1}x\left[e^{-x} - X_n(1 + e^{-x})\right].$$

We also need the expression for the neutron-to-proton conversion rate; following the details in Dodelson and Schmidt (2020), it can be evaluated as

$$\Gamma_n = \frac{255}{\tau_n}\frac{12 + 6x + x^2}{x^5},$$

where $\tau_n \simeq 887\,{\rm s}$ is the neutron decay time. One can see that the conversion time Γ_n is comparable to the age of the universe at a temperature $T \simeq 1\,{\rm MeV}$. At later times, $T \propto t^{-1/2}$ and thus $\Gamma_n \propto T^3 \propto t^{-3/2}$, so the neutron–proton conversion time $\Gamma^{-1} \propto t^{3/2}$ becomes longer than the age of the universe $t_{\rm to\,bigbang} \simeq H^{-1} \propto \rho_R^{-1/2} \propto a^2 \propto t$. Therefore we get *freezeout*, that is, the reaction rates become slow and the neutron/proton ratio approaches a constant.

Task at hand. Integrate the equation for dX_n/dx above to obtain the asymptotic value of the neutron fraction, X_n^∞. Plot X_n vs. x (or vs. T) for an interesting range of x (or T); make the x-axis logarithmic, and attach your code. *Hint:* Make sure you start with the right initial condition, that is, set $X_n^{\rm init} = X_n^{\rm eq}$.

8

Inflation

We now discuss **inflation**, a period of accelerated expansion that occurred in the very early history of the universe. Inflation is probably the most important idea in cosmology over the past 50 years (we discuss its history further below). Basic inflationary predictions have been confirmed by data, and there is great hope that upcoming cosmological observations from ground and space will shed further light on the physical processes during this period, which took place perhaps 10^{-35} seconds after the Big Bang. To start, we first outline what motivated inflation to be proposed by theorists in the first place.

8.1 Problems with the Standard Cosmological Model

The standard hot Big Bang model – the one without inflation – has a few problems that were recognized in the 1970s. These problems – the flatness, horizon, and monopole-abundance problem – each represent an observation which is very improbable in the standard cosmological model, and could only be explained as an extremely unlikely coincidence. We now describe these problems one at a time.

8.1.1 Flatness Problem

We know that the current *total* energy density of the universe is quite close to critical; recall from Eq. (2.55) that Planck satellite's measurements give

$$\Omega_{\mathrm{TOT}} = 0.999 \pm 0.002. \tag{8.1}$$

About 40 years ago, before the precision cosmology era, this number was not nearly as well measured, but it was nevertheless known that $|1 - \Omega_{\mathrm{TOT}}| \lesssim 0.5$. It turns out that even this much less precisely measured value makes Ω_{TOT} unexpectedly close to 1, as we now explain.

From Eq. (2.51), we get

$$1 - \Omega_{\mathrm{TOT}}(t) = -\frac{\kappa}{a(t)^2 H(t)^2}, \tag{8.2}$$

which is a rewriting of the Friedmann I equation. In particular, at the present time we can write (Eq. (2.52))

$$1 - \Omega_{\mathrm{TOT}} = -\frac{\kappa}{H_0^2}. \tag{8.3}$$

Dividing Eq. (8.2) by Eq. (8.3), we get

$$1 - \Omega_{\mathrm{TOT}}(t) = (1 - \Omega_{\mathrm{TOT}}) \left(\frac{H_0^2}{a(t)^2 H(t)^2} \right). \tag{8.4}$$

Let us for simplicity assume that the universe is dominated by a single component with an equation of state w, then we have

$$\frac{H^2}{H_0^2} = \frac{\rho(z)}{\rho_0} = a^{-3(1+w)}, \tag{8.5}$$

while the scale factor itself evolves as

$$a(t) = \left(\frac{t}{t_0} \right)^{\frac{2}{3(1+w)}}. \tag{8.6}$$

Therefore, we now have

$$1 - \Omega_{\mathrm{TOT}}(t) = (1 - \Omega_{\mathrm{TOT}}) \left(\frac{a^{3(1+w)}}{a(t)^2} \right)$$

$$= (1 - \Omega_{\mathrm{TOT}}) \, a^{1+3w} \tag{8.7}$$

$$= (1 - \Omega_{\mathrm{TOT}}) \left(\frac{t}{t_0} \right)^{\frac{2}{3} \frac{1+3w}{1+w}},$$

so that $1 - \Omega_{\mathrm{TOT}}(t)$ scales as a *power law in time*. So in the matter-dominated era

$$1 - \Omega_{\mathrm{TOT}}(t) \propto t^{\frac{2}{3}} \quad \text{(matter dominated)}, \tag{8.8}$$

while in the radiation-dominated era

$$1 - \Omega_{\mathrm{TOT}}(t) \propto t \quad \text{(radiation dominated)}. \tag{8.9}$$

Therefore,

> If the total energy density is close to critical today, it had to be *extremely* close to critical in the early universe.

Consider some examples. Let us assume that the universe is purely matter dominated between now and recombination ($z \simeq 1000$ or $t \simeq 300,000$ years after the Big Bang). This is not a very accurate approximation since the universe today is 70 percent dark energy dominated and was, at $z \simeq 1000$, about 10 percent radiation dominated (and 90 percent matter), but we just want a rough estimate. Then

$$|1 - \Omega_{\mathrm{TOT}}(t_{\mathrm{rec}})| \approx (1 - \Omega_{\mathrm{TOT}}) \left(\frac{300,000 \, \mathrm{yr}}{14 \times 10^9 \, \mathrm{yr}} \right)^{2/3} \lesssim 10^{-5}. \tag{8.10}$$

Next, consider the time of Big Bang nucleosynthesis, about 3 minutes after the Big Bang. It is now a better approximation to assume that the universe has been radiation dominated (not super accurate again since the universe, in the latter stages of the period since the BBN, is matter and then dark-energy dominated, but never mind). Then

$$|1 - \Omega_{\text{TOT}}(t_{\text{BBN}})| \approx (1 - \Omega_{\text{TOT}}) \left(\frac{180\,\text{s}}{14 \times 10^9\,\text{yr}} \right) \lesssim 10^{-18}. \qquad (8.11)$$

Finally, let us dare to consider how close the universe is to flat at the Planck epoch, when $t \approx 10^{-43}\,\text{s}$. The physics at the Planck time is not well understood, and the following part of the calculation is necessarily more speculative than that for the earlier epochs – yet it is nevertheless instructive. Assuming radiation domination throughout again for simplicity:

$$|1 - \Omega_{\text{TOT}}(t_{\text{Planck}})| \approx (1 - \Omega_{\text{TOT}}) \left(\frac{10^{-43}\,\text{s}}{14 \times 10^9\,\text{yr}} \right) \lesssim 10^{-63}. \qquad (8.12)$$

So at the Planck epoch the universe had to be flat to better than one part in 10^{60}! What is the physical origin of initial conditions that appear so finely tuned? Figure 8.1 illustrates how $\Omega_{\text{TOT}}(t) = 1$ is an unstable point in the evolution of the universe – as long as the universe starts out not exactly flat, Ω_{TOT} rapidly diverges from unity as the universe ages.

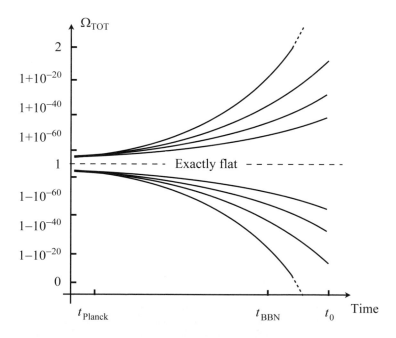

Fig. 8.1 Sketch illustrating how the flat limit, $\Omega_{\text{TOT}}(t) = 1$, is an unstable point in the evolution of the universe. To be anywhere close to flat today, the universe (without inflation) would have to have started incredibly close to flat, indicating a fine-tuning in the initial conditions.

8.1.2 Horizon Problem

Next in line is the horizon problem – the fact that, on large spatial scales, the universe looks statistically remarkably uniform, which is unexpected in a model without inflation.

Consider, for example, the cosmic microwave background radiation, which we will cover at length in Chapter 13. The most important (and remarkable!) thing about the CMB is that its anisotropies – direction-dependent fluctuations – are, on average, $\delta T/T \sim 10^{-5}$. Thus the CMB sky is uniform across the sky to one part in 100,000.

However, consider now the particle horizon distance *at the last scattering surface* (i.e., at recombination). Recall, this is given by Eqs. (3.37) and (3.38):[1]

$$d_{\mathrm{hor}}(t_{LS}) = a(t_{LS}) \int_0^{t_{LS}} \frac{dt}{a(t)} = \frac{2}{1+3w} \, H(t_{LS})^{-1}, \qquad (8.13)$$

where we have again for simplicity assumed a single-component universe with an equation of state w. Approximating the universe to be matter dominated ($w = 0$) since recombination, we have

$$d_{\mathrm{hor}}(t_{LS}) \approx 2H(t_{LS})^{-1} \approx 0.4\,\mathrm{Mpc}. \qquad (8.14)$$

In other words, the particle horizon is (approximately) twice the Hubble radius at that time.

To calculate the angle at which we "see" the particle horizon distance, we have to compute the angular diameter distance to the last scattering surface:

$$d_A(z_{LS}) = \frac{1}{1+z_{LS}} \int_0^{z_{LS}} \frac{dz}{H(z)} \approx 13\,\mathrm{Mpc}, \qquad (8.15)$$

where we evaluated the integral numerically (which you can check, e.g. by extending your angular diameter distance plot from Problem 3.5 in Chapter 3 to higher redshift).[2] Therefore, the angle subtended on the sky today by the particle horizon-sized ruler at the last-scattering surface is

$$\theta_{\mathrm{hor}} = \frac{d_{\mathrm{hor}}(t_{LS})}{d_A(z_{LS})} \approx \frac{0.4\,\mathrm{Mpc}}{13\,\mathrm{Mpc}} \approx 0.031\,\mathrm{rad} \approx 1.8\,\mathrm{deg}. \qquad (8.16)$$

So, regions of diameter only 2° on the CMB sky had a chance to be in causal contact with each other. Remember that the CMB is remarkably uniform over the whole sky, that is, over about 40,000 square degrees. Hence there are roughly $40,000/(2° \times 2°) = 10,000$ patches on the sky that were causally disconnected – yet

---

[1] The particle horizon in Eq. (3.37) was defined at the present time, so there is an additional $a(t_{\mathrm{LS}})$ factor when calculating it at time t_{LS} as we do here. Remember, this is the proper distance evaluated at the corresponding time.

[2] Don't get confused that this is apparently 1000 smaller than the well-known Hubble radius today – the latter is a proper distance, while what we have here is an angular diameter distance that has a factor $1/(1+z) \approx 1/1000$ in front. In fact, the whole calculation of θ_{hor} could be carried out in comoving coordinates; there would be no factor $a(t_{LS})$ in either Eq. (8.13) or Eq. (8.15), and the result would be the same.

all have the same temperature tuned to one part in 100,000. This is very much like seeing a group of 10,000 people wearing t-shirts of the same color; you *know* they must've been in causal contact in this regard – for example, maybe they are all fans of a soccer team and are going to a game – otherwise, such an event would be a coincidence which is simply too unlikely. The uniform CMB sky also requires such a causal explanation. This is the horizon problem.

8.1.3 The Magnetic Monopole Problem

Magnetic monopoles are putative objects which have *magnetic* charge, much like the electric charge possessed by electrons and protons for example. In undergraduate physics courses we are taught that magnetic monopoles do not exist, but known physical principles do not forbid the existence of such monopoles. If magnetic monopoles exist, the relevant Maxwell's equation simply becomes $\nabla \cdot \boldsymbol{B} = 4\pi\rho_{\mathrm{mon}}$, where $\boldsymbol{B}$ is the magnetic field and ρ_{mon} the magnetic monopole energy density.

The existence of magnetic monopoles is predicted by the Grand Unified Theory (GUT), a framework whose goal is to unify electroweak and nuclear forces. Note that gravity is conspicuously absent from this unification – in fact, combining the GUT and gravity would presumably give the full model of particles and fields (the "quantum gravity"). The GUT framework was very popular in the 1980s, as it offered the best promise toward a unified theory. The GUT is somewhat less favored now as it predicts proton decay with a lifetime of $\sim 10^{30}$ years, in apparent conflict with the current experimental lower limit of 10^{33} years.

The characteristic energy scale of the GUT is $10^{16}\,\mathrm{GeV}$. This is also the mass of magnetic monopoles that would be expected from the GUT, $m_{\mathrm{mon}} \sim 10^{16}\,\mathrm{GeV} \sim 1$ nanogram, which is very large for a particle. Moreover, we expect about one magnetic monopole per Hubble radius at the time they were created. Because monopoles are non-relativistic, they rapidly come to dominate the energy density in the universe because

$$\rho_{\mathrm{monopoles}} \propto a^{-3}, \tag{8.17}$$

and the radiation density falls off faster, as a^{-4}.

Clearly our universe is not dominated by magnetic monopoles – in fact, we did not even detect a single monopole so far! Nevertheless, if we take the prediction of their existence and abundance seriously, we have a problem, as they are overproduced. As we will see below, inflation comes to the rescue. By dramatically increasing the size of the universe, inflation simply dilutes the number and energy density of magnetic monopoles.[3]

---

[3] For this to work, inflation needs to take place *after* the GUT epoch.

8.2 Inflationary Solution

Inflation has been proposed specifically to solve the problems outlined in the previous section. We now describe how it does that, proceed with a more precise description of what inflation actually is, and end with some history of how it was discovered.

8.2.1 Solving the Coincidence Problems

What if, at some point in the early history of the universe, the energy density was dominated by a component which has $P \approx -\rho$, or $w \approx -1$? Then, the Friedmann II equation tells us that

$$\frac{\ddot{a}}{a} = -\frac{4\pi G}{3}(\rho + 3P) \approx \frac{8\pi G}{3}\rho. \tag{8.18}$$

Assuming a flat universe for a moment (in Problem 8.1 you will show that the argument remains unchanged in non-flat cases), the Friedmann I equation says

$$\left(\frac{\dot{a}}{a}\right)^2 = \frac{8\pi G}{3}\rho, \tag{8.19}$$

so that

$$\frac{\ddot{a}}{a} = \left(\frac{\dot{a}}{a}\right)^2, \tag{8.20}$$

or

$$\frac{d}{dt}\left(\frac{\dot{a}}{a}\right) = 0$$

$$\left(\frac{\dot{a}}{a}\right) = \text{const.} \equiv H \tag{8.21}$$

$$a \propto e^{Ht}.$$

Therefore

- the scale factor increases nearly exponentially with time, and
- the Hubble parameter is nearly constant during this (near-)exponential expansion.

[The word "nearly" features here because, recall, we started talking about the $\rho \approx -3P$ case, but to simplify notation and our arguments we replaced the approximate sign with an equality in the equations.] Let us further simplify our notation and introduce the **number of e-folds**, N:

$$a \propto e^N; \qquad N \equiv \ln a = \int \frac{d\ln a}{dt}\, dt \equiv \int H dt. \tag{8.22}$$

Since $H \approx$ const. during inflation, $N \simeq H\Delta t$, where Δt is the duration of inflation.

Let us now see how inflation resolves the problems of standard cosmology that we talked about at the beginning of this chapter.

• **Solution to the flatness problem.** Recalling Eq. (8.4), we can write

$$
1 - \Omega_{\mathrm{TOT}}(t_i) = [1 - \Omega_{\mathrm{TOT}}(t_f)] \left(\frac{a(t_f)^2 H(t_f)^2}{a(t_i)^2 H(t_i)^2} \right)
$$

$$
\approx [1 - \Omega_{\mathrm{TOT}}(t_f)] \left(\frac{a(t_f)^2}{a(t_i)^2} \right) \qquad (8.23)
$$

$$
\approx [1 - \Omega_{\mathrm{TOT}}(t_f)] e^{2N},
$$

where t_i is the time at the beginning of the expansion and t_f at the end and we used the fact that $H(t_i) \simeq H(t_f)$. Now, note that the most extreme fine-tuning was found by taking $t_i = t_{\mathrm{PL}}$; then $1 - \Omega_{\mathrm{TOT}}(t_i)$ had to be close to zero to one part in 10^{60}. Therefore, to solve the flatness problem we need $e^{2N} \simeq 10^N \gtrsim 10^{60}$, or

$$
N \gtrsim 60 \qquad \text{(required to solve flatness problem)}, \qquad (8.24)
$$

or at least 60 e-folds. Essentially, inflation dramatically increases the size of space, making any local patch look spatially flat. This is illustrated for a 2D space – surface of a sphere – in Fig. 8.2.

• **Solution to the horizon problem.** The proper distance prior to the exponential expansion is

$$
d_p(t_i) = a(t_i) \int_0^{t_i} \frac{dt}{a(t)}, \qquad (8.25)
$$

while the distance after inflation is

$$
d_p(t_f) = a(t_f) \int_0^{t_f} \frac{dt}{a(t)}
$$

$$
= e^N a(t_i) \left[\int_0^{t_i} \frac{dt}{a(t)} + \int_{t_i}^{t_f} \frac{dt}{a(t)} \right] \qquad (8.26)
$$

$$
\simeq e^N a(t_i) \left[\int_0^{t_i} \frac{dt}{a(t)} \right]
$$

$$
= e^N d_p(t_i),
$$

where we made an approximation by ignoring the second term in parentheses in the second line (which is actually equal to $1/H_i$). Therefore, the proper distance also increases by the factor of e^N. Recall from Eq. (8.14) that the standard cosmology predicts that the physical proper distance at recombination is $0.4\,\mathrm{Mpc}$.

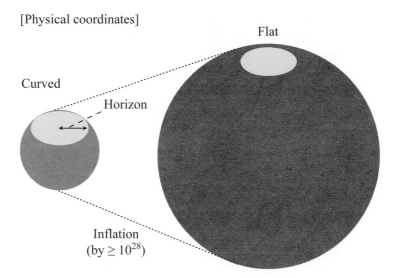

[Physical coordinates]

Flat

Curved

Horizon

Inflation
(by $\geq 10^{28}$)

Fig. 8.2 Physical-coordinates illustration of inflation. Inflation blows up the size of the universe (large dark hypersurface) to a huge size, so that the observable patch (light area), which is roughly unchanged during inflation in physical coordinates, looks flat after inflation. This is how inflation resolves the flatness problem.

With our new paradigm, the proper distance becomes at least e^{60} times that, or $\gtrsim 10^{30}$ Mpc. This is more than enough to ensure that all of the observed sky (which, remember, is only about 13 Mpc "far away" as measured by the angular diameter distance) has had a chance to come in causal contact. Therefore, inflation easily solves the horizon problem.

- **Solution to the monopole problem.** Finally, the monopole over-abundance problem is also solved, basically by diluting the number density of monopoles by $V_f/V_i = e^{3N}$. In Problem 8.2 you will have a chance to estimate the necessary number of e-folds N to achieve dilution of the monopole abundance to insignificant levels consistent with the current experimental upper bound.

We recapitulate the benefits of inflation as follows:

> An early accelerated expansion of the universe, provided it lasts for $N \gtrsim 60$ e-folds, makes the universe naturally flat, solves the horizon problem, and exponentially dilutes the abundance of any relic particles that originated before inflation.

8.2.2 Definition of Inflation

We started talking about inflation by proposing a scenario in which $P \simeq -\rho$ (or in terms of the equation of state, $w \equiv P/\rho = -1$), so that $\ddot{a}/a \simeq (8\pi G/3)\rho$. This is optimal for inflation to be effective, but a weaker condition (e.g., $w = -0.8$) may still lead to some inflation. We now discuss the precise condition for inflation.

The condition that the universe accelerates can be written in four equivalent ways:

$$\ddot{a} > 0 \iff \frac{d}{dt}\left(\frac{H^{-1}}{a}\right) < 0 \iff \rho + 3P < 0 \iff w < -\frac{1}{3}. \quad (8.27)$$

You can easily check that these conditions are equivalent. The second condition, in particular, says the following:

> The comoving Hubble radius, $(aH)^{-1}$, is shrinking during inflation.

Therefore our Hubble sphere encompasses an increasingly smaller chunk of the universe, so that our observable volume is "zeroing in" to a small part of the space that we started with; see Fig. 8.3. After the end of inflation, the universe reverts to the standard decelerating expansion, and the comoving horizon again increases in time.[4]

8.2.3 Pre-inflationary Inhomogeneities are "Inflated Away"

Consider some patch of the universe that initially (before inflation) had a *large* density fluctuation $(\delta\rho/\rho)_i$. Note that

$$\left(\frac{\delta\rho}{\rho}\right)_i \simeq \left(\frac{\nabla\rho}{\rho}\right)\left(\frac{H_i^{-1}}{a_i}\right), \quad (8.28)$$

where ∇ is the gradient that involves the comoving distance. It is easy to see what the *final* (i.e., after inflation) density fluctuation is:

$$\left(\frac{\delta\rho}{\rho}\right)_f = \left(\frac{\nabla\rho}{\rho}\right)\left(\frac{H_f^{-1}}{a_f}\right) \simeq \left(\frac{\delta\rho}{\rho}\right)_i \times e^{-N}, \quad (8.29)$$

since $a_f = a_i\, e^N$ and $H_f \simeq H_i$. Therefore, any pre-inflationary inhomogeneities within the Hubble patch are exponentially suppressed, or "inflated away" in the parlance common in cosmology. We can intuitively understand this as follows: the inhomogeneities are stretched exponentially while the (physical) Hubble radius is nearly constant. Equivalently, the *comoving* perturbation length is fixed, while the comoving Hubble radius shrinks exponentially. Either way, the effect is the same as that of zooming in to a small patch of a geographical landscape, when the bumpy mountains, fields, and canyons give way to one zoomed-in, smaller patch that is much smoother in comparison.

---

[4] Much later in the history of the universe, at $z \sim 0.5$ when dark energy starts to dominate the expansion, the universe commences a new epoch of acceleration, and the comoving horizon starts to shrink again. We will discuss dark energy in Chapter 12.

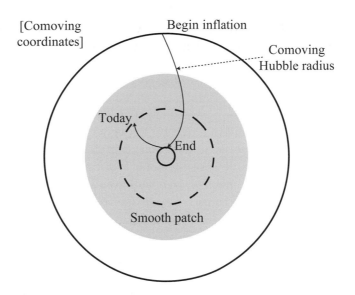

[Comoving coordinates]

Begin inflation

Comoving Hubble radius

Today

End

Smooth patch

Fig. 8.3 Evolution of the comoving Hubble radius, $(aH)^{-1}$, during and after inflation. The comoving Hubble sphere shrinks during inflation and expands after inflation. Inflation is therefore a mechanism to "zoom in" on a smooth sub-horizon patch, which is shaded in gray here. This comoving-coordinates illustration complements the physical-coordinate (that is, non-comoving) picture in Fig. 8.2. [Inspired by a similar figure in Baumann (2011).]

8.2.4 Early History of Inflation

Inflation has an interesting history. It was first proposed by a young cosmologist Alan Guth in 1981 as a possible solution to the flatness and horizon problem (Guth, 1981). Guth was, at the time, a postdoc working at the interface of particle physics and cosmology, and he had heard a talk about the flatness and horizon problems. Then he had a "spectacular realization" (according to his own notes – now at the Adler Planetarium in Chicago) that an exponential expansion would solve the problems that we just described. Upon the publication of his groundbreaking paper, Guth's idea was recognized as important almost immediately, and was a shot heard around the world (as a bonus, it got Guth multiple faculty job offers). Guth named the proposed process "inflation."

Guth's idea was based on scalar fields. A classical example of a scalar field is temperature, which is a scalar (number) and varies in space and time, $T = T(x, y, z, t)$. Gravitational potential is also an example of a scalar field, but the gravitational force, on the other hand, is a vector and not a scalar. In quantum field theory, a scalar field corresponds to space filled with particles – in fact, field excitations (locations where the field is "large") are places where we would find particles. A putative particle corresponding to the scalar field responsible for inflation is called an **inflaton**, and usually labeled with the Greek letter ϕ. The inflaton (particle) has not yet been unambiguously identified, let alone directly detected; it may well be a particle in one of the extensions of the Standard Model of particle physics.

Guth's idea was that inflation proceeds via a **first-order phase transition**, where a scalar field tunnels through a potential barrier separating the "old phase" from a "new phase," leading to exponential expansion. You can think of this as the standard tunneling of a particle through a potential barrier in elementary quantum mechanics. However, it was soon realized that such a first-order phase transition would not work because the bubbles of the old phase percolate and expand. This is similar to air bubbles in boiling water that expand and come to the surface; water (old phase) turns to vapor (new phase) via these bubbles. In cosmology, the bubbles expand at the speed of light and, if this really happened, would make the universe extremely inhomogeneous (and potentially dangerous for life!). We will not discuss the physical details of these processes, but rather just emphasize that the original Guth inflation was a brilliant idea realized in an imperfect physical scenario. In common parlance, the original Guth inflation suffers from a "**graceful exit problem**," that is, the inability to end inflation resulting in a smooth, largely uniform universe.

We now discuss the modern theory of inflation (called "new inflation" in the early 1980s, now just referred to as inflation) that does not have the graceful exit problem.

8.3 Inflation: Modern Picture

A new physical mechanism for inflation which is favored to this day was proposed soon after Guth's revolutionary paper. Linde (1982) and, independently, Albrecht and Steinhardt (1982), argued that a **second-order phase transition** would work better than Guth's first-order transition. In this scheme, there is no percolation of bubbles or the old/new phase; instead, the inflaton ϕ smoothly evolves according to its potential $V(\phi)$, as we now describe (and illustrate in Fig. 8.4).

8.3.1 Scalar-Field Physics

A **scalar field** has the kinetic energy (in natural units with $\hbar = c = 1$) of $\dot{\phi}^2/2$ and potential energy $V(\phi)$. This is pretty much equivalent to a classical particle with mass that has some kinetic and potential energy. The energy density and pressure of the scalar field are given by

$$\rho = \frac{\dot{\phi}^2}{2} + V(\phi)$$

$$P = \frac{\dot{\phi}^2}{2} - V(\phi).$$

(8.30)

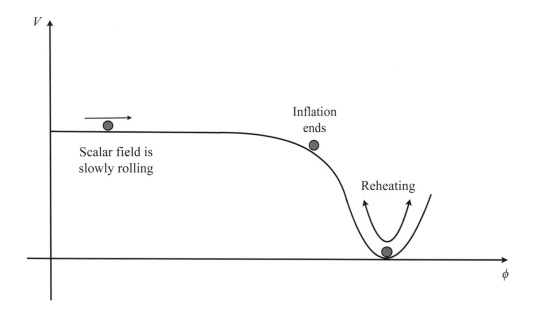

Fig. 8.4 Schematic representation of a typical inflationary scalar field potential, and the important stages of inflation.

There are also terms corresponding to spatial gradients of the scalar field, $(\nabla \phi)^2$, but we ignore those; this turns out to be a very good approximation (we will return to inhomogeneities later). Further, the continuity equation

$$\dot{\rho} + 3H(P + \rho) = 0 \tag{8.31}$$

can be rewritten as (plugging in P and ρ from Eq. (8.30) and dividing all terms by $\dot{\phi}$)

$$\ddot{\phi} + 3H\dot{\phi} + V'(\phi) = 0, \tag{8.32}$$

where dots are derivatives with respect to physical time and prime is the derivative with respect to ϕ. Using the language of quantum field theory, this is the Klein–Gordon equation for the field ϕ – a relativistic version of the Schrödinger equation that describes the behavior of scalar (spin-zero) particles.

Equation (8.32) therefore describes the temporal evolution of the inflaton field. Note that this is the same as the equation that governs the motion of a ball rolling with friction down a hill:

$$\ddot{x} + C\dot{x}/m + V'(x)/m = 0, \tag{8.33}$$

where x is the displacement measured down the slope, $V' = d(mgh)/dx = mg\sin(\theta)$ is the change of potential energy per x, θ is the angle that the hill forms with respect to the horizontal, and C/m is the coefficient multiplying the velocity $\dot{x}$ that gives the friction force. Setting $m = 1$ and $C = 3H$, we can see that Eq. (8.33) is equivalent to Eq. (8.32).

This is why cosmologists often talk about inflaton "rolling down its potential" – the scalar field is actually not rolling down anything, it's just that the evolution of inflaton is *equivalent* to a ball rolling down an inclined plane from your freshman physics course.

8.3.2 Reheating after Inflation

Inflationary's "graceful exit" implies that, in addition to inflation ending, the energy density in the inflaton needs to disappear, so $V(\phi)$ has to go to zero. However, the universe also needs to give birth to the particles of the Standard Model (quarks, gluons, leptons, photons, etc.) from what is, at the time, a very empty space with negligible energy density of any species except for the inflaton.

This conversion of the inflaton field to Standard-Model particles goes under the name of **reheating**. In the standard reheating picture, $V(\phi)$ goes to zero smoothly, and has some curvature V'' around the minimum. The field oscillates rapidly around the minimum, and in the process decays to particles.

Let us investigate the reheating process in a little more detail. During the rapid oscillations of the inflaton, equipartition of energy tells us that the time averages of the kinetic and potential energy are equal to each other:

$$\left\langle \frac{\dot{\phi}^2}{2} \right\rangle = \langle V(\phi) \rangle, \tag{8.34}$$

meaning that the time-averaged pressure, $P = \dot{\phi}^2/2 - V$, is zero. Then the Klein–Gordon equation, Eq. (8.32), becomes $\dot{\rho}_\phi + 3H\rho_\phi = 0$. Adding the aforementioned coupling term for decay into particles, this becomes

$$\dot{\rho}_\phi + (3H + \Gamma_\phi)\rho_\phi = 0. \tag{8.35}$$

The coupling Γ_ϕ depends on potentially complicated physics, and can be calculated in a full model of inflation (i.e., in any reheating model that starts from the Lagrangian for all fields involved). In particular, Γ_ϕ is typically specified by the dominant Feynman diagrams for the decay of inflaton into Standard-Model particles.

After the decay of inflaton, a process of *thermalization* occurs, whereby the new (potentially heavy) particles may decay and interact. These particles had been created far out of thermal equilibrium, so during thermalization and these further decays/interactions, the final products finally settle into the thermal equilibrium, and the standard, hot Big Bang cosmology may begin.

It is perhaps not surprising that reheating (as well as preheating, see Box 8.1) is tough to test using currently available cosmological observations or laboratory

| Box 8.1 | Preheating |

The reheating process outlined in Sec. 8.3.2 is all that happens for fermions. For bosons, an interesting variant of reheating can be at work; it's called parametric-resonance reheating or **preheating**. [Parametric resonance generally refers to properties of some oscillating systems to exhibit oscillations at certain frequencies but instabilities or quasi-oscillations at others.]

In the preheating scenario, the fact that we are talking about the particles that obey the Bose–Einstein statistics allows their very high occupation number of quantum states. Assuming, for example, the coupling between inflaton ϕ and another field χ of the form $\mathcal{L} \propto \phi^2 \chi^2$, Fourier modes of particles χ can be shown to obey an equation of the form

$$\ddot{\chi}_k + \left[A + B \sin^2(Ct) \right] = 0, \tag{B1}$$

for some constants A, B, C which depend on the model. This is a Mathieu equation, well known to scientists studying chaotic systems; it exhibits instability strips in its parameter space. As a result, the occupation number of χ particles can be huge, and the inflaton's conversion to χ particles is rapid. Preheating is a nonlinear, non-perturbative process that occurs far out of equilibrium, and it leads to a non-trivial and interesting phenomenology.

experiments. Reheating is a process that almost certainly happened (given that inflation did), but its direct experimental verification is currently beyond our reach.

Summarizing, the elements of modern inflation are:

- The slow-roll regime, where the field is "slowly rolling down the hill" for at least 60 e-folds (and possibly many more).

- The reheating process, where the field comes to a concave part of the potential and oscillates rapidly, converting inflaton field into particles (i.e., into radiation). After this the universe is – again – radiation dominated, and its expansion decelerates.

See Fig. 8.4 for a schematic representation.

8.3.3 Conditions for Inflation and Slow-Roll Parameters

What is required for inflation to take place? As already discussed, clearly we need an inflaton. Aside from the field itself, we need its classical potential $V(\phi)$; note from Eq. (8.32) that the dynamics of the field is fully specified by this function. But what kind of potential $V(\phi)$ will support inflation?

It is clear that the field should be rolling slowly, since we need

$$\frac{P}{\rho} \simeq -1, \quad \text{or}$$

$$\frac{\frac{\dot{\phi}^2}{2} - V(\phi)}{\frac{\dot{\phi}^2}{2} + V(\phi)} \simeq -1, \quad \text{or} \tag{8.36}$$

$$\dot{\phi}^2/2 \ll V(\phi),$$

so that the potential energy of the field should dominate the kinetic energy. In other words:

> During inflation, the inflaton field evolves ("rolls down the potential") slowly.

After some initial transient time, the field will settle into the regime where the friction saturates and the second derivative is approximately zero, $\ddot{\phi} = 0$. This is called the **slow-roll approximation**. Then Eq. (8.32) simplifies to

$$3H\dot{\phi} + V'(\phi) \simeq 0 \qquad \text{(slow-roll approximation)}, \tag{8.37}$$

or

$$\dot{\phi} \simeq -\frac{V'(\phi)}{3H}. \tag{8.38}$$

In other words, defining the first **slow-roll parameter**[5]

$$\epsilon \equiv \frac{m_{\text{Pl}}{}^2}{16\pi} \left(\frac{V'}{V}\right)^2, \tag{8.39}$$

inflation requires (Problem 8.3)

$$\epsilon \ll 1. \tag{8.40}$$

That is, the slope of the potential needs to be suitably small.

It turns out that the *second* derivative of the potential also needs to be small. Let us define the second slow-roll parameter

$$\eta \equiv \frac{m_{\text{Pl}}{}^2}{8\pi} \frac{V''}{V}. \tag{8.41}$$

---

[5] There exist multiple definitions of slow-roll parameters in the literature. One notable alternative to our definition is that through derivatives of the Hubble parameter, $\epsilon_H \equiv (m_{\text{Pl}}{}^2/4\pi)(H'/H)^2$ and $\eta_H \equiv (m_{\text{Pl}}{}^2/4\pi)(H''/H)$.

Then the condition $|\ddot{\phi}| \ll |(3H\dot{\phi})|, |V'|$ also requires (see again Problem 8.3)

$$\eta \ll 1. \tag{8.42}$$

The end of inflation is most sensibly defined as the moment when $\ddot{a} = 0$, that is, when the universe goes from accelerating to decelerating. This corresponds to (Problem 8.4)

$$\epsilon = 1 \qquad \text{(end of inflation)}. \tag{8.43}$$

The slow-roll parameters ϵ and η are very useful in connecting the physics of inflation to observations, as we will demonstrate in the remainder of this chapter.

8.3.4 Last Out, First In

During inflation, physical Fourier modes of the density fluctuations are stretched enormously. When their wavelength exceeds the Hubble distance, we say that these modes "leave the horizon." We usually consider this situation in comoving coordinates, where the comoving wavelength of the modes is fixed, while the (comoving) horizon decreases exponentially. After inflation ends, the modes start re-entering the horizon. For example, the mode that enters the horizon at recombination has wavelength of about $H(t_{\rm rec})^{-1} \simeq 100\,{\rm Mpc}$ comoving, while that entering the horizon today has wavelength of $\sim 5\,{\rm Gpc}$ comoving. Figure 8.5 illustrates two modes that leave then re-enter the horizon.

Let us now consider different-wavelength modes leaving and re-entering the horizon. From Fig. 8.5, it is clear that the sequence is

$$\text{Last out} \Longleftrightarrow \text{First in}.$$

In other words, modes that leave the horizon last re-enter it first, and vice versa. For example, right after inflation ended, some mode that had just left the horizon re-entered it; that mode has a very short wavelength compared to cosmological distances today. And conversely, the mode that enters the horizon today (of size H_0^{-1}) left the horizon well before the end of inflation. This is the same rule as for passengers boarding a very busy bus: the poor last passsenger that boards – the one hanging off the door – will be the first one to leave. And the first passenger who boarded, the one stuck in a corner seat, will be the last one to leave.

Let us in fact estimate when the mode that corresponds to observable scales today exited the horizon. Assume for the moment a GUT-scale inflation with $T_{\rm end\,inf} \simeq 10^{15}\,{\rm GeV}$, and recall that the temperature today is $T_0 \simeq 10^{-3}\,{\rm eV}$. In the following, for simplicity we ignore the details of the reheating process. Since $aT \simeq {\rm const.}$

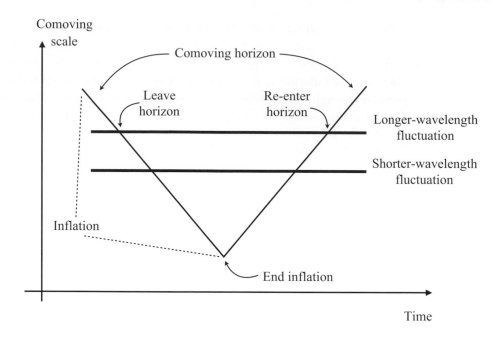

Fig. 8.5 Illustration of the evolution of the horizon and how perturbations leave and re-enter it, shown in comoving coordinates. The comoving horizon, $(aH)^{-1}$, shrinks during inflation and grows thereafter. A density fluctuation that was first inside the horizon leaves it during inflation, then re-enters at some point after the end of inflation. A shorter-wavelength perturbation leaves the horizon later and re-enters earlier than the longer-wavelength one, thus illustrating the "last out, first in" rule discussed in the text. For a perfectly scale-invariant spectrum with $n_s = 1$, the amplitude of each density-fluctuation mode upon horizon re-entry is the same, since $\delta_H^2 \propto \Delta_s^2 \propto k^{n_s-1} \propto \text{const.}$ [Inspired by a similar figure in Baumann (2011).]

during the adiabatic expansion of the universe (again for simplicity also ignoring any changes in the number of relativistic degrees of freedom g_{*S}), we have

$$\frac{a_0}{a_{\text{end inf}}} \simeq \frac{T_{\text{end inf}}}{T_0} \simeq \frac{10^{15}\,\text{GeV}}{10^{-3}\,\text{eV}} \simeq 10^{27}. \tag{8.44}$$

Therefore, the first mode to enter the horizon, right after inflation, has of order 10^{27} times shorter wavelength than the mode that is entering the horizon today. Hence the Hubble scale today left the horizon during inflation this many e-folds before the end of inflation:

$$N_{\text{obs}} = \ln\left(\frac{a_0}{a_{\text{end inf}}}\right) = \ln(10^{27}) \simeq 60. \tag{8.45}$$

Here we used the index "obs," to indicate that we observe the Hubble-scale physics today (in the CMB maps, for example) – hence we observe the conditions laid out by inflation about 60 e-folds before its end. In yet other words, we have observational access to physical conditions about 60 e-folds before the end of inflation.

We also observe scales smaller than the Hubble scale today, and hence some range of e-folds before the end of inflation. Let us now estimate this range. The largest scale we observe today is clearly the aforementioned $L_{\text{largest obs}} \simeq H_0^{-1} \simeq 5\,\text{Gpc}$. The *smallest* cosmologically observable scales which shed light on primordial physics are, very roughly, scales comparable to the size of the galaxy cluster, or $L_{\text{smallest obs}} \simeq 1\,\text{Mpc}$. [While we do observe smaller scales in general, they do not tell us about the primordial universe – think, for example, of the solar system which is very precisely measured but whose observed features have physical origin in the later-time evolution of the universe.] Hence the range of e-folds that corresponds to cosmologically observable scales today is

$$\Delta N_{\text{obs}} = \ln\left(\frac{L_{\text{largest obs}}}{L_{\text{smallest obs}}}\right) \simeq \ln\left(\frac{5\,\text{Gpc}}{1\,\text{Mpc}}\right) \simeq 8, \tag{8.46}$$

where, again, our estimates are very approximate but sufficient for our main take-aways. We conclude:

> Cosmologically observable structures were imprinted around 60 e-folds before the end of inflation, in a window lasting around 8 e-folds. This is the (only) epoch of inflation that we can directly probe with our observations.

Note that the number of e-folds that we estimated depends on the scale of inflation, on the reheat temperature, on the expansion history between the end of inflation and BBN, and on the number of relativistic species. As a result, the number of e-folds before the end of inflation from which we get our observables varies between about 40 and 70. We will not further pursue such details here.

8.4 Quantum Fluctuations of Inflation

One of the main triumphs of inflation is its success in laying out seeds of structure in the universe.

The source of these seeds are quantum fluctuations in the scalar field ϕ – the "quantum jiggles." The result that is derived much more rigorously using quantum field theory can also be roughly estimated by a very simple application of the Heisenberg uncertainty principle, $\delta E\,\delta t \approx \hbar/2 = 1/2$ (adopting natural units as usual). In natural units, ϕ has the units of energy so that $\delta E \simeq \delta\phi$, while the only relevant timescale in the problem is given by the Hubble time, $\delta t \sim 1/H$. Hence, the quantum fluctuation of inflaton is roughly equal to the Hubble parameter during inflation

$$\delta\phi \sim H. \tag{8.47}$$

[A slightly more sophisticated argument, often seen in the cosmology literature, gives $\delta\phi \simeq H/(2\pi)$.] At any rate:

> Quantum fluctuations of inflaton provide the seed for the origin of structure. From them arise the metric and matter perturbations and, eventually, the galaxies and galaxy clusters observed today.

We now talk about this process in a bit more detail.

8.4.1 Density Perturbations During Inflation: Gauge Issues

In order to theoretically predict the amplitude and statistics of the density fluctuations observed in the universe today, we need to perturb Einstein's equations of general relativity as follows:

$$\delta G_{\mu\nu} = 8\pi G \delta T_{\mu\nu}, \tag{8.48}$$

where $G_{\mu\nu}$ is the metric tensor, $T_{\mu\nu}$ is the stress-energy tensor, and G Newton's constant. This equation relates the inflaton perturbations (that enter $T_{\mu\nu}$) to ones in the metric (in $G_{\mu\nu}$). Because the primordial perturbation amplitude is small, $\delta \simeq 10^{-5}$, it is an excellent approximation to consider *linear* perturbation theory in all quantities of interest.

One issue that immediately comes up is that of **gauge freedom**, which is essentially freedom to select a set of coordinates. In a perfectly homogeneous universe without any perturbations, slices of constant time correspond to slices of constant density. In the presence of perturbations, however, it turns out that there is freedom to carry out the time slicing. For example, you could select the frame where $\rho(t, \mathbf{x})$ is uniform (the uniform-density gauge), or one where there exists a set of comoving observers who fall freely without changing their spatial coordinates (the synchronous gauge). The important point is that, when working in a specific gauge, one has to be careful how to interpret the results. Some results will be physical, observable effects, while others will be the so-called gauge artifacts. One can also work in a gauge-independent formalism, pioneered by Bardeen (1980).

The effect of gauge dependence can be seen in expressions for the metric of spacetime. For example, the familiar form for the metric

$$ds^2 = -(1 + 2\Psi)dt^2 + a^2(1 + 2\Phi)(dx^i)^2 \tag{8.49}$$

represents the metric in one particular gauge – the so-called Newtonian gauge. In other gauges the metric, as well as the matter perturbations, take different forms.

8.4.2 Curvature Perturbation

Speaking of the gauge-independent quantities, a very important one is the **comoving curvature perturbation** $\mathcal{R}$. Physically, this quantity measures the spatial curvature by local observers.[6] Referring to the flat-case ($\kappa = 0$) Friedmann I Eq. (2.46), one can introduce $\mathcal{R}$ as

$$\left(\frac{\dot{a}}{a}\right)^2 = \frac{8\pi G}{3}\rho + \frac{2}{3}\nabla^2\mathcal{R}. \tag{8.50}$$

We can see why the curvature perturbation deserves its name – it is directly related to the perturbations in curvature: $(2/3)a^2\nabla^2\mathcal{R} = \delta k_{\mathrm{curv}}$, where we temporarily label the curvature parameter k_{curv} to distinguish it from the wavenumber k. In fact, $-(2/3)a^2k^2\mathcal{R}_{\mathbf{k}}$ can be thought of as a Fourier component of the locally measured curvature.

We shall not rigorously treat $\mathcal{R}$ here, but will rather motivate its relation to inflaton fluctuations $\delta\phi$. Curvature fluctuations can be thought of as fluctuations in the duration of inflation at different points in space that are caused by fluctuations in the field ϕ:

$$\mathcal{R} \approx H\delta t \simeq -H\frac{\delta\phi}{\dot{\phi}}, \tag{8.51}$$

where $\dot{\phi}$ is the inflaton speed. In other words, $\mathcal{R}$ is higher in parts of space where $\delta\phi$ fluctuates down ($\delta\phi < 0$), that is, where the inflation lasts a longer amount of time ($\delta t > 0$). And vice versa. The expression in Eq. (8.51), which can be rigorously derived using a *much* lengthier and more technical calculation than shown here, will be important in what follows.

The comoving curvature perturbation $\mathcal{R}$ has a crucial property that it is conserved on superhorizon scales. That is, between the time that some fluctuation mode exits and re-enters the horizon, $\mathcal{R}$ doesn't change. This makes $\mathcal{R}$ a key quantity to bridge the time from when some quantum fluctuation $\delta\phi$ of some wavelength leaves the horizon, and the time when it re-enters as a classical density fluctuation. In particular, $\mathcal{R}$ can be related to inflationary physics at the observationally relevant time, when modes leave the horizon, via Eq. (8.51). And, to obtain the matter (and radiation, etc.) perturbations observed in the universe, we simply evaluate the power spectrum of $\mathcal{R}$, as we explain next.

8.4.3 Quantum-Mechanical Treatment of Perturbations

Above we presented a slick argument to evaluate quantum fluctuations of inflaton, $\delta\phi \simeq H/(2\pi)$, using the Heisenberg uncertainty principle. To properly and robustly calculate the full spectrum of perturbations, a much more detailed analysis is warranted. It goes as follows:

1. Start with a free-field action for the inflaton scalar field.
2. Expand the action to second order in inflaton fluctuations $\delta\phi$.

---

[6] This statement of (gauge-independent) $\mathcal{R}$ is itself gauge dependent. In particular, $\mathcal{R}$ is identified as the curvature perturbation in comoving (or constant-ϕ) gauge, where the velocities vanish. Moreover, the expression for $\mathcal{R}$ in Eq. (8.51) holds only in the so-called spatially flat gauge. In the spirit of a simpler presentation, we sweep these details under the rug.

3. Using the classical (Euler–Lagrange) equations of motion applied to this action, derive the evolution equation for $\delta\phi$; it will be of the simple harmonic motion form.

4. Promote the classical field $\delta\phi$ to a quantum operator ("quantize it"), and calculate the zero-point quantum oscillations of this oscillator.

5. At horizon crossing (i.e., when $k = aH$), convert from fluctuations $\delta\phi$ to the conserved curvature perturbation $\mathcal{R}$. The curvature perturbation's key property is that it is constant at superhorizon scales. This fact enables us to connect the field fluctuations $\mathcal{R}$ at horizon exit to those in curvature at horizon entry.

6. Calculate the power spectrum of $\mathcal{R}$. This (modulo some conventional prefactors) is essentially the primordial matter power spectrum that will be discussed at length in Chapter 9.

7. If desired, repeat the whole procedure described above for the tensor fluctuations. This will give us a prediction for the power spectrum of primordial gravitational waves.

The calculation above is lengthy and technical. We therefore opt to quote its final result, then proceed to analyze and discuss it.

8.4.4 Power Spectrum of Curvature Perturbations

The calculation outlined above results in the **power spectrum of curvature perturbations** $\mathcal{R}$, also known as the **scalar power spectrum**. We will discuss the concept of the power spectrum from first statistical principles in Chapter 9, and here we outline it only briefly. First, the two-point correlation function of $\mathcal{R}$ is defined as

$$\langle \mathcal{R}_{\mathbf{k}} \mathcal{R}_{\mathbf{k}'}^* \rangle = (2\pi)^3 \, \delta^{(3)}(\mathbf{k} - \mathbf{k}') \, P_{\mathcal{R}}(k), \tag{8.52}$$

where the Dirac delta function signifies the fact that unequal modes are uncorrelated, while the fact that the power spectrum $P(k)$ is a function of $k = |\mathbf{k}|$ is due to homogeneity of the universe (again, please refer for details to Chapter 9). Next, we define the dimensionless scalar power spectrum as

$$\Delta_s^2 \equiv \Delta_{\mathcal{R}}^2(k) = \frac{k^3}{2\pi^2} \, P_{\mathcal{R}}(k). \tag{8.53}$$

The result of the lengthy computation outlined in Sec. 8.4.3 is the power spectrum of $\mathcal{R}$. This important result can actually be correctly surmised from our back-of-the-envelope arguments, by combining Eqs. (8.47) and (8.51):

$$\Delta_s^2(k) \equiv \Delta_{\mathcal{R}}^2(k) = \left(\frac{H_{60}}{2\pi} \right)^2 \left(\frac{H_{60}}{\dot{\phi}_{60}} \right)^2 \quad \text{(scalar power spectrum)}, \tag{8.54}$$

where H_{60} is the Hubble parameter evaluated when the desired mode leaves the horizon – so, $k = aH_{60}$.

8.4.5 Tensor Perturbations

As mentioned earlier, tensor perturbations generated during inflation also play an important role as they lead to the prediction of a stochastic background of gravitational waves. We will not review the theory behind gravitational waves, but rather briefly remind the reader that the tensor fluctuations are described by two functions, h_+ and $h_\times$ (see also Box 8.2). The analysis of tensor modes is actually simpler than for scalar perturbations due to the lack of gauge ambiguities.

Box 8.2 | **Scalar–Vector–Tensor Decomposition**

The treatment of metric and density perturbations in three-dimensional space is facilitated by the fact that they mathematically decompose into different "modes" which evolve independently. This is very similar to the statement in classical electromagnetism that any arbitrary force can be decomposed as a sum of the grad of some scalar potential Φ and the curl of some vector potential $\boldsymbol{A}$:

$$\boldsymbol{F} = \boldsymbol{\nabla}\Phi + \boldsymbol{\nabla} \times \boldsymbol{A}. \tag{B1}$$

Similarly, a traceless, symmetric stress tensor h_{ij} can be decomposed as

$$h_{ij} = h_{ij}^S + h_{ij}^V + h_{ij}^T, \tag{B2}$$

where the three tensors on the right denote the scalar (S), vector (V), and tensor (T) fluctuations. In particular:

- Scalar perturbations correspond to the honest-to-goodness fluctuations that are responsible for the structure observed today. In the process of reheating, these fluctuations are transferred to particles (photons, baryons, leptons, dark matter).
- Vector perturbations typically decay with time, and their effects are, at least in standard general relativity, negligible today.
- Tensor fluctuations manifest themselves as gravitational waves. Note that this is a stochastic (random) background of gravitational waves, not those found by LIGO that are due to mergers of black holes or neutron stars. The amplitude of this stochastic background is proportional to the energy scale of inflation (see Sec. 8.5.3).

For example, for a plane-wave density perturbation varying along the z-direction, we have

$$h_{ij}^S = \frac{1}{3}\begin{pmatrix} h & 0 & 0 \\ 0 & h & 0 \\ 0 & 0 & -2h \end{pmatrix}; \quad h_{ij}^V = -\frac{i}{2}\begin{pmatrix} 0 & 0 & h_{xz} \\ 0 & 0 & h_{yz} \\ h_{xz} & h_{yz} & 0 \end{pmatrix}; \quad h_{ij}^T = \begin{pmatrix} h_+ & h_\times & 0 \\ h_\times & -h_+ & 0 \\ 0 & 0 & 0 \end{pmatrix},$$

where h, $h_{+,\times}$, and $h_{xz,yz}$ are some quantities. Note the traceless nature of tensor fluctuations in the transverse directions (here, the x–y plane), a familiar mathematical property of gravitational waves.

The power spectrum for each of the two polarization modes, $h = \{h_+, h_\times\}$, is

$$\langle h_{\mathbf{k}} h_{\mathbf{k}'} \rangle = (2\pi)^3 \, \delta^{(3)}(\mathbf{k} - \mathbf{k}') \, P_h(k). \tag{8.55}$$

As we did with the scalar perturbations, we define Δ^2 in tensors:

$$\Delta_h^2 \equiv \frac{k^3}{2\pi^2} P_h(k). \tag{8.56}$$

The power spectrum of tensor perturbations is the sum of the power spectra for the two polarizations h_+ and $h_\times$, so simply

$$\Delta_t^2 = 2\Delta_h^2. \tag{8.57}$$

The inflationary prediction is

$$\Delta_h^2(k) = \frac{32\pi}{m_{\mathrm{Pl}}{}^2} \left(\frac{H_{60}}{2\pi} \right)^2, \tag{8.58}$$

so that

$$\Delta_t^2(k) \equiv 2\Delta_h^2(k) = \frac{64\pi}{m_{\mathrm{Pl}}{}^2} \left(\frac{H_{60}}{2\pi} \right)^2 \quad \text{(tensor power spectrum)}. \tag{8.59}$$

8.5 Connections to Observations

Inflationary theory has gained wide acceptance in cosmology because it is not only testable, but – thus far – in an excellent agreement with observations. We now review how the rather theoretical phenomena we discussed thus far connect to observations.

8.5.1 Scalar Spectral Index in Slow-Roll Parameters

Perhaps the most interesting thing about the scalar power spectrum is its shape in k. To study it, we make use of the **scalar spectral index** (or tilt), defined as

$$n_s - 1 \equiv \frac{d \ln \Delta_s^2}{d \ln k}. \tag{8.60}$$

We will return to the scalar spectral index in Chapter 9, where we will see that n_s was correctly postulated to be close to unity by Harrison, Zeldovich, and Peebles in 1969 – more than a decade before inflation was invented.[7]

---

[7] The Harrison–Zeldovich–Peebles conjecture refers to the matter power spectrum to be discussed in Chapter 9, roughly $P(k) = |\delta_k|^2 = Ak^{n_s}$, with A some constant. Equations (8.53) and (8.60), on the other hand, imply that $P_{\mathcal{R}} \propto k^{n_s - 4}$. Because $\mathcal{R}$ and δ_k are related by the Poisson equation, where $\delta_k \propto k^2 \mathcal{R}$, the two definitions of the spectral index, one via δ_k and one via $\mathcal{R}$, are consistent.

One can always define the derivative of a k-dependent function; the utility of Eq. (8.60) is that inflation predicts the scalar spectral index to be nearly k-independent – that is, the comoving curvature power spectrum is nearly a power law in k. In fact, inflation generically produces an *almost* scale-invariant spectral index, where scale invariance refers to the $n_s \to 1$ limit. Inflation predicts (Problem 8.5)

$$n_s = 1 - 6\epsilon_{60} + 2\eta_{60} + O(\epsilon_{60}^2, \eta_{60}^2), \qquad (8.61)$$

where the subscripts denote the fact that the slow-roll parameters are to be evaluated ~ 60 e-folds before the end of inflation. Of immediate note are the following facts:

- Inflation predicts n_s to be close to unity. For many inflationary models (though certainly not all), the term $-6\epsilon_{60}$ dominates over $+2\eta_{60}$, resulting in n_s slightly less than one – that is, a slightly red[8] spectrum. In another broad class of models (ones with a concave potential, $V'' < 0$), the second slow-roll parameter often dominates, but $\eta_{60} < 0$, leading again to a red spectrum.
- The generic inflationary prediction on Eq. (8.61) is in fabulous general agreement with modern measurements; $n_s = 0.966 \pm 0.005$ (see Table 3.1).

Inflation does predict small departures from scale independence of curvature power spectrum, that is, from the assumption that $n_s = \text{const}$. This is often encoded by the next parameter in the Taylor expansion in log wavenumber, the **running of the spectral index**:

$$\alpha_s \equiv \frac{d \ln n_s}{d \ln k} \propto O(\epsilon_{60}^2, \eta_{60}^2) \times (k\text{-dependent terms}). \qquad (8.62)$$

Current measurements already constrain the running to $\alpha_s = -0.005 \pm 0.007$. These measurements, while interesting, nevertheless have an order-of-magnitude larger error than the prediction of simple inflationary models, $\alpha_s \simeq (n_s - 1)^2 \simeq 0.001$. Therefore, inflationary prediction for the running of the spectral index will only be sharply tested with future surveys.

Summarizing, the scalar power spectrum can be written as:

$$\Delta_s^2(k) = A_s(k_{\text{piv}}) \left(\frac{k}{k_{\text{piv}}} \right)^{n_s - 1 + \alpha_s \ln(k/k_{\text{piv}})}, \qquad (8.63)$$

where A_s is its normalization and k_{piv} is an arbitrary pivot scale.

Let us finally run a small victory lap to celebrate the results that we obtained.

---

[8] Much like with the colors of a rainbow, a red spectrum is the one that favors large wavelengths, that is, small wavenumbers, and this is exactly what happens with k^{n_s} for $n_s < 1$. The converse holds for a blue spectrum, $n_s > 1$, which is now disfavored by data.

Equation (8.63), along the spectral index expression in Eq. (8.61), is a major prediction of inflation. This is *almost* what we measure with galaxy surveys today. The curvature power spectrum $\Delta_{\mathcal{R}}^2 \equiv \Delta_s^2$ determines the k-dependence and the amplitude of the primordial matter power spectrum. The additional things that the matter power spectrum requires – the growth rate and the transfer function, as well as any prescription for nonlinear clustering; see the master formula in Chapter 9, Eq. (9.61) – are all imprinted in the late evolution of the universe, and can be calculated in a given cosmological model.

8.5.2 Tensor Spectral Index

As in the scalar case, we define the **tensor spectral index** n_t, defined as

$$n_t \equiv \frac{d \ln \Delta_t^2}{d \ln k}, \tag{8.64}$$

which is given in terms of the slow-roll parameters as (Problem 8.5)

$$n_t = -2\epsilon_{60} + O(\epsilon_{60}^2, \eta_{60}^2). \tag{8.65}$$

The tensor power spectrum is then written as

$$\Delta_t^2(k) = A_t(k_*) \left(\frac{k}{k_*} \right)^{n_t(k_*)}, \tag{8.66}$$

where A_t is its normalization and k_* is an arbitrary pivot scale.

We typically talk about the possible future observation of tensor modes in terms of the ratio of their amplitude to that of the already observed scalar (curvature) perturbations. We now discuss what such an observation would imply for cosmology.

8.5.3 The Energy Scale of Inflation

We have seen in Eq. (8.54) above that the primordial power spectrum of scalar perturbations is proportional to $H_{60}^4/\dot{\phi}_{60}^2$; it is measured very well (e.g., by large-angle CMB anisotropy), and is known to have an amplitude of $\Delta_s^2 \simeq 10^{-9}$. Moreover, from Eq. (8.59) we see that the tensor power spectrum is proportional to the scale of inflation, $\Delta_t^2 \propto H_{60}^2 \propto V$. Hence, the **tensor-to-scalar ratio**

$$r \equiv \frac{\Delta_t^2(k)}{\Delta_s^2(k)}, \tag{8.67}$$

evaluated at some scale k, is a measure of the energy scale of inflation V. In particular, assuming $\Delta_s^2 \simeq 10^{-9}$, relating the Hubble parameter during inflation to the potential via $H^2 \simeq V/m_{\mathrm{Pl}}^2$, and plugging Eq. (8.59) into Eq. (8.67), we get

$$V^{1/4} \simeq \left(\frac{r}{0.01} \right)^{1/4} 10^{16} \, \mathrm{GeV}. \tag{8.68}$$

Therefore, in simple models of inflation:

> The tensor-to-scalar ratio r is directly related to the energy scale of inflation.

Hence, detecting a nonzero r would directly determine the Hubble parameter during inflation; more precisely, H_{60}. A measurement of r would also give us the evolution of the inflaton field $\Delta\phi$ during inflation (see Box 8.3).

Box 8.3   **The Lyth Bound**

In 1997, David Lyth showed that a detection of, or a limit on, the tensor-to-scalar ratio r has a direct consequence on the distance in the field space, $\Delta\phi$, that the inflaton covers during inflation (Lyth, 1997).

Deriving this relation – the Lyth bound – is remarkably simple. First, we will use the fact that

$$H = \frac{d\ln a}{dt} = \frac{dN}{dt} \implies dt = \frac{dN}{H}, \tag{B1}$$

where $N = \ln a$ is the number of e-folds. Then

$$|\Delta\phi| = \int |\dot\phi|\, dt \simeq \int \frac{|V'|}{3H}\frac{dN}{H} = \int \frac{m_{\rm Pl}^2}{8\pi}\frac{|V'|}{V}dN, \tag{B2}$$

using the slow-roll relations $\dot\phi = -V'/(3H)$ and $H^2 = 8\pi/(3m_{\rm Pl}^2)V$. Then, using the definition of the slow-roll parameter ϵ in Eq. (8.39), we have

$$\frac{|\Delta\phi|}{m_{\rm Pl}} = \int \frac{1}{2}\sqrt{\frac{\epsilon}{\pi}}\, dN \simeq 30\sqrt{\frac{\epsilon}{\pi}}\left(\frac{\Delta N}{60}\right) = \frac{15}{2}\sqrt{\frac{r}{\pi}}\left(\frac{\Delta N}{60}\right), \tag{B3}$$

where we used the fact that the slow-roll parameter ϵ varies slowly during inflation and can be pulled outside of the integral, and adopted the relation $r = 16\epsilon$ from Eq. (8.70). The Lyth bound is usually quoted in terms of the *reduced* Planck mass $m_{\rm Pl}^{\rm red} \equiv m_{\rm Pl}/\sqrt{8\pi}$ (see the end of Chapter 1); then, roughly

$$\frac{|\Delta\phi|}{m_{\rm Pl}^{\rm red}} \gtrsim \sqrt{\frac{r}{0.01}}, \tag{B4}$$

where the inequality sign comes from the requirement that inflation last *at least* 60 e-folds, $\Delta N \gtrsim 60$.

Equation (B4) is the Lyth bound. It indicates that, *if* the scalar-to-tensor parameter r is measured to be relatively large ($r \gtrsim 0.01$, so not much below the current upper limit $r \lesssim 0.06$), then the inflaton field evolved over at least a Planck-length distance in field space. This is interesting, because it means that the flatness of the inflaton potential had to be guaranteed over a super-Planckian field range – a very large $\Delta\phi$ from the quantum-field-theory point of view. This, in turn, may (or may not) be challenging to arrange for in specific classes of inflationary models.

Primordial tensor fluctuations manifest themselves as gravitational waves, and leave a signature in CMB polarization, as we will discuss in Chapter 13. CMB polarization measurements, therefore, are in principle sensitive to r. So far there has been no detection, despite the widely publicized claim in 2013 by the BICEP collaboration that would have implied $r = 0.2$ (Ade *et al.*, 2014a). Upon further analysis, it became clear that the BICEP result was due to systematic errors, specifically cosmic dust which was misinterpreted as a gravitational-wave signal. The currently accepted limit from Planck (plus some external data, particularly Keck Array and BICEP) is, at 95 percent confidence,

$$r < 0.06. \tag{8.69}$$

Constraints from Planck-2018 and some other data in the $n_s - r$ plane are shown in Fig. 8.6. It also shows lines or bands for a variety of inflationary models, with points indicating n_s and r evaluated 50 or 60 e-folds before the end of inflation. Figure 8.6 shows the very interesting interplay between inflationary theory and observations.

Finally, assuming single-field inflation, the tensor-to-scalar ratio can be expressed in terms of the slow-roll parameters. Dividing Eqs. (8.59) and (8.54) leads to

$$r = \frac{\Delta_t^2(k)}{\Delta_s^2(k)} = \frac{64\pi}{m_{\rm Pl}^2} \left(\frac{\dot{\phi}_{60}}{H_{60}} \right)^2 \simeq \frac{64\pi}{9m_{\rm Pl}^2} \left(\frac{V'}{H_{60}^2} \right)^2 \simeq \frac{m_{\rm Pl}^2}{\pi} \left(\frac{V'}{V} \right)^2 = 16\epsilon_{60}, \tag{8.70}$$

where we used the usual slow-roll expressions $\dot{\phi} \simeq -V'/(3H)$ and $H^2 \simeq 8\pi/(3m_{\rm Pl}^2)V$.

8.5.4 Single-Field Consistency Relation

For a single scalar field (and during slow-roll), there is an internal consistency relation between the tensor-to-scalar ratio r and the tensor spectral index n_t. From Eqs. (8.65) and (8.70), we see that

$$r = -8n_t. \tag{8.71}$$

This consistency relation can in principle be tested observationally, providing yet another direct window on the physics 10^{-35} seconds after the Big Bang. Unfortunately, in order to test this relation, we have to first (1) detect the signature of gravitational waves and hence r, and then (2) measure their full tensor power spectrum in order to determine the spectral index n_t. This is a challenging task, fit and potentially very rewarding for future generations of cosmologists.

8.5.5 Inflationary Models

Inflationary model-building has been a highly active area of research ever since Guth, Linde, and Albrecht and Steinhardt wrote their pioneering papers. The idea is to embed inflation in a realistic model of particle physics, in the same way that

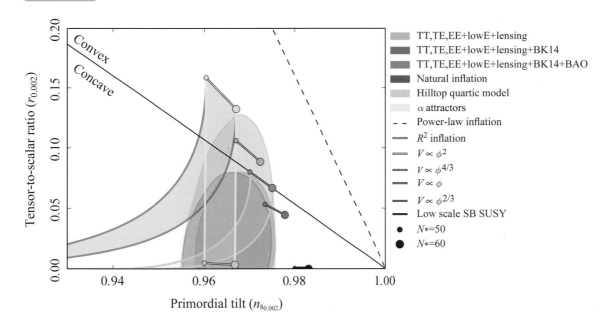

Fig. 8.6 Constraints in the $n_s - r$ plane from Planck and other data. The round filled contours show the following 95 percent credible interval constraints: the gray regions assume only the temperature, polarization, and lensing data from the 2018 release of Planck; the red contours additionally assume data from the BICEP-Keck array (BK); the blue contours additionally assume data from baryon acoustic oscillations (BAO), the probe that is described in Chapter 9. The figure also shows lines or bands for a variety of inflationary models, with points indicating n_s and r evaluated 50 or 60 e-folds before the end of inflation. Adopted from Planck Collaboration (Akrami *et al.*, 2020b). CC BY 4.0. A black and white version of this figure will appear in some formats. For the color version, please refer to the plate section.

we are able to explain various other physical phenomena in terms of exact, or sometimes approximate, physical laws.

If you are building your own inflationary model (and there have been literally thousands of papers on this), you should strive to achieve the following:

- Arrange that the inflationary potential be sufficiently flat to achieve at least 60 e-folds of inflation.
- Make sure that the spectral index comes out in the measured range close to unity (currently, $n_s = 0.966 \pm 0.005$ from Planck).
- Make sure that the amplitude of density perturbations comes out around the observed value of 10^{-5}.

These goals are often achieved by fine-tuning some free parameters in the inflationary potential. However, the requirements are stringent enough that it is not guaranteed that every potential will satisfy these requirements with some choice of the parameters. Let us list some popular models of inflation:

$$V(\phi) = \frac{m^2 \phi^2}{2} \qquad\qquad \text{(quadratic potential)}$$

$$V(\phi) = \lambda \phi^4 \qquad\qquad \text{(quartic potential)}$$

$$V(\phi) = V_0 \, e^{-\beta \phi} \qquad\qquad \text{(power-law inflation)}$$

$$V(\phi) = V_0 \left[1 + \cos\left(\frac{\phi}{f}\right) \right] \qquad \text{(natural inflation)}.$$

A worked example of one of these models, where we calculate the observable quantities starting from the potential $V(\phi)$, is given in Box 8.4.

8.6 Eternal Inflation and Multiverse

A fascinating possibility is that the standard inflationary framework may have implications for the universe far beyond our Hubble radius. It turns out that it is possible – maybe even plausible – that parts of the universe are undergoing inflation at any given time, including right now, resulting in the so-called eternal inflation. The argument goes as follows.

In the standard inflationary scenario described in this chapter, the universe undergoes inflation and exponentially expands as the field "rolls down the potential." Inflation stops when the potential of the scalar field becomes steep. But the field also fluctuates (quantum fluctuations), and does so randomly in different regions of space. Inevitably, somewhere in space the field makes a small jump *up* the potential, and there, inflation lasts a little longer. Now in that region of space where inflation lasts longer, there are sub-regions where the field again jumps up the potential. And so on – in the (huge!) inflated space, conditions are always ripe for a new bout of inflation to start somewhere anew. These newly created inflationary bubbles keep growing and self-reproducing (see Fig. 8.7 and Problem 8.7). They are mostly out of causal contact with each other, and their volume may be infinite. This scenario goes under the name of eternal inflation.

What is the probability that an observer is in a region of the universe that has inflated (in the eternal universe picture)? And how about the probability that our own patch will inflate in, say, the next 10 billion years? [Note, our patch *may* be inflating as we speak – dark energy behaves as the scalar field with $P \simeq -\rho$.]

The answer is – we don't know. It turns out that computing probabilities in an infinite universe is quite difficult. To see this, consider the following simple question: what fraction of integers are odd? Easy peasy you would say, if we count all integers as follows:

$$\mathbf{1}, 2, \mathbf{3}, 4, \ldots \tag{8.72}$$

(where boldface denotes odd numbers), then clearly the answer is one-half. However,

Box 8.4   **Worked Example: Quadratic Potential**

We demonstrate on a worked example that it is fairly straightforward to compare inflationary theory to observations. Consider the quadratic potential (Linde, 1982)

$$V(\phi) = \frac{1}{2}m^2\phi^2, \tag{B1}$$

where m is a constant with units of mass. The slow-roll parameters are

$$\epsilon = \frac{m_{\rm Pl}^2}{16\pi}\left(\frac{V'}{V}\right)^2 = \frac{m_{\rm Pl}^2}{16\pi}\left(\frac{2}{\phi}\right)^2 \equiv \frac{1}{4\pi\tilde{\phi}^2}$$

$$\eta = \frac{m_{\rm Pl}^2}{8\pi}\frac{V''}{V} = \frac{m_{\rm Pl}^2}{8\pi}\frac{2}{\phi^2} \equiv \frac{1}{4\pi\tilde{\phi}^2}, \tag{B2}$$

where we defined $\tilde{\phi} \equiv \phi/m_{\rm Pl}$. Note thus that $\epsilon = \eta$ for this potential. The spectral index is

$$n_s \simeq 1 - 6\epsilon_{60} + 2\eta_{60} = 1 - \frac{1}{\pi\tilde{\phi}_{60}^2}. \tag{B3}$$

Requiring that $n_s \simeq 0.966$ (see Table 3.1) gives $\epsilon_{60} = \eta_{60} \simeq 0.0085$, and thus $\tilde{\phi}_{60} \simeq 3.0$. Moreover, the end of inflation, $\epsilon_{\rm end} = 1$, implies $\tilde{\phi}_{\rm end} \simeq 0.3$. Thus (quoting the field displacement in terms of the reduced Planck mass $m_{\rm Pl}^{\rm red} \equiv m_{\rm Pl}/\sqrt{8\pi}$ as is the convention in this corner of cosmology; see Box 8.3)

$$\frac{|\Delta\phi|}{m_{\rm Pl}^{\rm red}} = \sqrt{8\pi}(\tilde{\phi}_{60} - \tilde{\phi}_{\rm end}) \simeq 14. \tag{B4}$$

So the field moves a super-Planckian distance. Next, we calculate the tensor-to-scalar ratio r:

$$r \simeq 16\epsilon_{60} \simeq 0.14. \tag{B5}$$

Because the current upper bound from the CMB experiments is $r \lesssim 0.06$ (see Eq. (8.69)), we find that *this model is ruled out by the data*. This showcases the impressive ability of data to constrain inflation. In truth, such a strong constraint is also made possible by the simplicity of this particular model which only has a single free parameter, m.

Finally, we can calculate the total number of e-folds between when the observables are imprinted and the end of inflation:

$$N = \int_{60}^{\rm end} H dt = \int_{60}^{\rm end} H\frac{d\phi}{\dot{\phi}} = -\int_{60}^{\rm end}\frac{3H^2}{V'}d\phi$$

$$= -4\pi\int_{60}^{\rm end}\phi\,d\phi = 2\pi(\tilde{\phi}_{60}^2 - \tilde{\phi}_{\rm end}^2) \simeq 58, \tag{B6}$$

where we used the slow-roll relation $\dot{\phi} \simeq -V'/(3H)$ and also $H^2 \simeq 8\pi/(3m_{\rm Pl}^2)V$. This number of e-folds is typical of inflationary models.

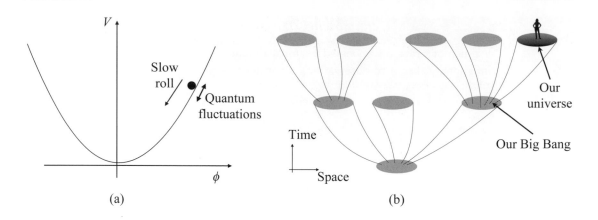

(a) Illustration of how fluctuations in the scalar field lead to prolonged inflation in some regions of space, eventually leading to eternal inflation. (b) Sketch of an eternally inflating universe.

what if we go through all integers as follows:

$$1, \mathbf{3}, 2, \mathbf{5}, \mathbf{7}, 4, \ldots . \tag{8.73}$$

While this counting scheme may seem somehow "not right," it is valid as we *will* get to every integer eventually. However, the fraction of odd integers now comes out to be two-thirds! The arbitrariness of the final answer is due to the fact that the set we are counting – integers – is infinite.

Eternal inflation suffers from similar problems. Because its space is infinite, computing probabilities in eternal inflation is tricky, and is in fact a subject of modern research in cosmology.

Finally, the eternal-inflation picture (illustrated in Fig. 8.7) leads to one of the most sensible definitions[9] of the **multiverse**. In this picture, each inflating pocket in the multiverse is what we usually refer to as the universe that started with its own Big Bang. The multiverse picture is speculative. Remarkably, however, there may be ways to test this framework with observations; for example, if the different pockets of the multiverse intersect, then they may leave signatures in the CMB pattern on the sky observed from within pockets. No such signatures have been found to date, but the quest to observationally test this fascinating scenario continues.

[9] Here, the multiverse refers to an eternally inflating universe, with different "pockets." Other contexts where this term is used are for different branches in the many-world interpretation of quantum mechanics, or else generally conjectured (but often vaguely defined) structures whose different pockets may have different laws of physics.

Bibliographical Notes

Inflation is reviewed in many excellent articles and a few books. The original paper by Guth (1981) is remarkably clear, and constitutes essential reading on the topic. Very clear expositions are given in the review article by Baumann (2011) and the books by Ryden (2016) and Liddle and Lyth (2000). More advanced coverage is in Kolb and Turner (1994) and Mukhanov (2005). A superb popular book on the history of the discovery of inflation is the one by Guth (1998). Finally, the literature on the multiverse is, much like its subject, not yet very mature, but the popular article by Tegmark (2003) is very informative and entertaining.

Problems

8.1 Exponential expansion in non-flat universes. [Adapted from Guth (2013).]
Recall that the evolution of a Robertson–Walker universe is described by the equation

$$\left(\frac{\dot{a}}{a}\right)^2 = \frac{8\pi G}{3}\rho(t) - \frac{\kappa}{R_0^2 a(t)^2}.$$

Suppose that the mass density ρ is given by the constant mass density ρ_{vac} of the false vacuum. For the case $\kappa = 0$, the solution is given simply by

$$a(t) \propto e^{Ht},$$

where $H^2 = 8\pi G\rho/3$. Find the solution to this equation for an arbitrary value of κ. Be sure to consider both possibilities for the sign of κ. You may find the following integrals useful:

$$\int \frac{dx}{\sqrt{1+x^2}} = \sinh^{-1} x$$

$$\int \frac{dx}{\sqrt{1-x^2}} = \sin^{-1} x$$

$$\int \frac{dx}{\sqrt{x^2-1}} = \cosh^{-1} x.$$

Show that for large times one has $a(t) \propto e^{Ht}$ for all choices of κ.

8.2 Inflation and the density of monopoles. [Adapted from Ryden (2016).]
Observational limits on the abundance of magnetic monopoles tell us that their abundance is currently $\Omega_{\mathrm{mon},0} < 10^{-6}$. If monopoles formed at the GUT

time ($a_{\text{GUT}} = 10^{-27}$), with energy density $\rho_{\text{mon,GUT}} \sim 10^{94}\,\text{TeV}\,\text{m}^{-3}$, how many e-folds of inflation would be required to drive the current density of monopoles below the bound $\Omega_{\text{mon,0}} < 10^{-6}$? Assume also that inflation took place immediately after the formation of monopoles.

8.3 **Slow-roll parameters are small.** Prove that the slow-roll parameters

$$\epsilon = \frac{m_{\text{Pl}}^2}{16\pi} \left(\frac{V'}{V} \right)^2, \qquad \eta = \frac{m_{\text{Pl}}^2}{8\pi} \frac{V''}{V}$$

satisfy

$$\epsilon, |\eta| \ll 1.$$

Do not worry if you get them to be $\ll X$, where X is a small integer, and not one. *Hint:* There are several ways to go about doing this. One way is to get a condition on ϵ using $\dot{\phi}^2/2 \ll V$, and a condition on a linear combination of ϵ and η using $|\ddot{\phi}| \ll 3\dot{\phi}$. For the latter, start by evaluating $\ddot{\phi}$ as a time derivative of $\dot{\phi} \simeq -V'/3H$ and remember that, at this level of approximation, you cannot assume that H is constant.

8.4 **End of inflation.** Prove that inflation ends ($\ddot{a} = 0$) when $\epsilon = 1$. *Hint:* Find the relation between the kinetic and potential energy of inflaton when inflation ends, then rewrite the second-order ODE for ϕ evolution. Note that you cannot assume the slow-roll approximation as the field is not slowly rolling near the end of inflation.

8.5 **Spectral indices in terms of slow-roll parameters.** Prove the relations between the scalar and tensor spectral indices, n_s and n_t, in terms of slow-roll parameters given respectively in Eqs. (8.61) and (8.65).

(a) First prove this relation for the tensor spectral index (Eq. (8.65)). Start by writing

$$n_t \equiv \frac{d\Delta_t^2(k)}{d\ln k} = \frac{d\Delta_t^2(k)}{dt} \frac{dt}{d\ln k}$$

and work out $dt/d\ln k$ using the horizon-exit relation $k = aH$. For the first term in the chain rule above, remember that the tensor power spectrum is given in terms of the Hubble constant at horizon exit, Eq. (8.59).

(b) Then, prove this relation for the scalar spectral index (Eq. (8.61)). Start by writing

$$n_s - 1 \equiv \frac{d\Delta_s^2(k)}{d\ln k} = \frac{d\Delta_s^2(k)}{dt} \frac{dt}{d\ln k}.$$

The derivation is similar to that for the tensor index in part (a). The added complication is that the scalar power spectrum is given in terms of the Hubble constant at horizon exit and the field velocity, Eq. (8.54). To address this, use the slow-roll relation $\dot{\phi} \simeq -V'/(3H)$.

8.6 **Power-law inflation.** Consider one of the early – and exactly solvable – models proposed for inflation: the power-law inflation, named because the scale factor $a(t)$ is a power law

$$a(t) \propto t^p,$$

where p is some constant. The corresponding potential $V(\phi)$ is actually an exponential

$$V(\phi) = V_0 \exp\left(-\beta \, \frac{\phi}{m_{\mathrm{Pl}}}\right),$$

where $\beta = \sqrt{16\pi/p}$.

(a) By solving the (full, second-order) Klein–Gordon differential equation that governs the temporal evolution of ϕ, determine the solution for $\phi(t)$. Search for the solution of the form $\phi(t)/m_{\mathrm{Pl}} = c_0 \ln(c_1 t)$, and determine the two constants in terms of p.

(b) Determine the slow-roll parameters ϵ and η in terms of p and any other parameters you may need. What does the condition for inflation ($\epsilon < 1$) imply for p?

(c) Given that at least about 60 e-folds ($N \gtrsim 60$) are required for inflation, what does that imply for the ratio of the final-to-initial proper time (t_f/t_i) for this model? And what does it imply for the change in the field value during inflation $\Delta\phi \equiv |\phi_f - \phi_i|$?

(d) Strictly speaking, is it possible that this – without any additional features – is the true model for inflation? *Hint:* Can inflation end in this model as written? What qualitative feature (bump? steepening? something else?) do we need to add to the potential to end inflation?

8.7 **Eternally inflating, self-reproducing universe.** Let us investigate the conditions for eternal inflation, discussed in Sec. 8.6. Consider the quadratic inflaton potential

$$V(\phi) = \frac{1}{2}m^2\phi^2,$$

where m is the inflaton mass. Over the characteristic time $\Delta t \simeq H^{-1}$, the field will classically "roll down the potential" the distance

$$\Delta\phi = |\dot{\phi}|H^{-1}$$

(where we take the absolute sign since, in this inflationary model, $\dot{\phi} < 0$ and we care about the absolute field shift only).

Meanwhile, there are also quantum fluctuations of the inflaton field, which can be shown to be of size

$$\delta\phi \simeq \frac{H}{2\pi}.$$

When the quantum "jumps" dominate the classical evolution of the field

$$\frac{\delta\phi}{\Delta\phi} \gg 1,$$

then the field will jump to a higher value ($\sim 1/2$ of the time) than where it is at a given moment. In turn, that chunk of space will undergo inflation a little longer and, because the inflating volume increases exponentially ($V \propto a^3 \propto e^{3N} = e^{3H\Delta t}$), starts to dominate – so, for example, over one e-fold with $\Delta t \simeq H^{-1}$, V increases a factor $e^3 \simeq 20$. Hence, inflation spawns new space without end. The condition in the equation above ($\delta\phi/\Delta\phi \gg 1$) is satisfied when

$$\phi \gg \phi_*,$$

where ϕ_* is some critical value of the field.

Find ϕ_* in terms of the inflationary potential mass m and Planck mass m_{Pl}. You may ignore all order-unity numerical prefactors, as we are only interested in the functional dependence on key physical parameters. *Hint:* You will want to use the Friedmann I equation, and the Klein–Gordon equation (the one with $\ddot{\phi}$). You may assume the slow-roll approximation throughout.

8.8 **The supercooled, then reheated, universe.** Entropy in the universe behaves in a wild way during and after inflation. The universe is supercooled during inflation, resulting in low entropy. Then, during the process of reheating, entropy increases by a massive amount, generating the present-day, high-entropy-type universe.

When quoting your entropy numbers, set to unity all prefactors to powers of 10 (i.e., report "round numbers").

(a) As a warm-up, calculate the entropy of the million-solar-mass black hole at the center of our galaxy. Consider that Bekenstein and Hawking told us that the mass of the black hole is roughly its area, $S_{\mathrm{BH}} = A/4$, where the natural-unit lengths are implied in Planck length, $\ell_{\mathrm{Planck}} = 1.61 \times 10^{-35}$ m (so area is in units of ℓ_{Planck}^2). Also remember the relation between the Schwarzschild radius (i.e., event horizon) and the mass of a black hole, $R = 2GM/c^2 \to 2M/m_{\mathrm{Pl}}^2$. You can quickly get the black hole event horizon radius by scaling it to the corresponding (Schwarzschild) radius of the Sun, which is $R_{event,\odot} \simeq 3\,\mathrm{km}$.

(b) Now consider a typical GUT-scale inflationary scenario, with the temperature at the start of inflation $T_i \simeq 10^{14}\,\mathrm{GeV}$. Let the inflating patch have initial physical size $H_i^{-1} \simeq 10^{-25}\,\mathrm{m}$. During inflation the temperature gets exponentially suppressed (as the scale factor grows exponentially), but after inflation the universe reheats to the temperature $T_{\mathrm{RH}} = T_i$. Calculate the entropy in the initial patch, S_i. To do that, recall that the entropy *density* is $s \simeq T^3$ (remembering that we are ignoring all numerical prefactors of order unity in this problem).

(c) Assuming inflation lasts a total of $N = 100$ e-folds, calculate the entropy in the same patch after reheating, S_{RH}. To do that, first ask yourself what the (physical) size of our patch is relative to its value at the beginning of inflation. [You should be finding that the result in part (b) is, to put it mildly, a "little" smaller than that in part (c).]

8.9 [**Computational**] **Numerical evolution of inflationary dynamics.** In this problem you will explore the dynamics of scalar fields that power inflation. The equation that describes temporal evolution of the scalar field, the inflaton ϕ, is

$$\ddot{\phi} + 3H\dot{\phi} + dV/d\phi = 0,$$

where the dots are derivatives with respect to physical (not conformal) time. Assume that there is a single field, and that it is homogeneous (i.e., only depends on time, not space). Also assume that the universe's energy density is entirely inflaton dominated (what does that imply for H?).

(a) First, rewrite the equation above in terms of the derivative with respect to the field, where $' \equiv d/d\phi$; you should get

$$\phi'' + \left(3 + \frac{H'}{H}\right)\phi' + \frac{1}{H^2}\frac{dV}{d\phi} = 0.$$

Further, write the Hubble parameter in terms of ϕ and ϕ'; you should get

$$H^2 = \frac{CV}{1 - C(\phi')^2/2},$$

where $C \equiv 8\pi/(3m_{\mathrm{Pl}}^2)$.

Now you are ready to put the ϕ'' equation on the computer. For this, you could use ready-made integrators such as `odeint` in Python, but I instead recommend using a ground-up implementation of the fourth-order Runge–Kutta method; see Numerical Recipes (Press *et al.*, 2007). This will allow you to update not only ϕ and ϕ', but also H and H' (which are required to calculate the ϕ and ϕ' updates), as well as any other parameters that you are asked about below. You can adopt the stepsize $d\ln a = 0.1$, or else experiment with it as you wish. Note that the absolute scale factor a never enters the calculation, so you don't need to specify its initial value or track it.

Assume the quadratic potential

$$V(\phi) = \frac{m^2\phi^2}{2}$$

with $m = 10^{-6}m_{\mathrm{Pl}}$. You might want to start with the following initial conditions:

$$\phi_0 = 10m_{\mathrm{Pl}}, \quad \phi_0' = 0.$$

Set all of the dimensionful quantities in units of Planck mass (i.e., set $m_{\rm Pl} = 1$ but remember that the field that you are reporting, for example, is not ϕ but rather $\phi/m_{\rm Pl}$). Then answer the following questions:

(b) Plot the potential as a function of the field value, $V(\phi)$, and the equation of state of the scalar field, $w(\phi)$, where ϕ starts at its initial value and ends when $\epsilon = 1$. In both cases, denote the point 60 e-folds before the end of inflation, ϕ_{60}.

(c) What is the total number of e-folds, N, between the beginning of the roll and the end of inflation (when $\epsilon(\phi) = 1$)? What is the value of the field ϕ at the end of inflation?

(d) Find the value of the scalar spectral index n_s, evaluated 60 e-folds before the end of inflation. Remember that $n_s = 1 - 6\epsilon + 2\eta$. How does this compare to Planck's constraint on n_s? [Note that the slow-roll parameters ϵ and η are fundamentally functions of the potential V and the Hubble parameter H, which in turn depend on ϕ and ϕ'. You will want to be updating all of these quantities at each step in your integrator.]

(e) Do the results in parts (b) and (c) change appreciably if you vary the initial conditions (ϕ_0 and ϕ_0')? Why do you think that is? What about if you make the initial field value *much* smaller?

(f) Evaluate the density fluctuation of modes leaving the horizon 60 e-folds before the end of inflation, δ_H. This quantity is given in terms of the curvature perturbation $\mathcal{R}$, and can be evaluated as

$$\delta_H = \frac{2}{5}|\mathcal{R}| = \frac{2}{5}\sqrt{\Delta_{\mathcal{R}}^2} = \frac{H_{60}^2}{5\pi|\dot{\phi}|_{60}},$$

where we used Eq. (8.54). Does δ_H that you find roughly agree with the size of the primordial density fluctuations measured by CMB experiments?

PART III

THE LATER UNIVERSE

9

Large-Scale Structure in the Universe

One of the most recognizable features of the maps of structure in the universe is that the distribution of galaxies is not random. Instead, galaxies form a network of sheets that meet in filaments, which in turn meet in clusters of galaxies; see Fig. 9.1. The goal of this chapter is to lay down the foundations for the study of large-scale structure, and specifically emphasize areas where theory and observations meet.

A central thing to settle early on is to decide which "test particles" we will use in order to connect theory to large-scale structure observations and thus test the laws of physics in the universe. Here, galaxies and their large cousins, clusters of galaxies, are the most natural choice, as both are easily identifiable in typical observations. However, the cosmological model most directly predicts the properties of **dark-matter halos** – massive bound structures largely made up of dark matter (but with an important baryonic component) that emerge in the cosmic web of large-scale structure. We will start out this chapter referring to halos and galaxies/clusters interchangeably, as most of our formalism applies to both of these classes of objects. We will finally address the relation between the two near the end of the chapter.

First, however, we will lay out the foundations of the study of cosmic structure. We start with the definition of overdensity, and then describe how the density fluctuations evolve in an expanding universe. We continue with the statistical description of cosmic structure, and end by discussing some of the cosmological probes that enable cosmologists to connect the observations of large-scale structure to the properties of the cosmological model.

9.1 Density Perturbations

A cosmological model predicts statistical properties of the distribution of any set of objects on the sky. For example, a model does *not* predict the number density of galaxies in some direction on the sky, but rather it predicts their statistical distribution – say, the mean and variance of the galaxy number density in some volume.

The spatial distribution of objects is most generally described in terms of an infinite hierarchy of correlation functions of object positions, which are equivalent to an infinite hierarchy of counts in cells of finite volume. The lowest-order and most useful ones are:

1. The **number density** of objects, which is formally just the one-point correla-

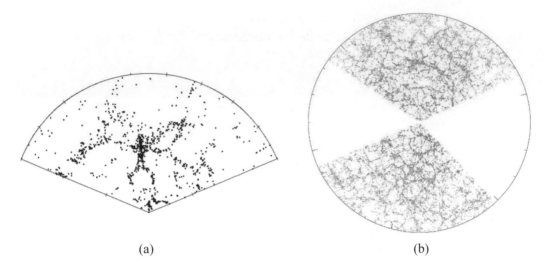

(a)   (b)

Fig. 9.1 Historical and modern maps of large-scale structure in the universe. (a) Distribution of galaxies in the Cfa (Harvard Center for Astrophysics) galaxy redshift survey, with about 1100 galaxies. Notice the "stick-man" structure, which is due to a modest number of galaxies in the survey and not the work of aliens. [The stick-man structure reflects a complex local velocity field, generated by large density fluctuations, in which clusters are radially stretched and filaments radially compressed, helping thin the arms and elongate the body.] The Cfa map was arguably the first clear image of the large-scale structure. Reproduced with permission from de Lapparent *et al.* (1986), © AAS. (b) Distribution of galaxies in the complete Baryon Oscillation Sky Survey (BOSS; extension of SDSS). Each point represents one of roughly a million galaxies with accurate redshift measurements. We are at the center of the diagram. Adopted from www.sdss3.org/science. Sloan Digital Sky Survey, CC BY 4.0.

tion function. While counting galaxies or dark-matter halos on the sky seems an obvious way to compare theory to observations, it turns out that the theory prediction is reliable only for the most massive objects – galaxy clusters. Near the end of this chapter, we will cover the so-called *mass function*, $dn/d\ln M$, that quantifies how the number density of galaxy clusters can be predicted from theory.

2. The **two-point correlation function** (of galaxies, clusters, etc.) $\xi(r)$ – and its Fourier-space cousin, the power spectrum $P(k)$ – turn out to be a most useful meeting point between observations and theory. As we will discuss at length in what follows, the two-point function and/or the power spectrum can both be accurately predicted from theory and readily measured.

Higher-point correlation functions (especially the three-point function, or its Fourier transform, the bispectrum) are also very useful and the subject of much research; we briefly review them below.

We next introduce the fundamental variable in structure formation – the overdensity, δ. The overdensity is defined as an excess mass density, evaluated over some region of space, relative to the mean cosmic density. At some location $\mathbf{x}$ and time t, the overdensity is thus defined as

$$\delta(\mathbf{x}, t) \equiv \frac{\rho(\mathbf{x}, t) - \bar{\rho}(t)}{\bar{\rho}(t)}, \qquad (9.1)$$

where $\rho(\mathbf{x}, t)$ is the mass density (mass per volume), and $\bar{\rho}(t)$ is the mean density of all space. While the equation above simply gives the overdensity as uniquely defined at a given location $\mathbf{x}$, for most practical applications the density is further *smoothed* over a region of space of some size that contains $\mathbf{x}$, something that we ignore for the moment but will explicitly cover in Sec. 9.6.2. In this chapter, we will largely specialize in small perturbations, where $|\delta(\mathbf{x}, t)| \ll 1$, whose evolution is tractable analytically.

Standard inflationary theory predicts that the distribution of the primordial density fluctuations is very nearly Gaussian. This means

$$P(\delta)d\delta = \frac{1}{\sqrt{2\pi}\sigma} e^{-\delta^2/(2\sigma^2)}d\delta, \qquad (9.2)$$

where $\sigma \sim 10^{-5}$ is the rms of primordial fluctuations predicted by inflation. Over time these fluctuations grow, and remain Gaussian until the onset of nonlinearity when $\delta \gtrsim 1$.

Because the mass density ρ can vary between zero and infinity, the overdensity $\delta \equiv (\rho - \bar{\rho})/\bar{\rho}$ can take values between -1 and $+\infty$. So, while the small perturbations $|\delta| \ll 1$ are symmetrical around zero (being nearly Gaussian!), we *know* that large fluctuations, $\delta \gtrsim 1$, cannot remain Gaussian, simply because large overdensities can have $\delta \gg 1$, while the underdensities can never have δ below -1 (because $\delta = -1$ is the absolute minimal value, and obtains when $\rho(\mathbf{x}, t) = 0$).

However, Gaussianity of the density fluctuations is not guaranteed in the inflationary hot Big Bang cosmological model, even for primordial fluctuations with $|\delta| \ll 1$, as there exist various inflationary models that predict primordial non-Gaussianity. Testing Gaussianity is therefore well worthwhile. Over the past couple of decades, tests of non-Gaussianity – departures from the Gaussian distribution in Eq. (9.2) – have become a hot topic in cosmology, both on the experimental front (measuring the statistics of the distribution of overdensities) and the theoretical one (identifying some, hopefully "natural," inflationary models predicts non-Gaussianities of measurable magnitude). In what follows, however, we assume the simplest scenario with Gaussian perturbations at early times (or Gaussian fluctuations for fluctuations filtered on large scales today[1] with $|\delta| \ll 1$) – the scenario that is thus far in agreement with the data. We will return to non-Gaussianity near

---

[1] In fact, as perturbations grow and are no longer linear, the field statistics transform toward a Gaussian form in the *logarithm* of the overdensity (more precisely, $\log(1 + \delta)$).

the end of Chapter 13, when we discuss its observability in the context of cosmic microwave background anisotropies.

9.2 Correlation Functions

Next, we introduce the most important statistical measure of the clustering of galaxies in the universe – the two-point correlation function. [Its equally important Fourier transform, the power spectrum, is covered in Sec. 9.5.] We also briefly touch upon the three-point and higher-order correlation functions.

9.2.1 From Probabilities to Correlations

Consider a point process (process with point particles in space), and let the mean density of points be n. Then the probability of finding a particle in an infinitesimal volume dV is[2]

$$dP = ndV. \tag{9.3}$$

Now consider the probability of finding two particles, one in volume dV_1 and another in volume dV_2; this is

$$dP = n^2(1 + \xi(r_{12}))\, dV_1 dV_2. \tag{9.4}$$

Here ξ is the *excess* probability of finding the second particle a distance r_{12} away (we assume isotropy and homogeneity here, so ξ can depend at most on distance, and not direction). In other words, *given* that we observe particle 1 in dV_1, the probability that we find the second particle in dV_2 is

$$dP(2|1) = n(1 + \xi(r_{12}))\, dV_2. \tag{9.5}$$

For a pure Poisson ("random") process, there is no correlation between counts in volumes dV_1 and dV_2, so that $\xi = 0$. Some of the earliest advances in the field of large-scale structure, made in the 1950s to 1970s, resulted in the measurements of the correlation function $\xi(r)$ of galaxies and clusters. A rough phenomenological approximation that did reasonably well in fitting the early clustering data is a pure power-law form

$$\xi(r) \simeq \left(\frac{r}{r_0}\right)^\gamma, \tag{9.6}$$

with $\gamma \approx -1.8$. The value of r_0 depends on the type of object we are talking about: for galaxies, $r_0 \simeq 5\,h^{-1}\mathrm{Mpc}$, while for clusters, $r_0 \simeq 20\,h^{-1}\mathrm{Mpc}$; we will see later how peaks in Gaussian random fields can explain this trend with the mass of the object. This power-law fit to the two-point correlation function is purely empirical

---

[2] Note that, strictly speaking, this is the number of points found in the volume. But since $dP \ll 1$, this is equivalent to the probability of finding a single particle.

and it is close to a pure power law mainly because, roughly speaking, "everything in astronomy is a power law." A more precise, ground-up theoretical prediction for $\xi(r)$ will be laid out further down in this chapter.

One can also now evaluate other, related quantities. Consider, for example, the expected number of particles in the space of volume V, around a particle centered within that volume (and excluding that particle). This evaluates to

$$\langle N \rangle = nV + n \int \xi(r) dV, \tag{9.7}$$

where the integral runs over the volume V and r is the distance from the central particle.

9.2.2 Two-Point Correlation Function

Consider now a continuous density field $\rho(\mathbf{x})$. To simplify notation, let us suppress the time dependence of quantities in what follows. The **two-point correlation function** can be defined as

$$\xi(r) = \frac{\langle \, [\, \rho(\mathbf{x} + \mathbf{r}) - \langle \rho \rangle \,]\,[\, \rho(\mathbf{x}) - \langle \rho \rangle \,]\,\rangle_\mathbf{x}}{\langle \rho \rangle^2} \equiv \langle \, \delta(\mathbf{x} + \mathbf{r})\, \delta(\mathbf{x})\, \rangle_\mathbf{x}, \tag{9.8}$$

where the subscript indicates averaging over all $\mathbf{x}$. Here, we picture taking the overdensity at position $\mathbf{x}$, multiplying it with an overdensity at location $\mathbf{x} + \mathbf{r}$, and averaging this over all positions in space $\mathbf{x}$ *and* over all the possible *directions* of $\mathbf{r}$, holding fixed the magnitude $r \equiv |\mathbf{r}|$. The two averagings are justified respectively by homogeneity of the universe as a whole (hence averaging over $\mathbf{x}$) and its isotropy (hence averaging over the direction of $\mathbf{r}$, holding $|\mathbf{r}|$ fixed). This equation can be rewritten as

$$\langle \, \rho(\mathbf{x} + \mathbf{r})\, \rho(\mathbf{x})\, \rangle_\mathbf{x} = \langle \rho \rangle^2 \, [1 + \xi(r)]. \tag{9.9}$$

Let us outline the big-picture summary straight away:

> The two-point correlation function is a workhorse of the practical cosmologist. It is relatively simple to both measure and theoretically predict, yet it contains a tremendous amount of information about the physical processes in both the early and late universe.

Much of the discussion in this chapter will be devoted to studying how to predict, measure, and utilize the two-point correlation function and its Fourier transform, the power spectrum.

9.2.3 Three-Point Correlation Function

Consider next the **three-point correlation function** in real space ζ_{123}. It can be defined via a probability of finding three particles in volumes dV_1, dV_2, and dV_3:

$$dP = n^3[1 + \xi(r_{12}) + \xi(r_{13}) + \xi(r_{23}) + \zeta(r_{123}))]\, dV_1 dV_2 dV_3, \qquad (9.10)$$

where $\zeta(r_{123})$ is the excess probability for a *triangle* configuration described by three sides (r_{12}, r_{13}, r_{23}), on top of that predicted by the number density n and the (necessarily pairwise) correlation functions $\xi(r_{12})$, $\xi(r_{13})$, and $\xi(r_{23})$.

Similar to the case of the two-point function, we can define the continuous version of the three-point correlation function:

$$\zeta(r, s, |\mathbf{r} - \mathbf{s}|) = \frac{\langle\, [\,\rho(\mathbf{x}+\mathbf{r}) - \langle\rho\rangle\,][\,\rho(\mathbf{x}+\mathbf{s}) - \langle\rho\rangle\,][\,\rho(\mathbf{x}) - \langle\rho\rangle\,]\,\rangle_{\mathbf{x}}}{\langle\rho\rangle^3}$$
$$\equiv \langle\, \delta(\mathbf{x}+\mathbf{r})\, \delta(\mathbf{x}+\mathbf{s})\, \delta(\mathbf{x})\,\rangle_{\mathbf{x}}, \qquad (9.11)$$

which can be rewritten as

$$\langle\, \rho(\mathbf{x}+\mathbf{r})\, \rho(\mathbf{x}+\mathbf{s})\, \rho(\mathbf{x})\,\rangle_{\mathbf{x}} = \langle\rho\rangle^3\, [1 + \xi(r) + \xi(s) + \xi(|\mathbf{r}-\mathbf{s}|) + \zeta(r, s, |\mathbf{r}-\mathbf{s}|). \quad (9.12)$$

Note the proliferation of dependent variables: while the two-point correlation function depends on a single variable (distance r), the three-point function depends on *three* variables – the distances $r_{12} \equiv r, r_{13} \equiv s$, and $r_{23} \equiv |\mathbf{r} - \mathbf{s}|$, which satisfy the condition that they form a triangle. This tripling of the quantities required to specify a configuration is the principal reason why both measuring and theoretically predicting the three-point correlation function is much more challenging than doing the same for the two-point function.

For a Gaussian distribution of overdensities, the three-point function is zero – to understand this qualitatively, consider that the third moment of a Gaussian-distributed variable X is zero, $\int_{-\infty}^{\infty} P_{\text{Gauss}}(X)X^3 dX = 0$. However, because the overdensities in the nonlinear regime are guaranteed to be non-Gaussian (see the discussion following Eq. (9.2)), the three-point function will also be nonzero in that regime.

Higher-order correlation functions can be defined in a similar way, but their practical importance in modern cosmology is less than that of the aforementioned two- and three-point functions and typically does not justify the far greater complexity in both measuring and predicting them.

9.2.4 Angular Two-Point Correlation Function

We readily observe angular positions of objects on the sky, but often do not have the means to measure radial distances to objects. We *can* measure the objects' redshifts in spectroscopic surveys, but photometric surveys, which provide us with

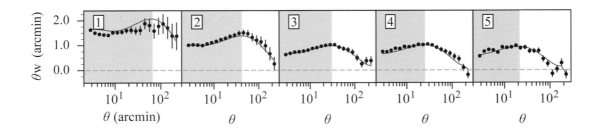

Fig. 9.2 Angular two-point correlation function $w(\theta)$ (multiplied by θ) as a function of galaxy separation θ as measured in five redshift bins (increasing in z from left to right) in the DES Year-1 analysis. The black data points are the DES measurements, the solid line is the best-fit ΛCDM theory, the horizontal dashed line is the zero-correlation value, and the gray shaded region shows scales *not* used because of concerns about accurately modeling nonlinear clustering at small spatial scales. Reproduced with permission from Abbott *et al.* (2018), copyright 2018 by the American Physical Society.

the largest and most complete samples of galaxies, typically only have limited per-object redshift precision.

It is thus of considerable interest to adopt and use a radially projected version of the two-point correlation function to quantify galaxy clustering. In the large-scale structure field, this statistic is called $w(\theta)$. [In the CMB, the equivalent quantity is called $C(\theta)$; see Chapter 13.] The angular two-point function is defined via

$$dP = \mathcal{N}^2 (1 + w(\theta)) \, d\Omega_1 d\Omega_2, \qquad (9.13)$$

where $\mathcal{N}$ is now the mean density of points/galaxies per solid angle, and Ω is the solid angle. As in the case of the full 3D correlation function $\xi(\mathbf{r}) \rightarrow \xi(r)$, the angular two-point function is a function of separation θ, rather than 2D sky coordinates $\hat{\theta}_{1,2} \equiv \hat{\mathbf{n}}_{1,2}$, due to homogeneity and isotropy.

Figure 9.2 shows the angular two-point correlation function $w(\theta)$ measured in five redshift bins in the Dark Energy Survey (DES) Year-1 data. The angular two-point function $w(\theta)$ by definition carries less information than the 3D correlation function $\xi(r)$, but is useful when we do not have excellent redshift information about the sources, and wish to purposefully integrate the density field along the radial direction.

We next turn to how to predict the temporal evolution of small density perturbations in the universe.

9.3 Density Perturbations: Basic Principles

Mathematical analysis of density perturbations in the expanding universe is *very* complex, as it involves both statistical/classical mechanics and general relativity. A full treatment that includes deriving and solving the coupled Einstein–Boltzmann equations for the perturbed quantities (dark matter, photons, etc.) has been described in book-length reviews of the subject. Here we adopt a much simplified, purely classical analysis which will give us sufficient insight into the resulting physics. Despite these simplifications, this material is still "dense" and heavy on the mathematics; readers interested in cosmological phenomenology and connections to data (rather than theory) can probably skip this section.

9.3.1 Fluid Equations

We will be modeling the density field of dark matter (but also, when needed, of radiation, baryons, neutrinos, or dark energy) as a fluid. Let us start with three fundamental equations that describe the evolution of a fluid. In what follows,

$$\frac{D}{Dt} \equiv \frac{\partial}{\partial t} + \mathbf{u} \cdot \nabla_{\mathbf{x}} \tag{9.14}$$

is the so-called **convective derivative** (or Lagrangian derivative, or total derivative), and measures what someone moving with the flow – whose velocity is $\mathbf{u}$ – would measure. In other words, the convective derivative is the derivative measured along the flow lines.

First, we have the continuity equation which encodes mass conservation:

$$\frac{D\rho}{Dt} + \rho(\nabla_{\mathbf{x}} \cdot \mathbf{u}) = 0 \quad \text{(continuity equation)}, \tag{9.15}$$

where ρ is mass density. Then, we have the Euler equation which quantifies conservation of momentum – these are really three equations of motion (one for each direction):

$$\frac{D\mathbf{u}}{Dt} = -\frac{\nabla_{\mathbf{x}} P}{\rho} - \nabla_{\mathbf{x}} \Phi \quad \text{(Euler equation)}, \tag{9.16}$$

where P is pressure and Φ is gravitational potential. Finally, the Poisson equation relates the gravitational potential Φ to the (matter) density ρ (and hence to matter overdensity δ):

$$\nabla^2 \Phi = 4\pi G \rho \quad \text{(Poisson equation)}. \tag{9.17}$$

We would like to consider these equations in comoving coordinates. To that effect, recall that the relation between a physical coordinate $\mathbf{x}$ and the corresponding comoving coordinate $\mathbf{r}$ is

$$\mathbf{x} = a(t)\mathbf{r}, \tag{9.18}$$

so that the physical and comoving velocities are related as

$$\dot{\mathbf{x}} \equiv \mathbf{u} = \dot{a}\mathbf{r} + a\dot{\mathbf{r}} = H\mathbf{x} + \mathbf{v}, \tag{9.19}$$

where $H \equiv \dot{a}/a$ is the Hubble parameter and $\mathbf{v} \equiv a\dot{\mathbf{r}}$ is the peculiar velocity which describes any motion beyond that due to the expansion of the universe. Equation (9.19) is just the familiar Hubble law, "corrected" for the peculiar velocity of the fluid element. Starting with the continuity, Euler, and Poisson equations, and making some approximations, one can derive a classical, linear-theory master perturbation equation (see Box 9.1 for the derivation):

$$\frac{\partial \mathbf{v}}{\partial t} + \frac{\dot{a}}{a}\mathbf{v} = -\frac{\nabla \Phi}{a} - \frac{c_s^2}{a}\nabla \delta - \frac{2T}{3a}\nabla S, \qquad (9.20)$$

where c_s is the speed of sound and S is the entropy of the system. We now investigate some important implications of this result.

Box 9.1   **Euler Equation for Perturbations**

Starting from Eq. (9.19), one can obtain the identities for going from physical to comoving time derivative and gradient:

$$\nabla_{\mathbf{x}} \rightarrow \frac{1}{a}\nabla_{\mathbf{r}}; \quad \frac{\partial}{\partial t} \rightarrow \frac{\partial}{\partial t} - \frac{\dot{a}}{a}\mathbf{r} \cdot \nabla_{\mathbf{r}}. \qquad (B1)$$

With the help of these, one can rewrite the continuity, Euler, and Poisson equations as

$$\frac{\partial \delta}{\partial t} + \frac{1}{a}\nabla \cdot [(1 + \delta)\mathbf{v}] = 0 \qquad (B2)$$

$$\frac{\partial \mathbf{v}}{\partial t} + \frac{\dot{a}}{a}\mathbf{v} + \frac{1}{a}(\mathbf{v} \cdot \nabla)\mathbf{v} = -\frac{\nabla \Phi}{a} - \frac{\nabla P}{a\bar{\rho}(1 + \delta)} \qquad (B3)$$

$$\nabla^2 \Phi = 4\pi G\bar{\rho}a^2\delta, \qquad (B4)$$

where in the last equation we made the redefinition $\Phi(r) \rightarrow \Phi(r) + a\dot{a}r^2/2$. Recall again that the partial derivative here is at a fixed comoving location $\mathbf{r}$, and that the gradient is also taken with respect to $\mathbf{r}$.

Our goal is to rewrite the gradient of pressure term, ∇P, on the right-hand side of Eq. (B3). We use the first law of thermodynamics, $TdS = dU + PdV$, where dS is the change in the entropy of the system, dU is the flow of energy into the system, and PdV is the work done on the system. Also, the pressure of an ideal gas is given by (recall that we work in units where $k_B = 1$)

$$P = nT = \frac{\rho T}{\langle m \rangle}, \qquad (B5)$$

where $\langle m \rangle$ is the mean mass which will cancel out below, so we just set it to unity now. Finally, for monoatomic gas, $U = (3/2)T = (3/2)(P/\rho)$, so that the first law becomes

$$TdS = d\left(\frac{3}{2}\frac{P}{\rho}\right) + Pd\left(\frac{1}{\rho}\right). \qquad (B6)$$

Substituting temperature from Eq. (B5), we have

$$d\ln P = \frac{5}{3}d\ln \rho + \frac{2}{3}Sd\ln S, \qquad (B7)$$

Box 9.1   **Euler Equation for Perturbations (continued)**

which can be integrated to give

$$P \propto \rho^{5/3} \exp\left(\frac{2}{3} S\right). \tag{B8}$$

Then we can express the gradient of the pressure as

$$\frac{\nabla P}{\rho} = \frac{1}{\rho}\left[\left(\frac{\partial P}{\partial \bar{\rho}}\right)_S \nabla\rho + \left(\frac{\partial P}{\partial S}\right)_\rho \nabla S\right] = c_s^2\,\nabla\delta + \frac{2}{3}(1+\delta)T\,\nabla S, \tag{B9}$$

where $c_s \equiv (\partial P/\partial\rho)_S^{1/2}$ is the (adiabatic) speed of sound. With the application of Eq. (B9), the Euler equation becomes

$$\frac{\partial\mathbf{v}}{\partial t} + \frac{\dot{a}}{a}\mathbf{v} + \frac{1}{a}(\mathbf{v}\cdot\nabla)\mathbf{v} = -\frac{\nabla\Phi}{a} - \frac{c_s^2}{a}\frac{\nabla\delta}{(1+\delta)} - \frac{2T}{3a}\nabla S. \tag{B10}$$

Finally, in linear theory we can ignore the terms quadratic in v and δ to get Eq. (9.20):

$$\frac{\partial\mathbf{v}}{\partial t} + \frac{\dot{a}}{a}\mathbf{v} = -\frac{\nabla\Phi}{a} - \frac{c_s^2}{a}\nabla\delta - \frac{2T}{3a}\nabla S. \tag{B11}$$

9.3.2 Curl Modes Decay

We are now ready for our first lesson from perturbation theory. In linear theory, curl modes in the velocity field decay, scaling inversely with the scale factor. This can be seen by operating with $\nabla\times$ on both sides of Eq. (9.20), and using the fact that the curl of the gradients on the right-hand side is zero. Then we get

$$\left[\frac{\partial}{\partial t} + \frac{\dot{a}}{a}\right](\nabla\times\mathbf{v}) = 0, \tag{9.21}$$

from which it follows that

$$(\nabla\times\mathbf{v}) \propto \frac{1}{a}. \tag{9.22}$$

Therefore, even if there are initial curl modes in the velocity distribution of baryonic or dark-matter particles in the universe, they decay in time, soon becoming negligible. This result can be loosely associated with, and interpreted as, conservation of the angular momentum. [Of course, the late-time gravitational collapse of small-scale structures regenerates angular momentum, as evidenced by the spins of galaxies observed today.]

9.3.3 Temporal Evolution of Fluctuations: General Case

Taking the derivative of the linear Euler equation (9.20) and combining with the continuity equation (9.15) and the Poisson equation (9.17), it is also possible to get the general equation for the evolution of fluctuations:

$$\frac{\partial^2 \delta}{\partial t^2} + 2\frac{\dot{a}}{a}\frac{\partial \delta}{\partial t} = 4\pi G\bar{\rho}\delta + \frac{c_s^2}{a^2}\nabla^2\delta + \frac{2}{3}\frac{T}{a^2}\nabla^2 S. \tag{9.23}$$

This is a second-order ordinary differential equation, which means that there is a growing and a decaying solution; of course, the former will dominate over time. The key source term is the first term on the right-hand side, which ensures that perturbations grow due to gravitational instability. The second term on the left-hand side is the friction term, which causes the fluctuations to grow slower than they would in a static universe: instead of exponentially, the perturbations grow as a power law in time.

To make even better progress, we expand the overdensity in the Fourier basis. We do so because it turns out that each Fourier mode $\delta_{\mathbf{k}}$ evolves independently (assuming linear perturbations and standard general relativity); we will demonstrate this in Sec. 9.4.2 below. The Fourier-space overdensity is

$$\delta_{\mathbf{k}}(t) = \frac{1}{\sqrt{V}} \int \delta(\mathbf{r}, t) e^{-i\mathbf{k}\cdot\mathbf{r}} d^3\mathbf{r}, \tag{9.24}$$

where V is the volume of the larger region over which the perturbations are assumed to be periodic. Note that $\mathbf{k}$ and $\mathbf{r}$ are both comoving quantities. After the Fourier transform, the time derivatives remain unchanged, while the gradients change as $\nabla \to -i\mathbf{k}$ and $\nabla^2 \to -k^2$. Then Eq. (9.23) becomes

$$\frac{\partial^2 \delta_{\mathbf{k}}}{\partial t^2} + 2\frac{\dot{a}}{a}\frac{\partial \delta_{\mathbf{k}}}{\partial t} = \left(4\pi G\bar{\rho} - \frac{k^2 c_s^2}{a^2}\right)\delta_{\mathbf{k}} - \frac{2}{3}\frac{T}{a^2}k^2 S_{\mathbf{k}}. \tag{9.25}$$

The overdensity $\delta_{\mathbf{k}}$ is "sourced" (generated to be nonzero) by the terms on the right-hand side, but its time evolution is slowed down by the Hubble term (proportional to $\dot{a}/a$) on the left-hand side. Intuitively, the expansion of the universe, represented by $\dot{a}/a > 0$, acts to slow down the growth of structure.

9.3.4 Initial Conditions

In principle, a universe can have either of the two general kinds of initial conditions (ICs). These are (see the illustration in Fig. 9.3):

1. **Isentropic ICs**, where there is no fluctuation in entropy in the initial conditions, so that $\nabla S = 0$ (hence, isentropic). Confusingly, these fluctuations are most often called **adiabatic ICs**, which strictly speaking is actually the $\dot{S} = 0$ condition, not $\nabla S = 0$; we henceforth adopt this imprecise but popular nomenclature. In the presence of adiabatic ICs, fluctuations in various components (matter, radiation, neutrinos, etc.) are proportional to each other, and the overall curvature fluctuation is nonzero. This is the kind of initial condition that inflation typically predicts, and that current data favor. Recall from Eq. (4.34) that the quantity

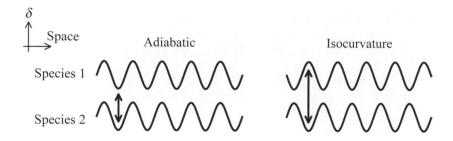

Fig. 9.3 Illustration of the adiabatic and isocurvature initial conditions, illustrated for two arbitrary species (e.g., cold dark matter and photons). For adiabatic initial conditions, the initial perturbations in all species (not just two as shown) are in phase. For isocurvature perturbations, the initial perturbations in a pair of species are out of phase, leading to the zero total curvature perturbation in these two.

n_i/s defines the number of particles of some species per comoving volume. For adiabatic perturbations, it is this comoving number density that is conserved, that is

$$0 = \delta\left(\frac{n_i}{s}\right) = \frac{\delta n_i}{s} - n_i \frac{\delta s}{s^2} \implies \frac{\delta n_i}{n_i} = \text{const (indep. of } i\text{)}, \qquad (9.26)$$

so that the adiabatic condition is that $\delta n/n$ be the same for each species. For baryons or CDM, $\delta n/n = \delta\rho/\rho$ since $\rho = nm$ (m being particle mass). For relativistic species such as photons or relativistic neutrinos, recall that $n \propto T^3$ and $\rho \propto T^4$, so that $\delta n/n = (3/4)(\delta\rho/\rho)$. Then the adiabatic initial condition implies the following relation between the energy density fluctuations for different species:

$$\frac{1}{3}\frac{\delta\rho_B}{\rho_B} = \frac{1}{3}\frac{\delta\rho_c}{\rho_c} = \frac{1}{4}\frac{\delta\rho_\gamma}{\rho_\gamma} = \frac{1}{4}\frac{\delta\rho_\nu}{\rho_\nu}, \qquad (9.27)$$

where the indices refer, respectively, to baryons, cold dark matter, photons, and relativistic neutrinos.

2. **Isocurvature ICs**, where there is identically zero curvature fluctuation, so that the fluctuations in different species effectively sum to zero. In this case, a nonzero entropy modulation is generated. There are a number of possibilities, which correspond to which of the equalities in Eq. (9.27) is violated. For example, a matter-radiation isocurvature mode corresponds to nonzero

$$S = \frac{\delta\rho_m}{\rho_m} - \frac{3}{4}\frac{\delta\rho_\gamma}{\rho_\gamma}, \qquad (9.28)$$

where $\rho_m = \rho_B + \rho_c$. Single-field inflation typically does not predict isocurvature perturbations but multi-field models, as well as other (non-inflationary) scenarios for the formation of structure, generally do.

Arbitrary initial conditions can be made up of a combination of the adiabatic and various isocurvature modes. In turn, the relative contributions from all of the aforementioned modes, adiabatic and isocurvature, can be constrained using data, as they lead to different predictions for galaxy clustering and CMB anisotropy.

Current cosmological constraints indicate that fluctuations are adiabatic and only a very small admixture of isocurvature fluctuations (upper limit of a few percent contribution to power, depending on which isocurvature mode is considered) is allowed by the data. In what follows, we assume adiabatic fluctuations only, and ignore the entropy term in Eq. (9.25).

9.4   Growth of Structure

Let us now study the gravitational collapse of structures. We first review the Jeans theory, which roughly implies: gravity wins, and matter collapses to form structures, on sufficently large scales; pressure wins, and prevents structure formation, on smaller scales. The mass/length scale separating these two regimes is called the Jeans scale. We now outline the derivation of these results.

9.4.1   The Jeans Length and Mass

Specializing in adiabatic density perturbations with $S_{\mathbf{k}} = 0$, and dropping the subscript $\mathbf{k}$ (remembering henceforth that δ refers to the Fourier component $\delta_{\mathbf{k}}$), Eq. (9.25) reads

$$\frac{\partial^2 \delta}{\partial t^2} + 2\frac{\dot{a}}{a}\frac{\partial \delta}{\partial t} = -\omega^2 \delta, \tag{9.29}$$

where

$$\omega^2 = \left(\frac{k^2 c_s^2}{a^2} - 4\pi G \bar{\rho}\right) \equiv (k^2 - k_J^2)\frac{c_s^2}{a^2}, \tag{9.30}$$

where $k_J = \sqrt{4\pi G\bar{\rho}}(a/c_s)$ is the Jeans wavenumber, which in turn defines the **Jeans length**

$$\lambda_J \equiv \frac{2\pi a}{k_J} \equiv c_s \sqrt{\frac{\pi}{G\bar{\rho}}}. \tag{9.31}$$

For simplicity, let us first consider the case of a *static universe* which does not contain the $\dot{\delta}$ term because $\dot{a}/a = 0$. Then the solution to Eq. (9.29) can be one of the following two:

- oscillating, $\delta(t) \propto \exp(\pm i\omega t)$ for $\lambda < \lambda_J$ (or $k > k_J$, i.e., when $\omega^2 > 0$), or
- exponentially growing, $\delta(t) \propto \exp(|\omega|t)$ for $\lambda > \lambda_J$ (or $k < k_J$, i.e., when $\omega^2 < 0$).

For the expanding universe, with the $\dot{\delta}$ term reinserted, qualitatively similar results hold: only Fourier modes larger than the Jeans length λ_J grow – though as a power law in time, not exponentially, as we prove further below.

The argument we just presented shows that only objects more massive than the Jeans mass can grow, where this mass is defined as

$$M_J \equiv \frac{4\pi}{3}\left(\frac{\lambda_J}{2}\right)^3 \rho_B \simeq c_s^3 \frac{1}{\sqrt{G^3 \rho_B}}, \tag{9.32}$$

where ρ_B is the baryonic energy density which is relevant here.

The Jeans mass is large before recombination, but drops like a rock at that time. The reason is that, once the electrons recombine with protons to create hydrogen atoms, the speed of sound goes from relativistic to highly sub-relativistic. In more detail, before recombination photons and baryons are tightly coupled, and the speed of sound is close to that of radiation alone, which is $c/\sqrt{3}$:

$$(c_s)^{\text{before recomb}} = \frac{c}{\sqrt{3\left(1 + \frac{3\rho_B(z)}{4\rho_\gamma(z)}\right)}} \simeq \frac{c}{\sqrt{3}}, \tag{9.33}$$

where the coefficient 3/4 comes about because under adiabatic compression, the energy densities change with volume as $\rho_M \propto V^{-1}$ and $\rho_\gamma \propto V^{-4/3}$. Then Eq. (9.32) implies

$$(M_J)^{\text{before recomb}} \simeq O(10^{19} M_\odot), \tag{9.34}$$

which corresponds to mass larger than that of any object in the universe. Before recombination, therefore, pressure prevents gravitational collapse of structures.

After recombination, however, baryons are not coupled to photons any more and they can be considered a non-relativistic monoatomic gas with the speed of sound

$$(c_s)^{\text{after recomb}} = \sqrt{\frac{5T}{3m_p}}, \tag{9.35}$$

which, just after recombination ($T \simeq 2\,\text{eV}$), evaluates to about 2×10^{-5} (in units of the speed of light), or about 6 km/s. Then Eq. (9.32) implies

$$(M_J)^{\text{after recomb}} \simeq O(10^5 M_\odot). \tag{9.36}$$

In the evolution of the universe post recombination, the Jeans mass continues to decrease further because $c_s \propto T^{1/2}$. Therefore, after recombination, the pressure support does not prevent the formation of cosmologically interesting structures.

Note that all of the aforementioned arguments apply to baryonic, "normal" matter. In contrast, *dark* matter has no pressure, and is not subject to the Jeans constraints. In fact, dark matter gets a head start (relative to baryonic matter), and its overdensities start to grow well before recombination. We return to this important point in Sec. 11.1.5 when we discuss how it constitutes a piece of evidence for the existence of dark matter.

9.4.2 Linear Growth of Cosmic Structure

On scales smaller than the horizon (so that the relativistic effects do not apply) and for matter at late times (so that the speed of sound is negligible), Eq. (9.23) reads

$$\ddot{\delta} + 2H\dot{\delta} - 4\pi G \rho_M(t)\delta = 0. \tag{9.37}$$

Because this is a second-order ordinary differential equation, it has two solutions, growing and decaying. We are interested in the growing solution, as the decaying one becomes unimportant quickly.

A most important feature of Eq. (9.37) is:

> Assuming standard general relativity, sub-horizon scales, and linear theory, all density-perturbation modes grow at the same rate – that is, independent of wavenumber k.

This simplifies our theory predictions for large-scale structure, since instead of computing $\delta(k, t)$, we need just $\delta(t)$.

9.4.3 Solutions to the Growth Equation

Let us work out solutions to the growth equation in a few simple cases. We will consider single-component scenarios, that is, periods in the history of the universe when a single type of energy density dominates the energy budget. We will also only consider a flat-universe scenario, so that $H^2 \propto \rho$, where ρ is the energy density of the dominant component. Then:

- **Radiation-dominated (RD) era.** In the RD era, recall that $a \propto t^{1/2}$, so that $H(t) \equiv \dot{a}/a \propto 1/(2t)$. Also, note that the last term in Eq. (9.37) is negligible because the Hubble parameter is dominated by radiation and not matter density, so that $4\pi G \rho_M \ll H^2$. Therefore, we need to solve the equation

$$\ddot{\delta} + 2H\dot{\delta} = 0 \tag{9.38}$$

  with $H = 1/(2t)$. Its solution is

$$\delta(t) = A_1 + A_2 \ln t \qquad \text{(radiation dominated)}. \tag{9.39}$$

Therefore, perturbations in the RD universe grow extremely slowly (logarithmically) – basically they do not grow much. This will become important in a bit, when we discuss the shape of the matter power spectrum.

- **Matter-dominated (MD) era.** In the MD era, recall that $a \propto t^{2/3}$ so that $H(t) \equiv \dot{a}/a \propto 2/(3t)$. Also, in the MD era the Hubble parameter is dominated by matter density, so that $4\pi G \rho_M = (3/2)H^2$. Let us assume that $\delta(t) \propto t^n$; then the growth equation simplifies to

$$n(n-1) + \frac{4}{3}n - \frac{2}{3} = 0, \tag{9.40}$$

whose solutions are easy to obtain: $n = +2/3$ and -1, so that

$$\delta(t) = B_1 t^{2/3} + B_2 t^{-1} \qquad \text{(matter dominated)}. \tag{9.41}$$

Since $a(t) \propto t^{2/3}$ in the MD era, the growing mode of the perturbations grows proportionally to the scale factor:

$$\delta(t) \propto a(t) \propto t^{2/3} \qquad \text{(matter dominated)}. \tag{9.42}$$

This scaling in the matter-dominated regime is of the utmost importance, as the universe spends of order 10 billion years in it – from the matter–radiation equality 50,000 years after the Big Bang, to the onset of dark energy a few billion years ago. During that time, structures in the universe grew appreciably, all thanks to the scaling in Eq. (9.42).

- **Lambda-dominated era.** In the cosmological constant-dominated era, which will presumably happen in our future when dark energy dominates completely, the scale factor will grow exponentially,[3] $a \propto e^{Ht}$, so that $H(t) \equiv H_\Lambda = $ const. Also, note that the last term in Eq. (9.37) is negligible since the matter density is negligible relative to vacuum energy in H. Therefore, we need to solve the equation

$$\ddot{\delta} + 2H_\Lambda \dot{\delta} = 0, \tag{9.43}$$

whose solution is

$$\delta(t) = C_1 + C_2 e^{-2H_\Lambda t} \simeq \text{const.} \qquad \text{(Lambda dominated)}, \tag{9.44}$$

where the exponentially decaying term becomes negligible quickly. Therefore, density perturbations do not grow at all in a Lambda-dominated universe. We are all witnesses to this effect today, as our universe with 70 percent dark energy and 30 percent matter displays a severely suppressed structure formation (relative to the matter-only Einstein–de Sitter scenario), something that is readily observed in cosmological data that probe $z \lesssim 1$.

Summarizing then: assuming adiabatic initial conditions, standard Einstein gravity, sub-horizon fluctuations ($k \gg aH$), and linear theory:

---

[3] The same happens during inflation, when the universe is completely vacuum-energy dominated.

> During radiation domination matter overdensities grow only logarithmically
> with time; during matter domination they grow robustly, proportionally to
> the scale factor; during the cosmological constant domination, they do not
> grow at all.

9.4.4 Growth of Perturbations: More General Solutions

We have just presented closed-form (and simple!) solutions for the growth of fluctuations in single-component cosmological models. For a multi-component model,
a closed-form solution can unfortunately not be found in general. For *special* multi-
component cases, such closed-form solutions can be found (see Problems 9.3 and
9.9), but cosmologists typically opt for numerically solving the second-order differential equation for growth. To do that, it is useful to cast Eq. (9.37) in a dimensionless form. Let us introduce the **linear growth function**

$$D(a) = \delta(a)/\delta(1), \tag{9.45}$$

where D at the present time is unity, $D(1) = 1$. The original equation determining
the growth of structure, (9.37), can now be set in a nice dimensionless form as
(Problem 9.2)

$$2\frac{d^2 g}{d\ln a^2} + [5 - 3w(a)\Omega_{\mathrm{DE}}(a)]\frac{dg}{d\ln a} + 3\left[1 - w(a)\right]\Omega_{\mathrm{DE}}(a)g = 0, \tag{9.46}$$

where $g(a)$ is the "growth suppression factor" – that is, growth relative to that in
an EdS universe. The growth suppression factor g is related to the growth factor
D implicitly via

$$D(a) \equiv \frac{ag(a)}{g(1)}. \tag{9.47}$$

Figure 9.4 shows the temporal evolution of $D(z)$ and $g(z)$ in our fiducial cosmological model. Note that the onset of dark energy at late times ($z \lesssim 1$) causes
these functions to deviate from their matter-domination behavior ($D(a) = a$ or
$D(z) = 1/(1 + z)$ and $g(z) = 1$). The late-time decrease in either $g(z)$ or $D(z)$
fully specifies the suppression in the growth of structure (in linear theory) due to
dark energy. In our fiducial ΛCDM cosmological model, the present-day value of
the suppression factor is $g(z = 0) \simeq 0.78$.

We now move from discussing the rate of growth of density perturbations in
cosmic time, to the shape of the *spectrum* of density perturbations.

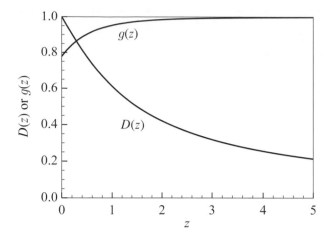

Fig. 9.4 Linear growth function $D(z)$ and the growth suppression factor $g(z)$, as a function of redshift z, in the fiducial cosmological model. Either $D(z)$ or $g(z)$ fully describes the linear growth of fluctuations; see text for details.

9.5 Power Spectrum

What is the spatial distribution of the overdensity δ in the universe? Does it have a lot of structure on small scales like the surface of a sandpaper, or on large scales like rolling hills in a large valley? At least on linear scales ($|\delta| \ll 1$), the standard cosmological model answers this question exactly, as we now discuss. On mildly and strongly nonlinear scales, numerical simulations complete the picture at late times.

It is most convenient to address this in Fourier space, looking at Fourier components of overdensity, $\delta_\mathbf{k}$. From here on, we will be making heavy use of the wavenumber k, defined in the usual way familiar from freshman physics:

$$k = \frac{2\pi}{\lambda},\tag{9.48}$$

where λ is the wavelength of the perturbation plane-wave. Large wavenumbers therefore correspond to short-wavelength modes of a perturbation, and vice versa. Because our favorite units of distance are $h^{-1}\mathrm{Mpc}$ (see Chapter 3, particularly near the end of Sec. 3.4.1), our favorite units of wavenumber are $h\,\mathrm{Mpc}^{-1}$.

The overdensity can be expressed in its Fourier components in some comoving volume V as

$$\delta(\mathbf{r}) = \frac{\sqrt{V}}{(2\pi)^3} \int \delta_\mathbf{k}\, e^{-i\mathbf{k}\cdot\mathbf{r}}\, d^3k,\tag{9.49}$$

and the Fourier components are, conversely,

$$\delta_\mathbf{k} = \frac{1}{\sqrt{V}} \int \delta(\mathbf{r})\, e^{i\mathbf{k}\cdot\mathbf{r}}\, d^3r.\tag{9.50}$$

We will continue to work in linear theory, that is, assume $|\delta| \ll 1$. Then each Fourier component satisfies the growth equation, Eq. (9.37). Note too that $\delta_{\mathbf{k}}$ are complex numbers in general, and that their units are $(\text{volume})^{1/2}$. [Recall that the real-space overdensity $\delta(\mathbf{r})$ is dimensionless, as is clear from the definition $\delta \equiv \rho/\bar{\rho} - 1$.]

If we shift each location in space $\mathbf{r}$ by some $\Delta\mathbf{r}$, Eq. (9.50) indicates that

$$\delta_{\mathbf{k}} \to \delta_{\mathbf{k}}\, e^{i\mathbf{k}\cdot\Delta\mathbf{r}}. \tag{9.51}$$

Then the two-point correlation function of Fourier-space overdensity transforms as

$$\langle \delta_{\mathbf{k}} \delta_{\mathbf{k}'}^* \rangle \to e^{i(\mathbf{k}-\mathbf{k}')\Delta\mathbf{r}} \langle \delta_{\mathbf{k}} \delta_{\mathbf{k}'}^* \rangle. \tag{9.52}$$

Assuming statistical homogeneity on large scales, we know that the right-hand side of Eq. (9.52) must not depend on $\Delta\mathbf{r}$, because the statistical properties do not change if we move some distance in space – it is useful here to recall the homogeneity illustration in Fig. 1.3 of Chapter 1. Therefore, the two-point function in Eq. (9.52) must be proportional to $\delta(\mathbf{k} - \mathbf{k}')$, and the remaining dependence is only on $\mathbf{k}$. Finally, due to statistical isotropy, only the magnitude of the wavenumber, $k \equiv |\mathbf{k}|$, enters, as clustering must be direction independent. Thus

$$\langle \delta_{\mathbf{k}}\, \delta_{\mathbf{k}'}^* \rangle = (2\pi)^3\, \delta^{(3)}(\mathbf{k} - \mathbf{k}')\, P(k), \tag{9.53}$$

where $P(k)$ is the **power spectrum** – the Fourier transform of the two-point function. Note that the factor $(2\pi)^3$ is due to conventions and not important.

The power spectrum is thus defined as the ensemble average, over all universes, of the square of the Fourier component $\delta_{\mathbf{k}}$. Since we are usually not able to average over different universes, we average over different locations in our universe.

The power spectrum tells us how much power there is on different scales – that is, different wavenumbers k. Consider the following gedankenexperiment: a sky that looks like a chessboard, with white and black pixels (see Fig. 9.5). Let the size of each pixel be L. Then the power spectrum will be smallish at all scales except at $k \sim \pi/L$, where it will peak (there is an additional $\sqrt{2}$ in Fig. 9.5 because that example is in two dimensions). At smaller scales, we are looking within a pixel where there is no variation in contrast, while at much larger scales, we are averaging over many pixels, and the signal gets washed out. It is only at scale $\sim L$ (or $k \sim \pi/L$, in Fourier space) that we see the black–white contrast and thus have nonzero power.

9.5.1 Shape of the Primordial $P(k)$

Inflation seeds power on all scales. But what is the relative power generated at different k? In the late 1960s, Harrison, Zeldovich, and Peebles put forth a conjecture that was far ahead of its time (well before inflation was invented!). They proposed the **Harrison–Zeldovich–Peebles spectrum**:

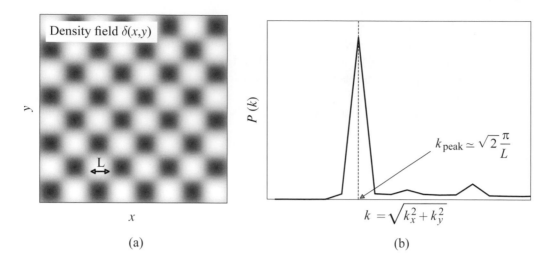

(a)   (b)

Fig. 9.5   Intuitive explanation of the meaning of the power spectrum. In this 2D example, the density field, shown in panel (a), resembles a chessboard, with black and white fields of size L. Panel (b) shows the power spectrum of this density field. Note that Fourier components in each direction peak when $k_x = k_y = \pi/L$, and hence the power spectrum in the total wavenumber $k = \sqrt{k_x^2 + k_y^2}$ peaks at $k = \sqrt{2}\,\pi/L$. The density field has been smoothed to avoid large higher harmonics, yet you can still see them as smaller bumps in $P(k)$.

$$P(k) \propto k^{n_s} \quad \text{with} \quad n_s \simeq 1. \tag{9.54}$$

The Harrison–Zeldovich–Peebles argument went like this. If n_s were much bigger than 1, there would be too much small-scale (large-k) power, and too many black holes would be created early on, in conflict with observations. If n_s were much less than one, there would be too much power on large scales (small k), and huge superclusters and voids would dominate the much smaller galaxies, which again is not what we observe. This goldilocks-type argument thus favors $n_s \simeq 1$.

Two remarkable things have happened since the Harrison–Zeldovich–Peebles conjecture ca. 1969:

- Inflation predicts that $n_s = 1 - 6\epsilon + 2\eta + O(\epsilon^2, \eta^2)$, where $\epsilon \ll 1$ and $|\eta| \ll 1$ are the so-called slow-roll parameters and are related to the first two derivatives of the effective potential of inflaton with respect to the field value (see Chapter 8). Thus, the inflationary prediction agrees very well with the Harrison–Zeldovich–Peebles conjecture.

- n_s has been measured by a number of modern experiments to be just below one; the Planck measurement is $n_s = 0.966 \pm 0.005$. This agrees with the inflationary prediction very well, as $-6\epsilon+2\eta$ is slightly negative for typical inflationary models, making the power spectrum slightly "red" (n_s smaller than one).

Therefore, the Harrison–Zeldovich–Peebles conjecture has proven to be both prescient and remarkably accurate.

The Harrison–Zeldovich–Peebles primordial power spectrum with $n_s = 1$ has another feature: all modes re-enter the horizon with the same amplitude, independently of their wavelength. We derive and explain this result in Box 9.2.

Box 9.2   **Specialness of the Harrison–Zeldovich–Peebles Spectrum**

Consider an object of (comoving) size r and mass $M \propto r^3$. One can approximately integrate Eq. (9.58), with limits from $k = 0$ to $k = 1/r$, at which point $\sin(kr)/kr$ begins to wildly oscillate. Doing this gives $\xi(r) \propto \int_0^{1/r} k^{n_s+2}dk \propto r^{-(n_s+3)}$. Then the mass rms fluctuation is, at a fixed epoch,

$$\sigma_{\rm rms}(M) \equiv \left\langle \left(\frac{\delta M}{M}\right)^2 \right\rangle^{1/2} \equiv \langle\delta^2\rangle^{1/2} \propto r^{-(n_s+3)/2} \propto M^{-(n_s+3)/6}, \qquad (\rm B1)$$

where the last proportionality follows from $M \propto r^3$. Note already one special result: for $n_s = 0$, we obtain the white-noise power spectrum. This follows because $\sigma_{\rm rms}(M) \propto M^{-1/2}$, similar to the familiar case of $\delta N/N \propto N^{-1/2}$, signifies a Poisson noise spectrum of fluctuations.

For $n_s = 1$ exactly, we have the so-called scale-invariant spectrum where every horizon-scale fluctuation mode has the same amplitude. We can see this as follows, by considering the temporal evolution of *super*-horizon fluctuations (in contrast to Sec. 9.4.3 where we considered *sub*-horizon fluctuations). Consider first the radiation-dominated regime. From the Poisson equation, $-k^2\Phi/a^2 \propto \rho\delta$, it follows that in the radiation-dominated epoch, when $\rho \propto a^{-4}$, keeping Φ constant at super-horizon scales[a] requires $\delta \propto a^2$. Similarly, in the matter-dominated epoch, $\rho \propto a^{-3}$ and then keeping Φ constant at super-horizon scales requires $\delta \propto a$ (which just happens to be identical to the scaling of matter perturbations on sub-horizon scales).

Then the growth of $\sigma_{\rm rms}(M)$ during the matter and radiation domination is the same in terms of the respective Hubble distance $R_H \simeq k_H^{-1}$. This follows from

$$\begin{cases} {\rm RD}\,(\delta \propto a^2):\ k_H = aH \propto a \times a^{-2}\ \propto a^{-1}\ \Rightarrow R_H \propto a\ \ \Rightarrow \delta \propto R_H^2 \\ {\rm MD}\,(\delta \propto a):\ k_H = aH \propto a \times a^{-3/2} \propto a^{-1/2} \Rightarrow R_H \propto a^{1/2} \Rightarrow \delta \propto R_H^2. \end{cases}$$
$$(\rm B2)$$

So in either case, $\delta \propto R_H^2$. The mass enclosed in the sphere of Hubble radius is $M \propto R_H^3$, and then the full temporal rms fluctuation is R_H^2 times what we found in Eq. (B1):

$$\sigma_{\rm rms}(M) \propto R_H^2 M^{-(n_s+3)/6} \propto M^{2/3} M^{-(n_s+3)/6} \xrightarrow{\text{for } n_s = 1} \text{const.} \qquad (\rm B3)$$

---

[a] The constancy of the so-called Bardeen potential on super-horizon scales is not treated explicitly in this book, but was briefly mentioned in Sec. 8.4.1, when we discussed the quantum-mechanical generation of fluctuations.

9.5.2 Relation Between $P(k)$ and $\xi(r)$

It is possible to relate the correlation function to the power spectrum. First, recall that the correlation function is given by

$$\langle \delta_{\mathbf{r}_1} \delta^*_{\mathbf{r}_2} \rangle = \xi(r_{12}) \qquad \text{where} \qquad r_{12} = |\mathbf{r}_1 - \mathbf{r}_2|, \tag{9.55}$$

where again, due to homogeneity, the correlation function must depend only on the distance between the two vectors. Now we can relate $P(k)$ to $\xi(r)$ via

$$
\begin{aligned}
P(k) &= \langle \delta_{\mathbf{k}} \delta^*_{\mathbf{k}} \rangle = \langle \delta_{\mathbf{k}} \delta_{-\mathbf{k}} \rangle \\
&= \frac{1}{V} \int \int \langle \delta(\mathbf{r}_1)\delta^*(\mathbf{r}_2) \rangle e^{-i\mathbf{k}\mathbf{r}_1} e^{i\mathbf{k}\mathbf{r}_2} d^3\mathbf{r}_1 d^3\mathbf{r}_2 \\
&= \int \xi(r_{12}) e^{-i\mathbf{k}\mathbf{r}_{12}} d^3\mathbf{r}_{12} \\
&\equiv \int \xi(r) e^{-i\mathbf{k}\mathbf{r}} d^3\mathbf{r},
\end{aligned}
\tag{9.56}
$$

where in going from the second to the third line we switched to $(\mathbf{r}_{12}, \mathbf{r}_2)$, trivially integrated over $\mathbf{r}_2$, and used $\int d^3\mathbf{r}_2 = V$. Notice a few nice things:

- The power spectrum is a Fourier transform of the two-point correlation function. This is actually guaranteed by the Wiener–Khinchin theorem in statistics, but we proved it explicitly here.

- The power spectrum does not depend on the (arbitrary) volume V.

- The units of $P(k)$ are just those of (comoving) volume, so that $P(k)/V$ is dimensionless, implying that $k^3 P(k)$ is also dimensionless. This agrees with our earlier statement that the dimensions of Fourier-space overdensity are $[\delta_{\mathbf{k}}] = (\text{volume})^{1/2}$.

Now let us inverse Fourier transform $P(k)$ to get (Problem 9.7)

$$\xi(r) = \frac{1}{(2\pi)^3} \int P(k) e^{i\mathbf{k}\mathbf{r}} d^3\mathbf{k} = \frac{1}{2\pi^2 r} \int_0^\infty P(k) \sin(kr) k \, dk. \tag{9.57}$$

Let us evaluate the **zero-lag** correlation function, $\xi(r = 0)$:

$$\xi(0) = \frac{1}{2\pi^2} \int_0^\infty P(k) \lim_{r \to 0} \frac{\sin(kr)}{kr} k^2 dk \tag{9.58}$$

$$\equiv \int_0^\infty \Delta^2(k) d\ln k, \tag{9.59}$$

where we have defined the **dimensionless power spectrum**

$$\Delta^2(k) \equiv \frac{k^3 P(k)}{2\pi^2}, \tag{9.60}$$

which is, mathematically, the contribution to variance per log wavenumber. If the peak of $\Delta^2(k)$ is at some k_*, then fluctuations in δ are dominated by wavelengths of order $2\pi/k_*$. Because $P(k)$ has units of k^{-3}, it follows that $\Delta^2(k)$ is dimensionless. Both $P(k)$ and $\Delta^2(k)$ are commonly referred to as the "power spectrum."

Note also that the integral in Eq. (9.59) is badly divergent in the ultraviolet (large-k limit). As an exercise, you can think about what the intuitive reason for this is. At any rate (and as a hint to the answer to the preceding question), this indicates that we will have to *smooth* the density field before calculating its variance; see Eq. (B1) and the upcoming discussion in Sec. 9.6.2. But first, we will declare victory on the complete expression for the power spectrum of density fluctuations.

9.5.3 Machine-Friendly Power Spectrum

Numerically evaluating the power spectrum, particularly the dimensionless form in Eq. (9.60), is essential in cosmology, as the power spectrum is a vital link between theory and observations of the galaxy distribution. Here we show the formula for the *linear* power spectrum of matter density perturbations in standard FLRW cosmology:

$$\Delta^2(k, a) = A_s \frac{4}{25} \frac{1}{\Omega_M^2} \left(\frac{k}{k_{\mathrm{piv}}}\right)^{n_s - 1} \left(\frac{k}{H_0}\right)^4 [ag(a)]^2 T^2(k), \tag{9.61}$$

where each individual parenthetical term is dimensionless. Here

- A_s is the normalization of the power spectrum (for the fiducial cosmology, $A_s = 2.1 \times 10^{-9}$) and n_s is the (scalar) spectral index.
- k_{piv} is the "pivot" wavenumber around which Δ^2 is expressed as a power law in k; Planck adopts $k_{\mathrm{piv}} = 0.05\,\mathrm{Mpc}^{-1}$, which is close to the wavenumber at which the primordial power is best constrained. Note that the measured value of A_s necessarily depends on the choice of k_{piv}.

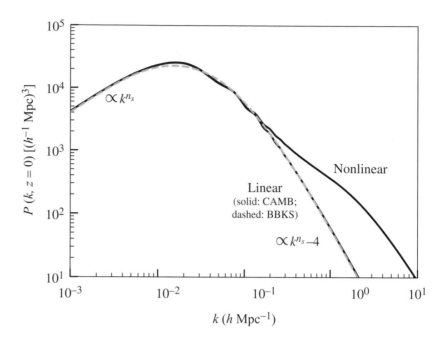

Linear and nonlinear power spectrum $P(k)$. The solid lines show the linear and nonlinear results obtained using the code CAMB (Lewis *et al.*, 2000), while the lighter, dashed line shows the linear-theory result obtained using our from-scratch code based on Eq. (9.61), including a simple fitting formula for the transfer function (BBKS). You will get a chance to obtain and check the latter result – see Problem 9.9.

- $[ag(a)]$ is the linear growth of perturbations. Note that in the Einstein–de Sitter model $g(a) = 1$ identically and at all times, and in the ΛCDM model $g(a)$ at recent times drops down to the value of ≈ 0.78 at $a = 1$. Recall that $ag(a) = D(a)g(1)$, as per Eq. (9.47).

- $T(k)$ is the linear transfer function (see Sec. 9.5.4 below).

Note again that this is the matter power spectrum evaluated in linear theory, $|\delta| \ll 1$. This assumption holds at present-day scales greater than about 10 Mpc. At smaller scales, nonlinearities are important, and they modify the power spectrum at these small scales (see Fig. 9.6), and can be calibrated using numerical simulations which we discuss in Sec. 9.6.1.

Figure 9.6 shows the numerically evaluated theoretical matter power spectrum – both the linear and the full, nonlinear $P(k)$. Note the characteristic scaling with wavenumber, going from $k^{n_s} \simeq k^1$ on large scales, to $k^{n_s-4} \simeq k^{-3}$ on small scales. We next discuss the reason for this turnover in the power spectrum, which occurs at $k \sim 0.02\,h\,\mathrm{Mpc}^{-1}$ in the standard cosmological model, and the transfer function that captures it.

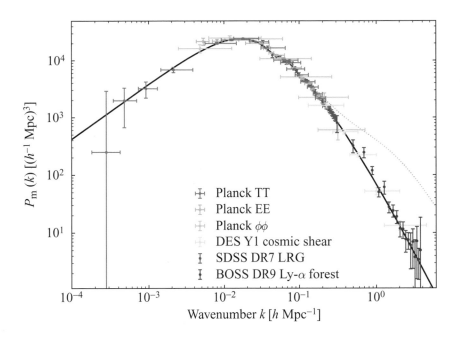

Fig. 9.7 Constraints on the linear power spectrum $P(k)$ at $z = 0$ from recent data, specifically Planck's CMB temperature (TT), polarization (EE), and gravitational lensing ($\phi\phi$); galaxy and Lyman-α feature clustering from the SDSS; and cosmic shear measurements from the DES. In all cases, the observable quantities were deconvolved in order to obtain error bars on $P(k, z = 0)$. The black curve shows the linear power spectrum predicted by the concordance ΛCDM cosmological model, while the dotted curve shows the impact of nonlinear clustering. Adopted from the Planck legacy paper (Aghanim *et al.*, 2020a), reproduced with permission © ESO. A black and white version of this figure will appear in some formats. For the color version, please refer to the plate section.

First, however, we show some *measurements* of $P(k)$ in Fig. 9.7. This shows the primordial (i.e., linear) power spectrum in the ΛCDM model, along with measurements deconvolved in a way to show their constraints on $P(k)$. The dashed line shows the effect of nonlinearities that enter on scales $k \gtrsim 0.1\, h\,\mathrm{Mpc}^{-1}$.

9.5.4 So What is the Transfer Function?

The (linear) **transfer function** mainly encodes the effect that the transition from radiation domination to matter domination (which happens at redshift $z \simeq 3500$) has on the growth of density fluctuations.

Inflation predicts that fluctuations enter the horizon with the amplitude that is independent of their wavelength. Larger-wavelength (smaller-k) fluctuations enter the horizon at later times, and vice versa. For example, fluctuations that are entering the horizon today have wavelength on the order of $\lambda \sim H_0^{-1} \simeq 5\,\mathrm{Gpc}$.

Consider the universe that was always matter dominated. Then, even though longer-wavelength fluctuations enter the horizon later than shorter-wavelength ones, their amplitudes upon re-entry would be the same[4] according to inflation. In that case, we would simply have $T(k) = 1$.

However, things are more complicated because of the existence of the radiation-dominated era. Recall that during radiation domination perturbations grow only logarithmically with time. So, the perturbation modes whose wavelength was small enough to have entered the horizon during the radiation-dominated era were "stunted," and these modes could not grow until the universe became matter dominated. Conversely, the perturbation mode whose wavelength was longer than the horizon at matter–radiation equality, and which entered the horizon during the matter-dominated era, never experienced stunting of the growth.

Following up on this intuitive picture, we now derive the most prominent feature of the transfer function – its turnover at $k \simeq 0.02\,h\,\mathrm{Mpc}^{-1}$. Recall from the discussion around Eq. (B2) in Box 9.2 that the super-horizon fluctuation δ is proportional to R_H^2 in both the radiation- and matter-dominated regime. Perturbation that enters the horizon at some scale $k_1 > k_{\mathrm{eq}}$, where k_{eq} is the wavenumber corresponding to the horizon scale at matter–radiation equality, will see its growth suppressed relative to the perturbation at some other scale $k_2 < k_{\mathrm{eq}}$ by the factor

$$\frac{\delta_{k_1}}{\delta_{k_2}} = \frac{\delta_{k_1}}{\delta_{k_{\mathrm{eq}}}} = \left(\frac{R_{\mathrm{H},k_1}}{R_{\mathrm{H,eq}}}\right)^2 \propto \left(\frac{k_1}{k_{\mathrm{eq}}}\right)^{-2}. \tag{9.62}$$

The growth therefore goes as the inverse square of the wavenumber; the higher the wavenumber is, the earlier the mode entered the horizon (before matter–radiation equality), and the more its growth was stunted. The transfer function simply summarizes the trend found in Eq. (9.62):

$$T(k) = \begin{cases} 1 & k \ll k_{\mathrm{eq}} \\ (k/k_{\mathrm{eq}})^{-2} & k \gg k_{\mathrm{eq}}, \end{cases} \tag{9.63}$$

where $1/k_{\mathrm{eq}}$ is the **(comoving) size of the horizon at matter–radiation equality**. The comoving wavenumber at matter–radiation equality (scale factor $a_{\mathrm{eq}} \simeq 1/3500$) is (Problem 9.4)

$$k_{\mathrm{eq}} = a_{\mathrm{eq}} H = H_0 \sqrt{2\frac{\Omega_M^2}{\Omega_R}} \simeq 0.02\,h\,\mathrm{Mpc}^{-1}. \tag{9.64}$$

Therefore, the power spectrum $P(k)$ has the following asymptotic behaviors:

$$P(k) \propto \begin{cases} k^{n_s} & k \ll k_{\mathrm{eq}} \\ k^{n_s - 4} & k \gg k_{\mathrm{eq}}, \end{cases} \tag{9.65}$$

---

[4] The amplitudes would be *nearly* the same; departures of this scale invariance are proportional to $n_s - 1$.

with $k_{\text{eq}} \simeq 0.02\,h\,\text{Mpc}^{-1}$.

Beyond the turnover in Eq. (9.63), there are more features in the transfer function that encode a variety of physical effects. Probably the *second* most prominent effect in $T(k)$ are the baryon acoustic oscillations, discussed in Sec. 9.5.5 just below. Accurate transfer functions, for a given cosmological model, can be inferred from the output of specialized computer codes that solve the full set of Einstein–Boltzmann equations for the evolution of density perturbations; examples of such codes are CAMB (Lewis *et al.*, 2000) or CLASS (Blas *et al.*, 2011; Lesgourgues, 2011). In Chapter 11 (Fig. 11.3), we show the transfer function for the standard CDM (cold dark matter) model, along with those for several more exotic dark-matter-model scenarios.

9.5.5 Baryon Acoustic Oscillations (BAO)

One especially important feature in the matter power spectrum are the **baryon acoustic oscillations**. This refers to a series of bumps and troughs in $P(k)$ on scales just smaller than the matter–radiation turnover feature. The series of BAO features in $P(k)$ corresponds to a *single*[5] peak in the two-point correlation function $\xi(r)$.

The BAO have the same origin as the oscillations in the CMB angular power spectrum, so we refer the reader to the more detailed discussion in Chapter 13, and only summarize the physics here. The story has its origin in the universe prior to recombination, when photons and baryons were tightly coupled. Competition between radiation pressure and gravity causes oscillations in the photon–baryon fluid. A single, spherical overdensity propagates outward with the speed of sound, $c_s \simeq 1/\sqrt{3(1+R)}$, where R is the ratio of baryon to total matter density. The (comoving) radius of the sphere – the distance that sound can travel since the Big Bang – is called the **sound horizon**. The sound horizon grows to a size of $\simeq 100\,h^{-1}\text{Mpc}$ comoving at recombination, and remains unchanged thereafter as, recall from Sec. 9.4, the speed of sound c_s rapidly drops after the universe becomes neutral.

After recombination, the newly freed baryons fall into the dark-matter potential wells. This happens at the end of the so-called **drag epoch**, which indicates the time when the baryons decoupled, and which occurs slightly later (at $z_d \simeq 1060$ in the Standard Model) than photon decoupling ($z_{\text{dec}} \sim 1090$) only because there are so few baryons relative to the number of photons.

The density field retains memory of this process by encoding excess probability of finding galaxy pairs at the distance corresponding to the sound horizon[6] at the end of the drag epoch – the **drag-epoch sound horizon**, $r_d \equiv r(z_d)$. It is this effectively geometrical feature in the clustering signal, corresponding to the distance r_d, that we call the BAO. Because the baryon density is subdominant to that of dark

---

[5] In mathematical treatment of processes in physics, a single feature in physical space corresponds to a harmonic series of features in Fourier space.

[6] See Sec. 13.3.2 for the mathematical expression for the sound horizon.

matter ($\Omega_B/\Omega_M \simeq 1/6$), the amplitude of the BAO feature(s) is only $\sim$10 percent of the total clustering amplitude, much smaller than the order-unity wiggles in the CMB angular power spectrum.

The BAO can be extracted statistically by measuring the two-point correlation function of galaxies and other tracers of large-scale structure such as quasars or Lyman-α absorbers. The BAO is instantly recognizable as a "bump" in the real-space correlation function, or multiple bumps in $P(k)$; see Fig. 9.8. The BAO feature was first detected in 2005, in data from the Sloan Digital Sky Survey (SDSS; Eisenstein *et al.*, 2005) and the 2dF Galaxy Redshift Survey (2dFGRS; Cole *et al.*, 2005). Since that time, the BAO have been detected in both real and harmonic space, and in both spectroscopic and photometric galaxy samples.

The BAO connects to theory in a remarkably simple way. By measuring the galaxy correlation function, we effectively detect the *angle* at which the BAO feature is seen from our vantage point. For galaxies in the plane of the sky, the angle is related to the **transverse** (perpendicular to the line of sight) separation $s_\perp$ as (see Fig. 9.9)

$$s_\perp(z) = d_A(z)\theta. \tag{9.66}$$

One could also look at the excess clustering in the **radial** (parallel to the line of sight) direction, and in that case one relates a radial feature $s_\parallel$ to the redshift

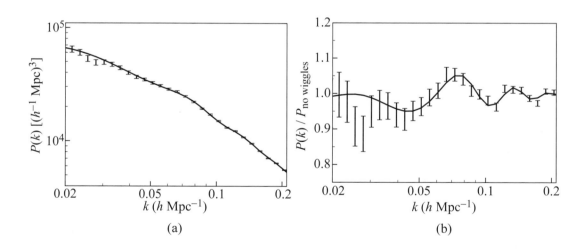

(a)   (b)

Baryon acoustic oscillation features measured in BOSS data (Gil-Marín *et al.*, 2016; data provided by Chris Blake). Both panels show the measurements from BOSS. Panel (a) shows the measured galaxy power spectrum $P(k)$, along with the best-fit ΛCDM theory model. Notice the barely perceptible BAO wiggles. Panel (b) also shows the measured and theoretical $P(k)$, but divided by a theory power spectrum that has the BAO artificially removed, $P_{\text{no wiggles}}(k)$, in order to emphasize the BAO features.

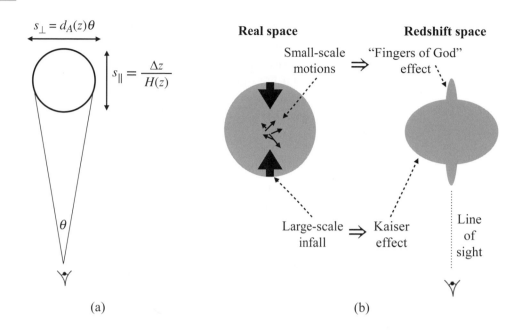

Fig. 9.9 (a) Illustration of the quantities measured parallel and perpendicular to the line of sight, assuming a perfectly spherical object (or feature in the distribution of galaxies). In the case of the BAO, both the parallel and the perpendicular distance can be measured in principle. In an Alcock–Paczynski test, one equates $s_\parallel = s_\perp$ in order to get a constraint on $H(z)d_A(z)$. (b) Illustration of the redshift-space distortions. There are two principal effects: (1) on large scales, velocity flows into large overdensities, "squishing" the appearance of the object along the line of sight – this is the Kaiser effect; (2) on smaller scales, random motions introduce apparent elongation along the line of sight – this is the somewhat hyperbolically called "fingers of God" effect.

extent:[7]

$$s_\parallel(z) = \frac{\Delta z}{H(z)}. \qquad (9.67)$$

For the BAO case, $s_\perp(z) = s_\parallel(z) = r_d$, a redshift-independent quantity. Hence the measurements θ and Δz respectively relate to theoretical quantities r_d, $d_A(z)$, and $H(z)$, via

$$\{\theta, \Delta z\}_{\mathrm{BAO}} = \left\{ \frac{r_d}{d_A(z)}, r_d H(z) \right\} \qquad \text{(from transverse, radial modes).} \qquad (9.68)$$

---

[7] For the radial-feature measurement in particular, accurate redshifts – typically meaning spectroscopic redshifts – of tracer objects are required in order to not wash out the observed BAO feature.

Thus we conclude:

> The angular (radial) extent of the BAO feature gives us the angular-diameter distance (the Hubble parameter), scaled to the comoving horizon at the end of the drag epoch, at the redshift of the source objects.

Separate constraints on the transverse and radial BAO are not always feasible due to imperfect redshifts, but also because of the inherently smaller signal-to-noise ratio in such a split which especially affects the radial modes. One thus often combines all the transverse and radial information into a single quantity:

$$D_V(z) \equiv \left(d_A^2(z) \frac{z}{H(z)} \right)^{1/3} , \tag{9.69}$$

where D_V has units of distance and is not a fundamental quantity but rather defined as above. We recognize two powers of the transverse modes (for two dimensions in the plane of the sky) and one power of the radial modes (for one dimension in the radial direction). Typically, galaxies in a survey are split in several redshift bins, and a value of $D_V(z)$, or else a pair $(r_d/d_A(z), r_d H(z))$, is reported from the measurement of the BAO feature in each bin.

9.5.6 Alcock–Paczynski (AP) Test

Let us look at the equations (9.66) and (9.67) again. Imagine a population of objects in the universe that are spherical ("spherical cows"). Then one would *know* that $s_\perp = s_\parallel$, and thus have the **Alcock–Paczynski** relation

$$\frac{\Delta z}{\theta} = H(z) d_A(z) \qquad \text{(Alcock–Paczynski test)}. \tag{9.70}$$

A comparison of the angular and radial extent of our spherical objects thus returns the product of the Hubble parameter and the angular diameter distance. This novel test of the cosmological model was proposed by Alcock and Paczynski (1979).

The universe is very sadly not full of perfectly spherical galaxies, nor are clusters of galaxies particularly round. Thus, it is a challenge to find a class of objects to which to apply the AP test. However, this is where the BAO feature proves to be useful again – while not an object per se, it is a well-defined and relatively easily measured physical feature imprinted in the galaxy correlation function, and it *is* perfectly spherical due to homogeneity and isotropy of space. Therefore, one can consider comparing the BAO location from the radial modes to that of the transverse modes to extract the combination $H(z) d_A(z)$.

The information in the Alcock–Paczynski quantity $H(z) d_A(z)$ is only a subset of the information content that is automatically extracted if we *separately* measure the transverse and radial modes as per Eq. (9.68). This is because we do have a very good theoretical prediction for the size of the feature in question – the sound horizon at the end of the drag epoch, r_d. Thus, literal application of the

AP test is a somewhat suboptimal way to extract the BAO information. However, AP has found applications for cosmological features which are (at least on average) spherical, but for whose radius we do not have a theory prediction. Notably, cosmic **voids** – regions largely devoid of galaxies that can be identified in galaxy maps – are perhaps a perfect subject for the application of the AP test. This is because voids are small enough (radius $\sim 10-50\,\mathrm{Mpc}$) to be numerous, but large enough that their effective boundary is not significantly distorted by peculiar motions of the galaxies that reside there.

Before we depart the observable features in the correlation function, let us mention redshift-space distortions (RSD), which enter at somewhat smaller angular scales than the BAO. The RSD have a particularly interesting property that they break the isotropy of the clustering signal; they are discussed in Box 9.3.

Box 9.3	Redshift-Space Distortions

The observed redshift of galaxies is affected by their gravitational infall into nearby large-scale structures, as well as the galaxies' own peculiar velocities. Because we typically measure the galaxies' redshift in order to get their radial location, clustering measurements in redshift space are subject to **redshift-space distortions** (RSD).

Figure 9.9(b) shows that there are two RSD effects. The first effect is the squishing of the apparent extent of a structure along the line of sight due to the mass infall. In this process, first discussed by Kaiser (1987), galaxies on the far side of the structure fall toward us and thus have a diminished observed redshift z_{obs}, and those on the near side are pulled outwards and have a higher z_{obs}. The second effect is called **fingers of God**, and corresponds to elongation of the observed structure along the line of sight due to peculiar velocities, which show up on very small scales.

We briefly review the first of these effects – the Kaiser effect. The observed redshift z_{obs} is related to the true redshift z by the special-relativistic formula

$$1 + z_{\mathrm{obs}} = (1 + z)\sqrt{\frac{1 + v_{\parallel}}{1 - v_{\parallel}}} \simeq (1 + z)(1 + v_{\parallel}), \tag{B1}$$

where $v_{\parallel}$ is the velocity of the galaxy along the line of sight in units of the speed of light, and for the second equality we assume non-relativistic motions with $v_{\parallel} \ll 1$. Next, if $\mathbf{r}$ is the real-space location of the galaxy, then the redshift-space location $\mathbf{s}$ is (choosing the z-axis to be the radial direction, and denoting it by r_z to distinguish it from the redshift)

$$\mathbf{s} = \mathbf{r} + \frac{(1 + z)v_{\parallel}}{H}\hat{r}_z, \tag{B2}$$

because $|\mathbf{s} - \mathbf{r}| = dz/H$ and $dz \equiv z_{\mathrm{obs}} - z = (1+z)v_{\parallel}$. With some additional steps – which you can complete in Problem 9.8 – we get the overdensity in redshift space, $\delta^{(s)}$, to be related to the overdensity in real space, $\delta \equiv \delta^{(r)}$, via

$$\delta^{(s)} = \left[1 + f\mu^2\right]\delta. \tag{B3}$$

Box 9.3   **Redshift-Space Distortions (continued)**

Here, $\mu = \hat{\mathbf{k}} \cdot \hat{r}_z = k_z/k$ is the cosine of the angle between the line-of-sight direction and the wavenumber, and $f \equiv d\ln\delta/d\ln a$ is the (linear) growth rate. Therefore, with redshift-space distortions taken into account, the clustering is **anisotropic** – it depends on the spatial direction of the wavenumber. To calculate the galaxy power spectrum, we simply square the expression for $\delta^{(s)}$. We also take into account galaxy bias b, discussed in Sec. 9.6.5, which however does not affect the μ^2 term which is due to velocities. The result is the **Kaiser RSD formula**

$$P(\mathbf{k})^{(s)} = b^2 \left[1 + \beta\mu^2\right]^2 P(k), \tag{B4}$$

where b is the galaxy bias, $P(k)$ is the usual isotropic power spectrum, and $\beta \equiv f/b$. The RSD effects enter on small scales ($k \gtrsim 0.1\,h\,\mathrm{Mpc}^{-1}$), where they can be readily measured in spectroscopic surveys where accurate galaxy redshifts are available. Because $P(\mathbf{k})^{(s)}$ has terms that go as μ^0, μ^2, and μ^4 (in linear theory), we can respectively measure the monopole, quadrupole, and hexadecapole of redshift-space galaxy clustering.

9.5.7 Estimators of the Correlation Function

Let us now find an *estimator* for the two-point function – a statistical operation that we can apply to the data and extract the two-point function and, ideally, the associated uncertainty. Any estimator (we discuss them further in Chapter 10) should have these desirable properties:

- The estimator should be *unbiased* – on average, it should return the correct, "true" result.

- The estimator should have *minimum variance* among all choices of estimators.

Historically, the first popular estimator for the correlation function $\xi(r)$ was the **Peebles–Hauser estimator** (Hauser and Peebles, 1973; Peebles, 1973)

$$\hat{\xi}_{\mathrm{PH}}(r) = \left(\frac{N_{\mathrm{rand}}}{N_{\mathrm{data}}}\right)^2 \frac{\mathrm{DD}(r)}{\mathrm{RR}(r)} - 1, \tag{9.71}$$

where $\mathrm{DD}(r)$ is the number of pairs in the catalog in the interval $r \pm \Delta r$ (where Δr is some width that determines the binning in r), while $\mathrm{RR}(r)$ is the number of pairs in a *random distribution*-generated catalog in the same distance interval. The numbers N_{rand} and N_{data} are the total numbers of points (say, galaxies) in the two catalogs, respectively. Note that the Peebles–Hauser estimator essentially mirrors the definition of the two-point correlation function – that is, it is the excess probability of clustering (in the data relative to a random distribution of objects).

Over time, estimators with better properties (smaller bias and variance) have been found. For practical purposes, it is sufficient to stick with the **Landy–Szalay estimator** (Landy and Szalay, 1993)

$$\hat{\xi}_{\mathrm{LS}}(r) = \left(\frac{N_{\mathrm{rand}}}{N_{\mathrm{data}}}\right)^2 \frac{\mathrm{DD}(r)}{\mathrm{RR}(r)} - 2\frac{N_{\mathrm{rand}}}{N_{\mathrm{data}}}\frac{\mathrm{DR}(r)}{\mathrm{RR}(r)} + 1, \tag{9.72}$$

where DD and RR are defined as before, while DR is the number of pairs where one galaxy is in the data and the other, a distance in the interval $r \pm \Delta r$ away, is in the random catalog. When $N_{\mathrm{rand}} = N_{\mathrm{data}}$, this takes a more memorable form:

$$\hat{\xi}_{\mathrm{LS}} = \frac{\mathrm{DD} - 2\mathrm{DR} + \mathrm{RR}}{\mathrm{RR}}. \tag{9.73}$$

The variance in these estimators, assuming we have a Poisson process and work in linear theory, is approximately

$$\sigma_{\xi}^2(r) \simeq \frac{1 + \xi(r)}{\mathrm{DD}(r)} \simeq \frac{1}{\mathrm{DD}(r)}. \tag{9.74}$$

Since a clustered field of galaxies for example is clearly not Poissonian, the actual variance can be bigger than this.

To estimate the *angular* correlation function from data, we can simply adopt the Landy–Szalay estimator in angular space:

$$\hat{w}_{\mathrm{LS}}(\theta) = \frac{\mathrm{DD}(\theta) - 2\mathrm{DR}(\theta) + \mathrm{RR}(\theta)}{\mathrm{RR}(\theta)}. \tag{9.75}$$

Note finally that the real-space estimators discussed in this section are relatively easy to implement because they deal effectively with the **survey mask** – the particular shape of the survey footprint on the sky, including any holes due to masked bright stars for example. Given a complicated survey mask, the only action required is to draw a random catalog (required for DR and RR terms) from a footprint with the same mask. This is in contrast to Fourier-space estimators of the power spectrum, for which the survey mask typically presents more of a challenge.

9.6 Structure Formation

Since the 1980s, it has been recognized that there are two possible, qualitatively different, structure formation histories, depending on the nature of dark matter.

- **Cold dark matter (CDM).** Here, dark matter is "cold," that is, *non-relativistic* at the time of matter–radiation equality (when, recall, the perturbations first get a chance to grow appreciably). An example of a CDM candidate is a weakly interacting massive particle (WIMP; see Chapter 11).

- **Hot dark matter (HDM).** Here, dark matter is hot, that is, *relativistic* at the time of matter–radiation equality. An example of an HDM candidate is a neutrino with mass of order a few electronvolts.

It turns out that hot dark matter does not clump very effectively, basically because the particles are relativistic, and all structures below the so-called free-streaming scale are washed out. This scale corresponds to roughly $10^{15} M_\odot$, the mass of a large galaxy cluster (free streaming is discussed further, in the context of dark matter, in Chapter 11). Therefore, the HDM scenario is **top-down**, since the largest objects (clusters) form first, and the smaller structures form later.

In contrast, in the CDM scenario the free-streaming scale is very small, and structures can grow just fine. The CDM scenario is **bottom-up** in that the less massive objects (stars and galaxies) form before the more massive clusters of galaxies.

Which of these paradigms, CDM or HDM, is realized in our universe? This question was conclusively answered by the comparison of observations to numerical "N-body" simulations, which we now introduce and then answer the question.

9.6.1 N-Body Simulations

Our understanding of many features of large-scale structure is informed by numerical N-**body simulations**. In these simulations, one tracks the motion of individual particles through cosmic time under the laws of physics. Simulating structure with dark-matter particles is relatively straightforward, as these simply follow Newton's law of gravity. The **dark-matter-only** simulations are therefore very reliable well into the nonlinear regime on smaller scales. Modeling of baryons is more challenging, as one has to keep track of electromagnetic forces between them as well, which lead to a variety of new effects such as gas cooling, collisional excitation and ionization, and eventually star formation. These (and many other) effects can be captured by treating the baryonic particles as a fluid, achieved **hydrodynamical** N-body simulations, which contain both the baryon and the dark-matter particles. The hydrodynamical simulations require considerably more inputs for the physical processes than the dark-matter-only simulations in order to track the interactions between the baryonic particles. The hydrodynamical simulations also require more computation to evolve the large-scale structure in cosmic time. Because the number of particles in any N-body simulation is finite (about a hundred in the earliest simulations around 1970, though up to a trillion particles per simulation today), each particle represents not a proton or a dark-matter particle, but a much larger mass (typically in the range between $10^4 M_\odot$ and $\sim 10^9 M_\odot$ in modern cosmological simulations, depending on how big a volume is simulated). You will have a chance to write your own N-body simulation in Problem 9.10.

The N-body simulations (as well as direct observations of cosmic structure) clearly favor the CDM paradigm; we illustrate this in Fig. 9.10, which shows results from an influential early N-body simulation which showed evidence against HDM and in favor of CDM. There is, however, an intermediate possibility that dark matter is neither cold but warm – hence WDM – with the dark-matter particle mass of order 1 keV (as opposed to MeV–GeV or higher for CDM, or else eV or smaller for HDM). In a universe with WDM, the power spectrum shows a telltale sign of

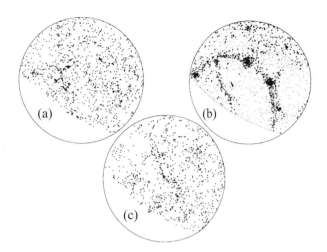

Fig. 9.10 Simulations that conclusively showed the structure in the cold-dark-matter-dominated universe (a) looks a lot more like actual measurements of galaxy distribution (c) than the neutrino-dominated universe (b). In other words, the observed structure is consistent with a bottom-up structure formation scenario where the largest structures form last (such as the CDM), rather than a top-down formation where the largest structures form first (such as the HDM). Reprinted by permission from Springer Nature Customer Service Centre GmbH: Davis *et al.* (1992).

suppression at small scales (high k). More discussion of this is in Chapter 11, where we cover dark matter.

A snapshot from a modern N-body simulation is shown in Fig. 9.11. Visually, the most striking feature in this map is the filamentary structure of matter distribution in the universe. The filamentary structure is a consequence of the force of gravity applied to the evolution of a multi-particle system. Because of this, large-scale structure in the universe is sometimes referred to as the **cosmic web**. The filamentary structure can be seen by eye even in the observed galaxy maps; see Fig. 9.1. Roughly speaking, at the intersection of dark-matter filaments reside the largest halos, whose baryonic component we observe as the clusters of galaxies. Individual galaxies reside within these clusters, but also along the dark-matter filaments and elsewhere in this cosmic web.

9.6.2 Smoothed Density Fields

An *unsmoothed* density field would have an infinite variance, which we already recognized in the apparent divergence of the real-space correlation function for a realistic input matter power spectrum in Eq. (9.59). This divergence corresponds to the possibility of the two-point averaging pointing to a point particle (of, say, baryons or dark matter) which has a near-inifinite overdensity. Realistic observations effec-

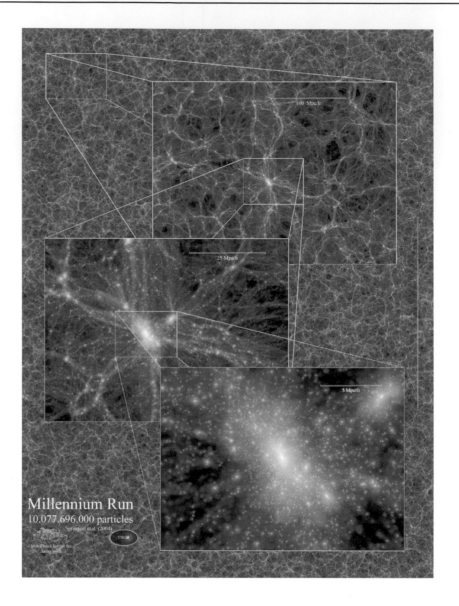

Fig. 9.11 A $z = 0$ snapshot with several levels of zoom-in, adopted from the Millennium simulation (Springel *et al.*, 2005). The starting area is several hundred Mpc on a side, and 15 Mpc thick. Each zoom-in is about a factor of four; the horizontal yardsticks shown are, top to bottom: 100 h^{-1}Mpc, 25 h^{-1}Mpc, 5 h^{-1}Mpc. A black and white version of this figure will appear in some formats. For the color version, please refer to the plate section.

tively smooth over regions of space, probing measurements of the coarse-grained density field averaged or smoothed on some characteristic spatial scale. This scale is typically the size of a galaxy or larger.

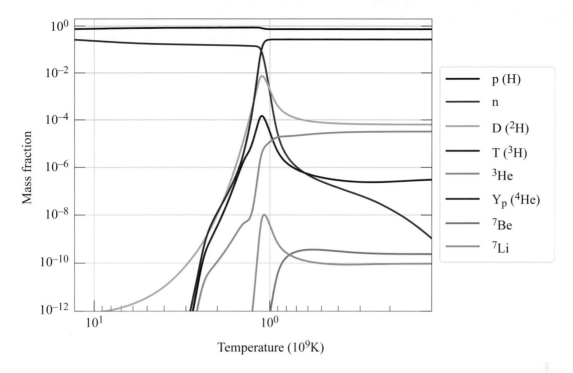

Elemental abundances (that is, mass fractions ρ/ρ_B) produced using a computer code that takes into account many nuclear reactions occurring during the BBN era. The abundances are shown as a function of decreasing temperature (or increasing time). Numerical calculation and figure courtesy of Aidan Meador-Woodruff.

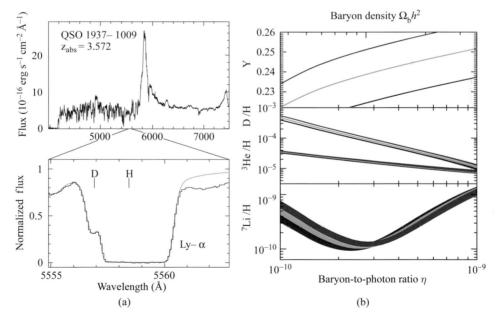

(a) The signature that deuterium leaves in the spectrum of a distant quasar. Note that the deuterium line is very slightly shifted relative to the dominant hydrogen line, and is correspondingly hard to measure (notice the hugely stretched wavelength scale). Figure courtesy of Ken Nollett; see also Burles *et al.* (2000). (b) Similar to Fig. 7.3, except shown for three values of number of neutrino species: $N_\nu = 2$ (blue), 3 (green), and 4 (red). The line thicknesses are equal to the current observational uncertainties. Reprinted with permission from Cyburt *et al.* (2016), copyright (2016) by American Physical Society.

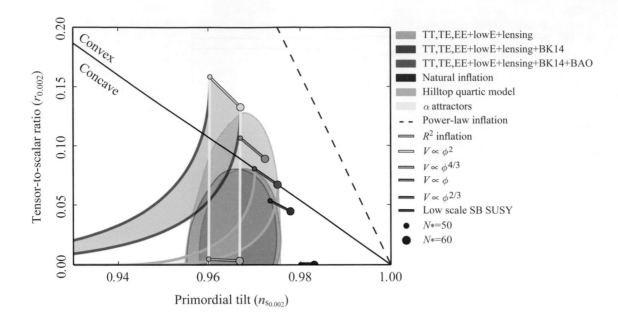

Fig. 8.6 Constraints in the $n_s - r$ plane from Planck and other data. The round filled contours show the following 95 percent credible interval constraints: the gray regions assume only the temperature, polarization, and lensing data from the 2018 release of Planck; the red contours additionally assume data from the BICEP-Keck array (BK); the blue contours additionally assume data from baryon acoustic oscillations (BAO), the probe that is described in Chapter 9. The figure also shows lines or bands for a variety of inflationary models, with points indicating n_s and r evaluated 50 or 60 e-folds before the end of inflation. Adopted from Planck Collaboration (Akrami *et al.*, 2020b). CC BY 4.0.

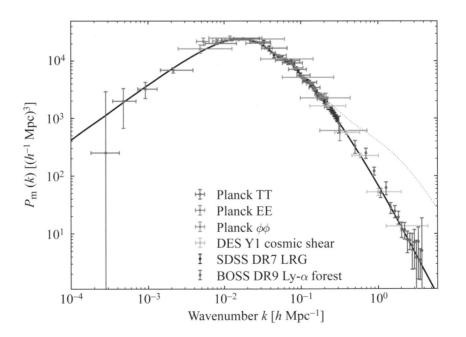

Fig. 9.7 Constraints on the linear power spectrum $P(k)$ at $z = 0$ from recent data, specifically Planck's CMB temperature (TT), polarization (EE), and gravitational lensing ($\phi\phi$); galaxy and Lyman-α feature clustering from the SDSS; and cosmic shear measurements from the DES. In all cases, the observable quantities were deconvolved in order to obtain error bars on $P(k, z = 0)$. The black curve shows the linear power spectrum predicted by the concordance ΛCDM cosmological model, while the dotted curve shows the impact of nonlinear clustering. Adopted from the Planck legacy paper (Aghanim *et al.*, 2020a), reproduced with permission © ESO.

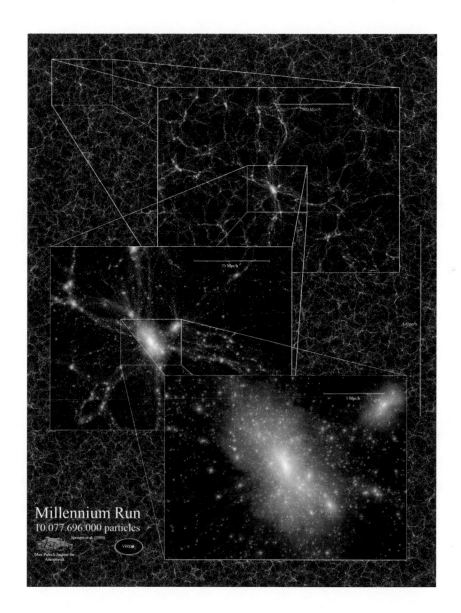

Fig. 9.11 A $z = 0$ snapshot with several levels of zoom-in, adopted from the Millennium simulation (Springel *et al.*, 2005). The starting area is several hundred Mpc on a side, and 15 Mpc thick. Each zoom-in is about a factor of four; the horizontal yardsticks shown are, top to bottom: 100 h^{-1}Mpc, 25 h^{-1}Mpc, 5 h^{-1}Mpc.

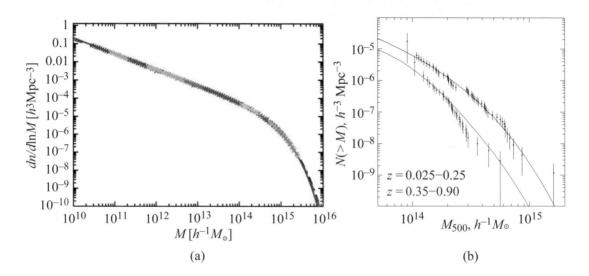

(a)   (b)

Fig. 9.12 (a) "Measurements" of the mass function from a suite of N-body simulations, together with an analytical fit (which is difficult to see as the curve is largely hidden by the simulations' data points which it matches well). Different colors denote different halo mass ranges, whose $dn/d\ln M$ is measured in simulations with sizes optimized for that mass range. The analytical fit, which assumes universality and hence depends on cosmology only via the rms mass fluctuation $\sigma(M)$ (see text), is accurate to about 5 percent. Reproduced with permission from Warren *et al.* (2006), © AAS.
(b) "Real" measurements of the mass function in two redshift bins from the 400-square-degree survey of ROSAT galaxy clusters followed up by Chandra Space Telescope. Points with error bars are data, and lines are fits to the theoretical mass function. Reproduced with permission from Vikhlinin *et al.* (2009), © AAS.

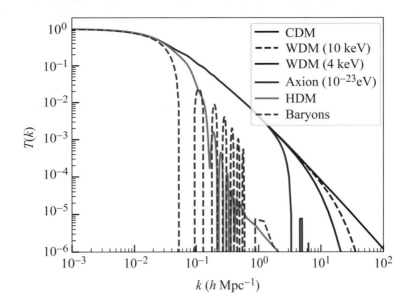

Fig. 11.3 Transfer functions for cold dark matter (CDM), warm dark matter (WDM) with two particle masses, hot dark matter (HDM; massive neutrinos), ultra-light axions, and a baryon-only case. In all cases we assume that all of the dark-matter content, $\Omega_{\mathrm{dm}}h^2 \simeq 0.112$, is made up of the particle in question (in addition to the usual $\Omega_B h^2 \simeq 0.022$). Warm and hot dark matter $T(k)$ are suppressed because of the free streaming of the dark-matter particle. Axion $T(k)$ is also suppressed, but for a different reason: because of the quantum pressure of these extremely light, long-wavelength particles. The baryonic $T(k)$ oscillates because baryons are coupled to radiation in the early universe. The plot was produced using CLASS (Blas *et al.*, 2011) and axionCAMB (Hlozek *et al.*, 2015).

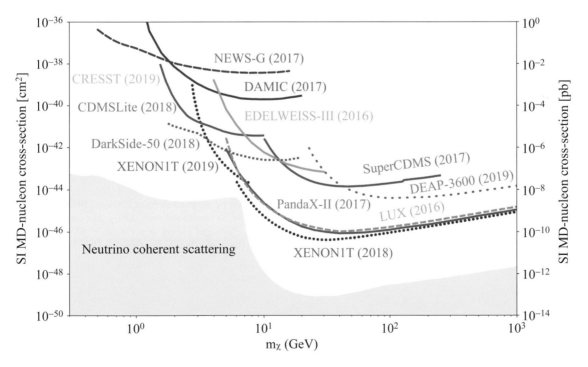

Fig. 11.6 Current limits on the dark-matter-nucleon cross-section as a function of dark-matter mass. The plot shows recent constraints; decades of improvements to the limits are not shown. Adopted from the Particle Data Group (Zyla *et al.*, 2020).

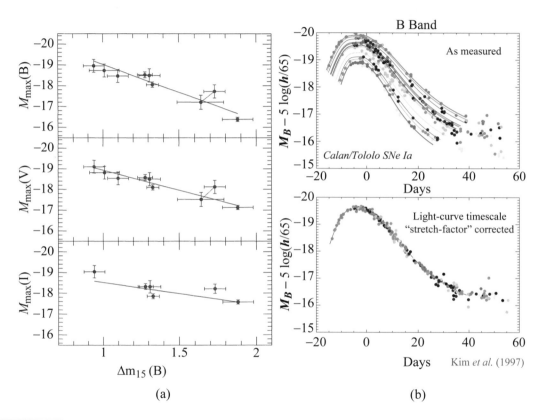

Fig. 12.1 (a) Phillips relation, adopted from his paper and reproduced by permission from Phillips (1993), © AAS. The absolute magnitude of type Ia supernovae, M, is correlated with Δm_{15}, the observed decay of the light-curve light 15 days after the maximum. (b) Light curves of a sample of SNe Ia before correction for the Phillips relation (top) and after (bottom). Figure courtesy of Alex Kim.

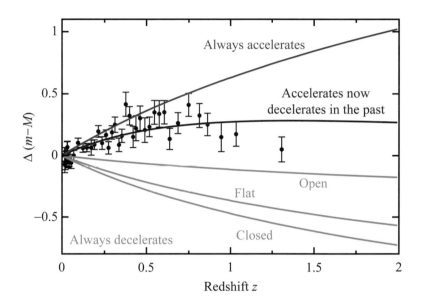

Fig. 12.2 Hubble diagram for the current SN Ia "Pantheon" dataset of 1048 SN Ia (Scolnic *et al.*, 2018). We have binned (grouped) SN Ia measurements in redshift for easier viewing; each datapoint shows the average of about 20 objects, with the error bar suitably scaled. We also show several theory models to guide the eye; the model favored by the data is the "accelerates now, decelerated in the past" model with $\Omega_M \simeq 1 - \Omega_\Lambda \simeq 0.3$. The vertical axis shows the distance modulus relative to some (unimportant) fiducial case. Remember that all theory curves are allowed to slide vertically (corresponding to an *a-priori* unknown $\mathcal{M}$ parameter); it is only the *shape* of the measured Hubble diagram that informs us about the cosmological model.

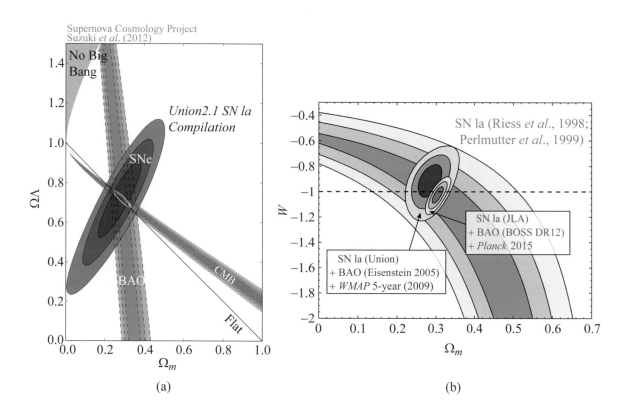

(a) Constraints on Ω_M and Ω_Λ in the consensus model using SN, baryon acoustic oscillation (BAO), and CMB measurements. Reprinted with permission from Suzuki *et al.* (2012), © AAS.
(b) Constraints on Ω_M and w in the wCDM cosmological model using the combination of SN Ia, BAO, and CMB. The plot shows the evolution of the constraints over the period of 20 years; constraints on panel (a) roughly correspond to data in the red contour on the right, while the blue contour is the more recent data. Reprinted with permission of IOP publishing, from Huterer and Shafer (2018); permission conveyed through Copyright Clearance Center Inc.

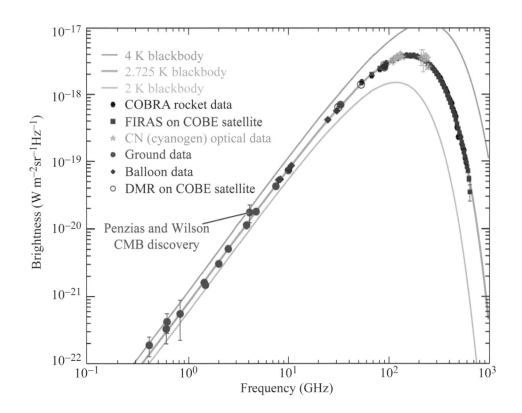

Fig. 13.1 CMB spectrum as measured by various early experiments. The best constraints to date have been provided by COBE FIRAS, whose actual error bars are much smaller than the size of the corresponding points in this figure. Reprinted with permission of *Annual Reviews*, from Samtleben *et al.* (2007); permission conveyed through Copyright Clearance Center Inc.

(a)   (b)

Fig. 13.2 (a) COBE DMR 90 GHz map of CMB fluctuations (credit: NASA / COBE Science Team; see also Bennett *et al.*, 1996). (b) WMAP's W-band (roughly also around 90 GHz) map (credit: NASA / LAMBDA Archive Team; see also Bennett *et al.*, 2013). Both maps show significant galactic contamination near the equator; the primordial CMB anisotropies are the features away from it. The resolution of the COBE map is about $7°$, while that of the WMAP map is about $15'$.

Planck experiment's full-sky map in Galactic coordinates. Galactic contamination has been cleaned using multi-frequency data using the so-called SMICA technique. The resolution of the map is approximately 5 arcmin. Adopted from Akrami *et al.* (2020a) and ESA and the Planck Collaboration.

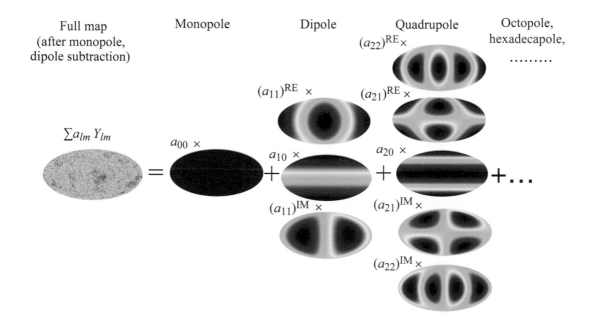

A sketch of how individual multipoles are combined to make the CMB map. In the sum, each $Y_{\ell m}$ pattern (individual ovals) is multiplied by a coefficient ($a_{\ell m}$). Note that the monopole and dipole are not used in the map of CMB anisotropy because the monopole is the mean temperature (around which we are expanding), while the dipole is due to our local motion through the CMB rest frame, and thus not cosmological. Note that the sum over the real and imaginary parts of $a_{\ell m}$ is a bit simplified in this sketch; see Eq. (13.68) for the exact expression.

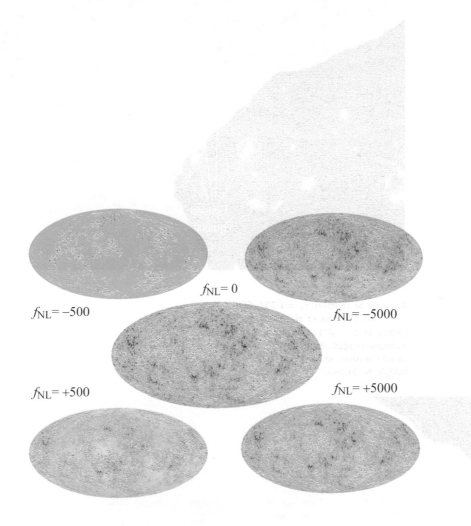

$f_{\mathrm{NL}}= 0$

$f_{\mathrm{NL}}= -500$

$f_{\mathrm{NL}}= -5000$

$f_{\mathrm{NL}}= +500$

$f_{\mathrm{NL}}= +5000$

Fig. 13.15 Synthetic maps of the Gaussian sky (center), and with non-Gaussianity of the local type for $f_{\mathrm{NL}} = \pm500$, and $f_{\mathrm{NL}} = \pm5000$. Current limit from Planck is $f_{\mathrm{NL}} = -0.9 \pm 5.1$.

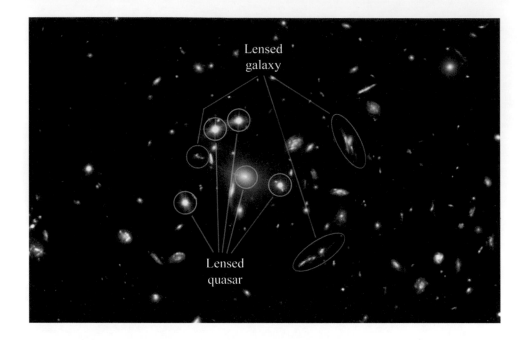

Fig. 14.4 A quasar at redshift $z_s = 1.734$, SDSS J1004+4112, is split into five images by a galaxy cluster at $z_d = 0.68$ (Sharon *et al.*, 2005). Further observations subsequently showed that a very distant source galaxy, at $z_s = 3.332$, is also multiply imaged. Therefore, there are at least two sources that are multiply imaged, in addition to distortions of numerous other sources (that are not strongly but only weakly lensed), showcasing the complexity of gravitational lensing in realistic situations. Credit: ESA, NASA, K. Sharon (Tel Aviv University) and E. Ofek (Caltech).

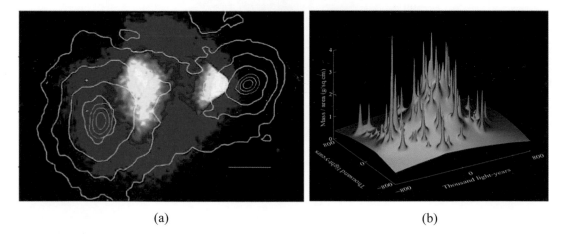

(a)   (b)

Fig. 14.8 (a) X-ray emission from the "Bullet" cluster of galaxies observed by ground and space telescopes. Colored features correspond to X-ray emission observed by Chandra Space Telescope, while the green contours correspond to the mass reconstruction from weak-lensing observations. Note that the X-ray features, which are dominated by baryons, are clearly separated from the green contours where most of the mass is located. Reprinted with permission from Clowe *et al.* (2006), © AAS.
(b) Reconstruction of the density profile of galaxy cluster CL 0024+1654 at $z = 0.39$ using strong gravitational lensing alone. The density peaks correspond to individual galaxies in the cluster, while the broad central feature is largely due to dark matter. Credit: Tony Tyson; see also Tyson *et al.* (1998).

Therefore, smoothing (or filtering) of the theoretically calculated density field is essential in order to connect observations to theory. This is accomplished via some window function. We now describe this process.

We first define a smoothing window function $W(r, R)$, where r is the dependent variable, and R is the characteristic smoothing scale. Popular choices are

$$W_{\mathrm{G}}(r, R) = \frac{1}{(2\pi)^{3/2} R^3} e^{-r^2/(2R^2)} \quad \text{(Gaussian smoothing)} \tag{9.76}$$

$$W_{\mathrm{TH}}(r, R) = \frac{1}{(4\pi/3) R^3} \mathcal{H}(R - r) \quad \text{(top-hat smoothing)}, \tag{9.77}$$

where $\mathcal{H}(x)$ is the Heaviside step function; $\mathcal{H}(x) = 1$ for $x > 0$ and $\mathcal{H}(x) = 0$ for $x < 0$.

Heuristically, the smoothed field at some point $\mathbf{r}$ is a weighted average over all points in space $\mathbf{r}'$, where the weight is specified by a function W that depends only on distance $|\mathbf{r} - \mathbf{r}'|$. Therefore, the smoothing operation is actually a **convolution** in real space, so that the smoothed density field becomes

$$\delta(\mathbf{r}, R) = \int W(|\mathbf{r} - \mathbf{r}'|) \delta(\mathbf{r}') d^3 \mathbf{r}'. \tag{9.78}$$

Very conveniently, convolution in real space corresponds to a *multiplication* in Fourier space by the Fourier transform of the window function, so that

$$\delta_{\mathbf{k}}(R) = W(k, R) \delta_{\mathbf{k}}, \tag{9.79}$$

and thus the smoothed power spectrum is

$$P(k, R) = |W(k, R)|^2 P(k). \tag{9.80}$$

The Fourier transforms of the window functions are easily computed:

$$W_{\mathrm{G}}(k, R) = e^{-k^2 R^2/2} \quad \text{(Gaussian)} \tag{9.81}$$

$$W_{\mathrm{TH}}(k, R) = 3 \frac{\sin(kR) - kR\cos(kR)}{(kR)^3} = \frac{3 j_1(kR)}{kR} \quad \text{(top hat)}, \tag{9.82}$$

where $j_1(x)$ is the spherical Bessel function of order one. Note that the top-hat window is conventionally accepted as a default in cosmology. This unfortunately means that our computer programs have to integrate wiggly Bessel functions, a task which cosmology students endure as a rite of passage – and so will you (Problem 9.9).

9.6.3 Amplitude of Mass Fluctuations

Let us henceforth adopt the top-hat smoothing window. What is the zero-lag auto-correlation function smoothed on some scale R, $\xi(0, R)$? Fixing up Eq. (9.59) to include smoothing *a lá* Eq. (9.80) leads to

$$\xi_{\mathrm{TH}}(0, R) = \int_0^\infty \Delta^2(k) |W_{\mathrm{TH}}(k, R)|^2 d\ln k, \tag{9.83}$$

or, renaming this quantity to agree with the literature, $\sigma^2(R) \equiv \xi_{\mathrm{TH}}(0, R)$, we have

$$\sigma^2(R) = \int_0^\infty \Delta^2(k) \left(\frac{3j_1(kR)}{kR}\right)^2 d\ln k. \tag{9.84}$$

This is the **rms amplitude** (squared) **of** (linear) **mass fluctuations** smoothed over scale R – a very important quantity in cosmology, as we now explain.

Cosmologists have historically quoted the rms amplitude of mass fluctuations $\sigma(R)$ on a conventionally chosen scale R that corresponds roughly to the size of a galaxy cluster $(5-10\,h^{-1}\mathrm{Mpc})$. The specific scale traditionally used is $R = 8\,h^{-1}\mathrm{Mpc}$, and thus the definition of the cosmological quantity σ_8

$$\sigma_8 \equiv \sigma(R = 8\,h^{-1}\mathrm{Mpc}, z = 0). \tag{9.85}$$

Here we explicitly indicate that σ_8 is conventionally defined at the present time.

The quantity σ_8 gives us one way to *normalize* the power spectrum: by measuring σ_8, from the distribution of galaxies for example, we can essentially determine the power spectrum normalization A_s from Eq. (9.61). It has only recently, with the precision of CMB experiments, been that we have measured A_s independently by studying the amplitude of fluctuations in the CMB. In a sense, A_s is a fundamental cosmological parameter, while σ_8 is a derived parameter, but more directly (than A_s) related to measurements from galaxy clustering and weak gravitational lensing. The measured value of σ_8 has – also historically – varied between about 0.6 and 1.1. Measurements of σ_8 from the abundance of clusters and weak lensing, and those of A_s from the CMB, both find $\sigma_8 \simeq 0.80$, so roughly in the middle of the range historically favored by different measurements. Very recent data show indication of some $(2-3\sigma)$ tension between the CMB and lensing constraints on σ_8, where the CMB comes out higher; the Planck result is $\sigma_8 = 0.810 \pm 0.007$ (see our Table 3.1), while the DES Year-3 constraint, for example, is $\sigma_8 = 0.74 \pm 0.04$ (Abbott *et al.*, 2022). The origin of this "sigma-8 tension" is currently being investigated.

The other major reason (i.e., beyond σ_8) for the utility of the amplitude of mass fluctuations $\sigma(R)$ is its central role in our understanding of the theory behind the abundance of collapsed structures in the universe. We now turn to discussing this in some detail.

9.6.4 Mass Function

Given the machinery we have developed, can we make an *ab initio* estimate of the so-called **mass function** – the number of collapsed objects as a function of their mass and redshift? This turns out to be typically too difficult for galaxies, as modeling their formation requires extensive use of N-body simulations and other tools that are motivated by the results of these simulations (such as the halo model mentioned at the end of this chapter). For the still smaller objects, such as stars and planets, a simple prediction of the mass function is clearly impossible due to the complicated physics of their birth and evolution. However, for the largest collapsed objects in the universe – clusters of galaxies – it turns out that we can obtain a reasonably good estimate of the mass function using essentially a back-of-the-envelope argument. Such an argument was first made by Press and Schechter (1974), and we now discuss it.

Press and Schechter stated that the likelihood for collapse of objects of specific size or mass ($R \propto M^{1/3}$) can be computed by examining the density fluctuations on the desired scale. They used a model for the collapse of a spherical top-hat over-density to argue that collapse on scale R should occur roughly when the smoothed density on that scale exceeds a critical value δ_c, of order unity, independent of R.

The mass within a region of size R is (implicitly assuming the top-hat window function that cuts off abruptly at R)

$$M = \frac{4\pi}{3}\rho_M R^3, \tag{9.86}$$

where as usual $\rho_M \equiv \rho_M(z) = \rho_{\text{crit}}\Omega_M(1+z)^3$.

Press and Schechter reasoned that, given the smoothing radius R, the fractional volume occupied by collapsed objects is proportional to regions whose overdensity is greater than some critical value. One can show that the critical value for collapse in an idealized spherical-overdensity model is (see Problem 9.6)

$$\delta_c \simeq 1.686 \qquad \text{(critical overdensity for collapse)}. \tag{9.87}$$

The quantity of interest is the fraction of the density field that has overdensity greater than the critical overdensity for collapse, δ_c. Assuming a Gaussian density field, this fraction is

$$F(M) \equiv \int_{\delta_c}^{\infty} P(\delta)d\delta = \frac{1}{\sqrt{2\pi}\sigma(M)}\int_{\delta_c}^{\infty} e^{-\delta^2/(2\sigma(M)^2)}d\delta$$

$$\equiv \frac{1}{2}\text{erfc}\left(\frac{\nu_c}{\sqrt{2}}\right), \tag{9.88}$$

where erfc is the complementary error function (see, e.g., Wikipedia) and

$$\nu_c \equiv \frac{\delta_c}{\sigma(M)} \tag{9.89}$$

is loosely called the **peak height**.

Note immediately one problem: as $\sigma(M) \to \infty$, $\nu_c \to 0$ and the collapsed fraction goes to 1/2, not 1. This intuitively corresponds to the fact that according to assumptions so far, only the overdensities, and not the underdensities, can lead to collapsed structures. Press and Schechter resolved this with an outrageously bold move – by simply multiplying the fraction of collapsed objects by a factor of 2!

We next relate the fraction of collapsed objects in the mass range $(M, M + dM)$ and the mass function $n(M)$; this is

$$dF(M) = -\frac{M \, dn(M)}{\rho_{M,0}}, \tag{9.90}$$

where the minus sign ensures positivity for $dn(M)$ (because there is less mass bound within high-mass halos than low-mass halos). Note that, since we are talking about a comoving density, we have $\rho_{M,0}$ evaluated at the present time [i.e., ρ_M at an arbitrary redshift divided by $(1 + z)^3$ gives just $\rho_{M,0}$]. Then it follows that the comoving number density of objects in an interval dM around a mass M is

$$\frac{dn}{d\ln M} d\ln M = \frac{\rho_{M,0}}{M} \left| \frac{dF(M)}{d\ln M} \right| d\ln M. \tag{9.91}$$

After taking the derivative of Eq. (9.88) analytically, and including the miraculous factor of 2, we get (see Problem 9.5) the **Press–Schechter mass function**

$$\frac{dn}{d\ln M} = \sqrt{\frac{2}{\pi}} \frac{\rho_{M,0}}{M} \frac{\delta_c}{\sigma} \left| \frac{d\ln \sigma}{d\ln M} \right| e^{-\delta_c^2/(2\sigma^2)}. \tag{9.92}$$

Note that $\sigma(M, z) = \sigma(M, 0)D(z)$, so that the mass–radius conversion from Eq. (9.86) only needs to be done at $z = 0$.

Note a few things:

- The fact that we have a formula that quite accurately describes the abundance of halos is fantastic, since halos are inherently *nonlinear* objects (i.e., it is not true that $|\delta| \ll 1$ at the scale of the halo).

- In particular, note that the assumption of Gaussianity was fishy, since we know that $\delta \geq -1$ by definition, and here we are talking about typical δs of order unity. Why does the Press–Schechter formula match the simulations reasonably well (to a few tens of a percent)? This is a subject of current research.

- The number density of objects falls exponentially with increasing mass. This is fundamentally a consequence of the nature of peaks in a Gaussian random field; the number density of high peaks is exponentially suppressed. There are fewer – much fewer – objects of high mass than of low mass.

- The Press–Schechter formula is "universal," since the halo abundance depends on the cosmological model only via the rms variance, σ. In other words, all of the dependence on the cosmological parameters (Ω_M, n_s, A_s, Ω_B, etc.) is channeled through the single cosmology-dependent function $\sigma(z, M)$. There is no fundamental reason why the mass function should be universal, however, and finding fundamental causes for the near universality is another subject of research. Recent work has found clear departures from universality, at $\sim$5 percent level in $dn/d\ln M$, by comparing theory to numerical simulations.

Precise determination of the mass function is done by analyzing the output of N-body simulations. Huge progress in this area has been made since the work of Press and Schechter. While the Press–Schechter mass function is only accurate at the $\sim$50 percent level, the most recent fit to simulations is accurate to better than 5 percent; see Fig. 9.12(a). Moreover, recent results from counting galaxy clusters effectively *measure* the mass function, and agree with numerical predictions; see Fig. 9.12(b).

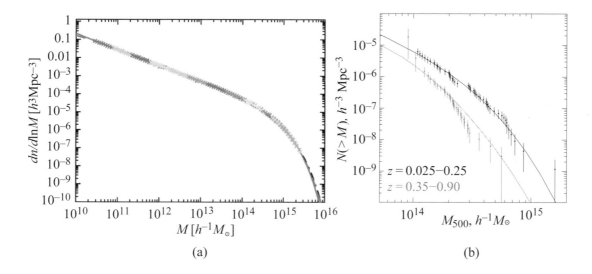

(a)   (b)

Fig. 9.12   (a) "Measurements" of the mass function from a suite of N-body simulations, together with an analytical fit (which is difficult to see as the curve is largely hidden by the simulations' data points which it matches well). Different colors denote different halo mass ranges, whose $dn/d\ln M$ is measured in simulations with sizes optimized for that mass range. The analytical fit, which assumes universality and hence depends on cosmology only via the rms mass fluctuation $\sigma(M)$ (see text), is accurate to about 5 percent. Reproduced with permission from Warren *et al.* (2006), © AAS. (b) "Real" measurements of the mass function in two redshift bins from the 400-square-degree survey of ROSAT galaxy clusters followed up by Chandra Space Telescope. Points with error bars are data, and lines are fits to the theoretical mass function. Reproduced with permission from Vikhlinin *et al.* (2009), © AAS. A black and white version of this figure will appear in some formats. For the color version, please refer to the plate section.

9.6.5 Bias of Halos and Galaxies

If you are standing at Mt. Everest, it is more likely that you will find another high peak near you than if you live in the Midwest of the United States. Similarly, peaks in the matter density field, which are the halos (i.e., galaxies and clusters with their dark-matter halos) that we observe on the sky, are more clustered than the density field as a whole. Schematically, the two are related with the factor b, which is called the bias of dark-matter halos (or just **bias**):

$$\delta_h = b\,\delta \qquad \text{(definition of halo/galaxy bias)}, \qquad (9.93)$$

where δ and δ_h are the overdensities in the matter field and dark-matter halos, respectively. Then the two corresponding power spectra are related by

$$P_h(k, z) = b^2(k, z)P(k, z), \qquad (9.94)$$

where we left the possibility that bias depends on scale k as well as redshift z. Early evidence for the presence of galaxy bias in the measurements of the correlation function is shown in Fig. 9.13(a). The correlation function of clusters of galaxies is larger than that of the galaxies themselves, because the bias of the former is larger than that of the latter; this was first pointed out by Kaiser (1984).

What we measure in cosmology is the clustering of halos (galaxies, clusters, etc.); this is represented by $P_h(k)$. What we can *predict* is the clustering of dark matter, $P(k)$ (see Eq. (9.61)). The ratio between the two is the bias squared.

It turns out that bias can be predicted, at least to a modest accuracy, using a clever trick called the peak-background split, covered in Box 9.4. The peak-background split, as well as its more sophisticated extensions, gives us a theoretical formula for the halo bias, which in principle sounds great. However, the extent to which we can trust such a formula when confronted with modeling real measurements is somewhat limited as we often measure the distribution of *galaxies*. Galaxies are also biased, but their bias is not necessarily the same as the halo bias (and even if it were, we typically do not have access to galaxies' mass, which is required to calculate the peak height ν_c in the peak-background formula).

We do know some trends about the galaxy bias. Most notably, bias is larger for more massive objects that correspond to higher peaks in the density field; see Eq. (B4) in Box 9.4. Galaxies are largely unbiased, with $b \sim 0.8-1.5$, while for clusters $b \sim 3-4$. Moreover, at least on large (linear) scales, bias is expected to be constant (as a function of scale k, not redshift) in standard ΛCDM with no primordial non-Gaussianities. However, analytical treatments of bias, such as the peak-background formalism outlined in Box 9.4, are not sufficiently accurate to impose meaningful priors on a typical analysis of galaxy-clustering observations. In such analyses, one therefore needs to marginalize over the bias parameter(s), which necessarily leads to a loss of information on the cosmological parameters of interest. Summarizing:

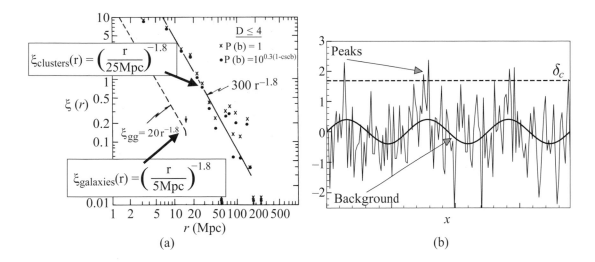

Fig. 9.13 (a) Early empirical evidence for the presence of bias, reproduced and adapted with permission from Bahcall and Soneira (1983), © AAS. The authors find that galaxies (lower points) and clusters of galaxies (higher points) have the respective correlation functions that scale identically with object separation, but have significantly different respective normalizations. The latter fact was soon thereafter explained by the fact that clusters have a larger bias than galaxies. (b) Illustration of the peak-background split. Short fluctuations (the "peaks") live on top of long fluctuations (the "background"), and the background raises or lowers the effective threshold for a halo to form.

Galaxy bias typically needs to be measured from the data. In general, it depends on the galaxy type and redshift, and thus contributes several nuisance parameters in a typical cosmological analysis.

Box 9.4   Peak-Background Split

Let us split the density fluctuations in the universe into ones of long and short wavelength. We will denote the long-wavelength perturbations as δ_b ("b" for background), and short-wavelength with δ_p ("p" for peaks):

$$\delta = \delta_b + \delta_p. \tag{B1}$$

We have already mentioned, while discussing the Press–Schechter formalism, that peaks form when the density exceeds some threshold ($\delta_c \simeq 1.686$ in the spherical collapse model). Note that the long-wavelength fluctuations form a background on which the peaks form, and therefore δ_b serves the role of changing the threshold from δ_c to $\delta_c - \delta_b$; see Fig. 9.13.

Let us now expand the number density in Taylor series, assuming the Press–Schechter mass function, $n(\nu_c) \propto \nu_c \exp(-\nu_c^2/2)$:

Box 9.4   Peak-Background Split (continued)

$$n(\nu_c + \delta\nu_c) = n\left(\frac{\delta_c - \delta_b}{\sigma}\right) \approx n(\nu_c) + \frac{dn}{d\nu_c}\frac{d\nu_c}{d\delta}(-\delta_b)$$

$$= n(\nu_c) + \left(\frac{1}{\nu_c} - \nu_c\right)n(\nu_c)\left(-\frac{\delta_b}{\sigma}\right) \quad \text{(B2)}$$

$$= n(\nu_c)\left[1 + \frac{\nu_c^2 - 1}{\nu_c\sigma}\delta_b\right].$$

Therefore, $\delta n/n = (\nu_c^2 - 1)/(\nu_c\sigma)\,\delta_b \equiv b_L\delta_b$. The quantity b_L is the **Lagrangian bias**, which is bias in coordinates moving with the expansion. We are ultimately interested in the **Eulerian bias** b_E, which is related to the Lagrangian bias via (e.g., Mo and White, 1996)

$$b_E(\equiv b) = b_L + 1. \quad \text{(B3)}$$

Therefore, the bias from the peak-background split is

$$b(M) \simeq 1 + \frac{\nu_c^2 - 1}{\delta_c}, \quad \text{(B4)}$$

where, recall, $\nu_c(M) = \delta_c/\sigma(M)$. This expression, which is in good agreement with the bias derived from both N-body simulations and observations, shows that the bias for more massive objects (higher ν_c) is larger.

The desire to avoid dealing with galaxy bias has historically motivated the development of weak gravitational lensing, discussed at length in Chapter 14.

9.7 Galaxy–Halo Connection

How is the distribution of galaxies related to the distribution of dark-matter halos? According to cosmological simulations, cosmic structure has a complex topology, with dark-matter sheets which surround voids (regions largely devoid of cosmic structures) and intersect at filaments. The filaments, in turn, intersect at dark-matter halos. The halos themselves have non-trivial structure; in particular, they contain localized dark-matter density peaks called **subhalos**, which are the remnants of halos accreted in the past. Massive dark-matter halos contain many galaxies: typically a bright central galaxy, along with a swarm of satellite systems. But how are the halos and subhalos occupied by galaxies? Do more massive halos contain more massive galaxies? Is there one, or can there be many, galaxies in a sub-halo? What determines the galaxies' properties, say their bias, morphology, or star-formation activity? Answering these questions is a task for **galaxy formation**, a large and notoriously complicated subject which occupies a large number of astrophysicists worldwide. The importance of answering the questions posed above can be appreciated by the fact that we typically observe galaxies, while we can accurately predict, using theory and simulations, the properties of dark matter and their halos.

9.7.1 Halo Model and the HOD

Modeling of the formation of cosmic structure that includes the physics of galaxy formation is in principle available through the results obtained by running hydro-dynamical simulations (see Sec. 9.6.1). While not a definitive model, the hydro simulations represent a first-principles approach to galaxy formulation which is steadily improving in its accuracy. However, there are many free parameters in such hydro simulations, and running a simulation for each combination of those parameters requires significant computing resources. To streamline the implementation of galaxy formation, the community additionally relies on various empirical models, where a combination of physical reasoning, direct observations, and results of the aforementioned simulations is used to connect the statistical properties of halos to those of galaxies.

One such semi-analytic model is the **halo model** of large-scale structure. The halo model, first formulated more than half a century ago, aims to describe the clustering of dark-matter halos and their resident galaxies on both large and small scales. On large scales, the typical distance between halos is greater than the halo size, and in this so-called **two-halo** regime the predictions are accurate and largely equivalent to our prescription in Eq. (9.61).

Modeling of the clustering of subhalos within a single halo – in the so-called **one-halo** regime – is more challenging. Nothing in our model for the power spectrum in Eq. (9.61), or the bias prescription in Box 9.4, contains information about how many galaxies, N, occupy a halo of a given mass, M.

One avenue to theoretically model the probability $P(N|M)$ within a halo is to adopt the **halo occupation distribution (HOD)** modeling. A typical HOD makes use of the fact that the description of the structure in a halo is simplified if we assume one **central galaxy**, along with a number of **satellite galaxies**; this also agrees with observations made in clusters of galaxies. The expected number of central galaxies $\langle N_{\rm cen}\rangle_M$ in a halo of mass M is modeled as

$$\langle N_{\rm cen}\rangle_M = \frac{1}{2}\left[1 + {\rm erf}\left(\frac{\log M - \log M_{\rm min}}{\sigma_{\log M}}\right)\right], \qquad (9.95)$$

where erf is the error function, while the number of satelltes $\langle N_{\text{sat}} \rangle_M$ is modeled by

$$\langle N_{\text{sat}} \rangle_M = \left(\frac{M - M_0}{M_1} \right)^{\alpha}. \tag{9.96}$$

Here $M_{\text{min}}, \sigma_{\log M}, M_0, M_1$, and α are free parameters which are typically calibrated by N-body simulations. Therefore, the expected number of central galaxies in a halo increases from zero to one around the halo mass M_{min} (typically $M_{\text{min}} \simeq 10^{13} h^{-1} \text{M}_\odot$ at $z = 0$), while the number of satellite galaxies increases as a power law.

To compute various statistical quantities from a HOD and use them to compare theory with observations, further steps may be required. The *distribution* of satellite galaxies in a halo may need to be assumed (a common choice is the Poisson distribution). More ambitiously, we may want to model the distribution of galaxy luminosities, or their morphological properties. A number of empirical methods exist to do just this; we will not discuss them here but rather point the reader to the review of the subject by Wechsler and Tinker (2018).

Wrapping up this chapter, let us mention that some of the most exciting and powerful upcoming cosmological measurements are precisely those that will map the large-scale structure of the universe in far greater detail than previously possible. These surveys include Vera C. Rubin Observatory, Dark Energy Spectroscopic Instrument (DESI), Euclid Space Telescope, and Nancy Grace Roman Space Telescope, as well as a number of other smaller but no less promising surveys and experiments.

Bibliographical Notes

The field of large-scale structure is reviewed in a number of books and a large number of review articles. Nevertheless, it can be challenging to find material that is easily accessible, and not overly technical or detailed.

For the basic statistics of large-scale structure and an introduction to the two-point correlation function, Jim Peebles' (1980) pioneering book on the subject is a very good starting place. So is Peacock (1998), which covers all of cosmology and is especially strong on cosmic structure. Cosmological perturbation theory is covered by many review articles, but most of them take no prisoners in terms of the mathematical level; a good starting point is Bertschinger (1993), while our analysis in Sec. 9.3 follows Mo *et al.* (2010), a wide-ranging book covering both galaxy formation and large-scale structure. Galaxy clusters from the point of view of cosmology are lucidly reviewed in Allen *et al.* (2011), while the baryon acoustic oscillations are compactly and clearly covered in Bassett and Hlozek (2009). End-to-end coverage of the whole subject, both theory and connection to observations, is provided by Dodelson and Schmidt (2020).

Problems

9.1 Developing intuition about the size and growth of density perturbations. Consider the density perturbation mode that enters the horizon at matter–radiation equality. Recall that the amplitude of this fluctuation is, in the scale-invariant ($n = 1$) primordial power spectrum limit, the same as the amplitude of any mode just entering the horizon – take that to be $\delta_H \simeq 2 \times 10^{-5}$.

(a) What is the size R of that mode? In other words, what is the comoving sound horizon at matter–radiation equality? Give the answer in $h^{-1}\text{Mpc}$ comoving, and remember that $H_0^{-1} \simeq 3000 \, h^{-1}\text{Mpc}$.

(b) What amplitude does that mode have *today*? Assume the linear growth and, for simplicity (so that you won't even need a calculator), assume that the universe has been matter dominated between MR equality and today.

(c) Using your code from Problem 9.9, numerically calculate the linear rms of mass fluctuations on scales corresponding to that mode. That is, evaluate $\sigma(R, z = 0)$, where R is the value you found in part (a). *Very* roughly, is it comparable to the (analytically calculated) answer in part (b)?

(d) If the answer in part (c) is somewhat different from that in part (b), can you explain why the difference is in that direction (larger or smaller)? *Hint:* Think about which mode(s) $\sigma(R)$ averages.

9.2 Growth equation in terms of the growth suppression factor g. Starting from the standard differential equation for the linear growth of fluctuations D, Eq. (9.37), derive the equivalent equation in the growth suppression factor g, Eq. (9.46). *Hints:* You will want to make use of Eq. (3.11). Also, express the ratio $4\pi G\rho_M/H^2$ as well as H'/H (where $' = d/d\ln a$) in terms of the time-dependent matter density relative to critical, $\Omega_M(a)$.

9.3 Closed-form solution for the growth equation? The growth of linear density perturbations in a general cosmological model with multiple components is described by a second-order ordinary differential equation (so Eq. (9.37) for $D(a)$, or else Eq. (9.46) for $g(a)$). However, it would be nice to have a closed-form, integral solution. Such solutions exist only for some special cases. Here, consider the Heath (1977) ansatz

$$D(a) = \frac{5}{2}\Omega_M H(a) \int_0^a \frac{da'}{[a'H(a')]^3},$$

where the constant of proportionality ($5/2$) is determined by requiring that $D(a = 1) = 1$.

Assume a flat wCDM model (so a flat model with matter density Ω_M and dark-energy density Ω_{DE} and constant equation of state w). Plug the expression above into the growth equation (9.37). For which value(s) of w does the Heath expression satisfy the growth equation? *Hints:* You will need the time derivatives of the Hubble parameter, $\dot{H}$ and $\ddot{H}$, so you might want to work them out first. For a further small shortcut, you will get terms that contain the integral in the Heath expression above, and terms that contain derivatives of the integral (so no integral itself), and both kinds of terms need to separately sum to zero.

9.4 Value for k_{eq}. Prove the analytical relation for the comoving wavenumber corresponding to the horizon at matter–radiation equality given in Eq. (9.64):

$$k_{\mathrm{eq}} = H_0 \sqrt{2 \frac{\Omega_M^2}{\Omega_R}}.$$

Plug in the numbers for our fiducial cosmological model and evaluate k_{eq}.

9.5 Fill in the steps: Press–Schechter mass function. Starting from Eq. (9.91), derive Eq. (9.92).

9.6 Spherical collapse model. [Adopted from Gus Evrard.] The spherical collapse (SC) model is a simplified framework in which one can analytically track the evolution of an overdense region well into the nonlinear regime. The model adopts a number of simplifying assumptions, the main one being that the perturbation is perfectly spherical. Despite the obvious mismatch between this assumption and reality, the SC model's results – for example, the critical density for collapse, δ_c, which is worked out below – are a very useful input to various analytic results in the field of large-scale structure.

Let us start with some preliminaries. Assume an EdS universe, with $a(t) \propto t^{2/3}$. Then the mean (background) mass density evolves as

$$\bar{\rho}(t) = \frac{1}{6\pi G t^2},$$

and the Hubble parameter is $H(t) = (2/3t)^2$.

Now consider a spherical patch of radius R_i within which the interior density is slightly higher, $\rho(<R_i, t_i) = \bar{\rho}(1 + \delta_i)$. Also imagine the boundary of this patch to be expanding outward with initial velocity $v_i = H_i R_i$, with H_i a constant. Repurposing almost exactly the similar Newtonian derivation of the Friedmann I equation in Chapter 2, the radius of the overdensity is governed by Newton's law of gravity:

$$\frac{d^2 R}{dt^2} = -\frac{GM}{R^2},$$

the first integral of which gives us a specific energy equation

$$\frac{1}{2}\left(\frac{dR}{dt}\right)^2 - \frac{GM}{R} = E_i,$$

where E_i is a constant of integration. At the initial time, the specific kinetic and potential energies of the outer shell of the perturbation are

$$K_i = \frac{1}{2}\left(\frac{dR}{dt}\right)^2 = \frac{1}{2}(H_i R_i)^2$$

$$|U_i| = \frac{GM}{R_i} = G\frac{4\pi}{3}\bar{\rho}(t_i)(1+\delta_i)R_i^2 = \frac{1}{2}(H_i R_i)^2(1+\delta_i) = K_i(1+\delta_i).$$

The total specific energy of the shell is then

$$E_i = K_i - K_i(1+\delta_i) = -K_i\delta_i.$$

Since $E_i < 0$ for $\delta_i > 0$, a locally overdense patch is bound. Like a rocket launched with sub-escape velocity, the shell expands to a maximum radius, turns around (at the turnaround time $t = t_{\mathrm{ta}}$), and then collapses (at time $t = t_c$).

(a) **SC model: turnaround radius.** Determine the maximum radius of the shell, R_{ta}, which is reached at turnaround. Using conservation of mechanical energy of the shell, show that

$$\frac{R_{\mathrm{ta}}}{R_i} = \frac{1+\delta_i}{\delta_i} \simeq \delta_i^{-1}.$$

(b) **SC model: parametric solution.** There is a parametric solution for the bound trajectory of the shell that almost exactly parallels that for the expansion of the closed universe discussed in Chapter 2. This solution takes the mathematical form of the equation describing a cycloid (see Box 2.3):

$$R(\theta) = A(1-\cos\theta)$$
$$t(\theta) = B(\theta - \sin\theta),$$

where $0 \leq \theta \leq 2\pi$ is a dummy variable, and the two constants are

$$A = \frac{R_i}{2}\frac{1+\delta_i}{\delta_i}; \qquad B = \frac{1}{2H_i}\frac{1+\delta_i}{\delta_i^{3/2}}.$$

The two constants are related by an expression $A^3 = GMB^2$ that looks a lot like Kepler's third law of planetary motion (Newtonian gravity is the basis for both, after all). Show that the expression for the ratio of the perturbation's mean interior density, $\rho(\theta)$, to that of the background universe is

$$\Delta(\theta) \equiv \frac{\rho(\theta)}{\bar{\rho}(\theta)} \equiv 1 + \delta(\theta) = \frac{9}{2}\frac{(\theta - \sin\theta)^2}{(1-\cos\theta)^3}.$$

(c) **SC model: overdensity at turnaround.** Numerically evaluate this expression for Δ at the moment of the shell's maximum expansion (i.e., at the turnaround).

(d) **SC model: early-time expansion.** Now Taylor-expand the expression for $1 + \delta$ that you found in part (b) for $\theta \ll 1$, and define $\delta(\theta \ll 1) \equiv \delta_i$. Show that

$$\delta_i = \frac{3}{20} \left(6\pi\right)^{2/3} \left(\frac{t_i}{t_{\mathrm{ta}}}\right)^{2/3}.$$

(e) **Linearly extrapolated SC model: overdensity at collapse.** The spherical model predicts that, at the collapse time ($\theta = 2\pi$), the overdensity of the perturbation becomes infinite because its size goes to zero. A better approximation is that of a *linearly extrapolated* evolution of the overdensity, which assumes that $\delta(t)$, which you found in part (d) to scale like density perturbations in the EdS model

$$\delta(a) \propto D(a) \propto a \propto t^{2/3},$$

scales like this not only at early times, but at *all* times.

Evaluate your result from part (d) at the time of collapse ($t = t_c = 2t_{\mathrm{ta}}$) to find the linearly extrapolated overdensity at collapse, $\delta_c \equiv \delta(t_c)$. You should find that $\delta_c \simeq 1.686$. This quantity is prominently featured in the derivation of the Press–Schechter mass function in Sec. 9.6.4.

9.7 Single-integral relation between $P(k)$ and $\xi(r)$. Consider Eq. (9.56):

$$P(k) = \int \xi(r) e^{-i\mathbf{kr}} d^3\mathbf{r}.$$

Solve the angular integrals to simplify the equation and be left with a 1D integral over (scalar) dr on the right-hand side. *Hint:* Work in a coordinate system where the angle θ is measured with respect to the $\mathbf{k}$ direction, so that $\mathbf{kr} = kr\cos\theta$. Then the volume element is $2\pi d(\cos\theta) r^2 dr$.

9.8 Fill in the steps: redshift-space distortions. Complete the derivation of the Kaiser formula in Box 9.3 following these steps:

- The Jacobian for going from real space $\mathbf{r}$ to redshift space $\mathbf{s}$ is $J \equiv d^3r/d^3s$. Make use of Eq. (B2) and make use of the so-called distant-observer approximation, where $\partial v_\parallel/\partial r \equiv k v_\parallel \gg v_\parallel/r$ (so that the observed modes are well within the typical size of the survey, $kr \gg 1$). Show that

$$J \simeq \left(1 - \frac{1+z}{H}\frac{dv_\parallel}{dr}\right).$$

- Use the conservation of mass density, $\rho(\mathbf{s})d^3s = \rho(\mathbf{r})d^3r$, along with the Jacobian above, to get the relation between the overdensities $\delta^{(\mathbf{s})}$ and $\delta \equiv \delta^{(\mathbf{r})}$:

$$\delta^{(s)} = \delta - \frac{1}{aH}\frac{dv_{\parallel}}{dr}.$$

- Use the continuity equation in the comoving coordinates ($\nabla_{\text{comov}} = a^{-1}\nabla$)

$$\dot{\delta} = -a^{-1}\nabla \cdot \mathbf{v},$$

along with the relation between the velocity and its potential, $\mathbf{v} = \nabla\Phi_v$, as well as $v_{\parallel} = \partial\Phi_v/\partial r_z$, to get

$$\frac{\partial v_{\parallel}}{\partial r_z} = -faH\frac{\partial^2}{\partial r_z^2}\left(\nabla^{-2}\delta\right),$$

where $f(a) = d\ln\delta/d\ln a$ is the growth rate and $H(a) = d\ln a/dt$ is the Hubble parameter.

- Combine the results of the previous two bullet points to obtain the Kaiser formula for the density perturbation with RSD:

$$\delta^{(s)} = (1 + f\mu^2)\delta,$$

where $\mu \equiv \hat{\mathbf{k}}\cdot\hat{r}_z = k_z/k$ is the cosine of the line-of-sight to the wavenumber describing the feature that we observe. From here, the final formula for the redshift-space power spectrum follows as explained in Box 9.3.

9.9 **[Computational] Calculate the mass function "from scratch."** In this multi-part problem, you will produce the Press–Schechter mass function numerically and from scratch.

Assume the usual fiducial model from Table 3.1. Make sure you use the convenient units of $h^{-1}M_{\odot}$ for the mass of halos, and h^{-1}Mpc for distance (and its inverse $h\,$Mpc^{-1} for wavenumber). You might find it useful that $\rho_{\text{crit},0} = 2.775 \times 10^{11}(h^{-1}M_{\odot})/(h^{-1}\text{Mpc})^3$.

(a) **Linear growth rate.** Calculate and plot the linear growth suppression factor $g(z)$ vs. redshift. You will want to utilize Eq. (9.46). Also plot the linear growth rate $D(z) \equiv g(z)/[(1+z)g(0)]$ vs. z. Make the x-axis (i.e., redshift z) range from 0 to 5 and the y-axis from 0 to 1. Both axes should be linear.

[Plan B for those who want a faster way to the growth function, and a useful check for everyone regardless: It happens that for the ΛCDM model with Ω_M and Ω_{Λ} (doesn't even have to be flat), there is a simple and accurate analytic fitting formula (Carroll *et al.*, 1992). It gives the suppression rate as

$$g(z) = \frac{5}{2}\frac{\Omega_M(z)}{\Omega_M(z)^{4/7} - \Omega_{\Lambda}(z) + \left[1 + \frac{1}{2}\Omega_M(z)\right]\left[1 + \frac{1}{70}\Omega_{\Lambda}(z)\right]},$$

where, notice, you need to calculate the z-dependent omegas. You may use this $g(z)$-fitting formula instead if solving the second-order ordinary differential equation is not working out.]

(b) **Power spectrum.** Plot the dimensionless linear power spectrum, as given in Eq. (9.61), at the present time (so $a = 1$). Plot it for the range of scales $0.001 < k/(h\,\mathrm{Mpc}^{-1}) < 10$, and make the y-axis go from 0.001 to 10 as well. Both axes should be log.

For the power spectrum, you have everything you need except for the transfer function. Here I would like you to use the old but simple (and famous) BBKS transfer function (Bardeen *et al.*, 1986):

$$T(k) = \frac{\ln(1 + 2.34q)}{2.34q} \left[1 + 3.89q + (16.1q)^2 + (5.46q)^3 + (6.71q)^4\right]^{-1/4}$$

$$q \equiv \frac{1}{\Gamma}\frac{k}{h\,\mathrm{Mpc}^{-1}}; \qquad \Gamma \equiv \Omega_M h \exp(-\Omega_B - 1.3\Omega_B/\Omega_M).$$

Plot the transfer function $T(k)$ in the same plot as $\Delta^2(k)$.

(c) **Amplitude of mass fluctuations.** Calculate the amplitude of mass fluctuations, $\sigma(z, M)$, for the assumed cosmological model:

$$\sigma^2(z, R) = \int_{k=0}^{\infty} \Delta^2(k, z) \left(\frac{3j_1(kR)}{kR}\right)^2 d\ln k,$$

where $j_1(x) = \sin x / x^2 - \cos x / x$. Be *very careful* to check convergence of the integral, you can see from above that it rings in a nasty manner. Make a plot of $\sigma(0, M)$ for $10^{12} < M/h^{-1}\mathrm{M}_\odot < 10^{15}$ (and don't forget to take the square root of σ^2 in the above equation). The x-axis should be log, and the y-axis linear. Recall that the conversion between the radius R and mass M is $M = (4\pi/3)R^3\rho_M$, where here you will be evaluating everything (σ, ρ_M, Δ^2) at $z = 0$.

(d) **Mass function and halo counts.** Now you are ready to plot the Press–Schechter mass function, given in Eq. (9.92). Plot the mass function $(dn/d\ln M)$ vs. redshift z for fixed-mass halos of three discrete values: $M = 10^{13}\,h^{-1}\mathrm{M}_\odot$, $M = 10^{14}\,h^{-1}\mathrm{M}_\odot$, and $M = 10^{15}\,h^{-1}\mathrm{M}_\odot$. (So you will have three curves.) Note that the units of $dn/d\ln M$ are $(h^{-1}\mathrm{Mpc})^{-3}$. Do not forget that $\sigma = \sigma(z, M) = D(z)\sigma(z = 0, M)$ in the Press–Schechter formula. Make the redshift axis go from zero to five, $0 < z < 5$, and make the y-axis go from 10^{-10} to 10^{-3}.

(e) **How many halos?** What is the total number of halos expected above $M_{\min} = 10^{14}\,h^{-1}\mathrm{M}_\odot$ in a survey that covers 5000 square degrees, and covers the redshift range $0 < z < 1.0$? Assume the Press–Schechter mass function you just calculated, and remember that the comoving volume element is $dV/(d\Omega dz) = r^2(z)/H(z)$, where $r(z)$ is the comoving distance and $H(z)$ is the Hubble parameter. Be careful about units; for example, the solid angle is measured in steradians, while the volume should be in units that are inverse of those for the number density.

Repeat the calculation above but now assuming the EdS universe with $\Omega_M = 1.0$, and lower the Hubble constant to $h = 0.50$ (to make it a better fit to the data, which are mainly sensitive to $\Omega_M h$ and $\Omega_M h^2$). How many clusters is our assumed survey finding now? Intuitively, why has the number changed so much in the direction (up or down) that you find?

9.10 [**Computational**] N-**body simulation.** Create your own N-body simulation! You will calculate (and visualize) the motion of N massive bodies which mutually interact via gravity. N-body simulations play an important role in cosmology. To understand the creation of structure in the universe, cosmologists evolve the gravitational interactions of particles in time. Here, a "particle" would ideally represent say a proton, or at the very least an Earth-sized body. Because the universe is big, even in a simulation with $N = 10^{11}$ objects, you can only afford each such object to represent something like several thousand Sun masses. Nevertheless, this is sufficient to give us an idea how large structures in the universe – galaxies and clusters of galaxies – form in cosmic time.

Start with a modest number of particles $N \sim 5$ and, when the code is polished to work well, try $N \sim 50$. The goal is to evolve each particle according to Newton's law of gravity. In this brute-force evaluation, for N particles you will have $\binom{N}{2}$ force evaluations at each step – most modern methods avoid this $\propto N^2$ scaling using some tricks, but we will not.

You first have to decide what mass objects you will have. You can have Earth mass bodies, or galaxy mass, or anything in between. While this is up to you, you have to make sure to use time steps and initial placement (size of box and particle separations) that are appropriate for your choice. If you place 1000 Earth masses in a Hubble-sized volume, not much will happen!

You also have to figure out how to set initial positions and initial velocities of the particles. First you can do something simple, but later, when your code is working, you should place them more realistically – maybe randomly over some region of space for example. You may also want to impose periodic boundary conditions (once your basic code is working properly).

The pseudo-code may look something like this:

```
# relevant pseudo-code; note, it's incomplete!
N = 5 # number of objects

populate objects in the volume
initialize their positions and velocities/momenta

loop over time
```

```
loop over all objects
    update force
    update momentum/velocity
    update position
```

Finally, if everything is working fine, you should still be seeing that close encounters between particles lead to wild ejections from the system. This is a problem well known to N-body simulators, and requires some kind of **force softening**. The simplest way to implement it is to set the gravitational force to zero when the two masses are closer than some small distance (think about what distance makes sense here on physical grounds).

It is best if you write your program using an environment that allows visualization, such as Visual Python (VPython). Then you can directly observe what happens to your particles as they interact.

10 Statistical Methods in Cosmology and Astrophysics

The "official" starting date for when cosmology became truly data-driven science is arguably the year 1992, when the COBE satellite released full-sky maps of the CMB radiation pattern on the sky, and effectively measured the amplitude of primordial density fluctuations. Ever since, cosmology's progress has relied on an increasingly more sophisticated – and stunningly diverse – set of tools borrowed or adopted from statistics. In this chapter, we review some of the most important such tools, with the specific emphasis on how they are used in cosmology and astrophysics.

10.1 Introduction to Statistics

Before we apply statistics to cosmology, we first review some basic concepts and tools in statistics.

10.1.1 Random Variables and Probability

A random (or stochastic) variable X is a parameter whose value is subject to variations due to chance. A random variable's possible values are the possible outcomes of an experiment that measures those values. Examples of random variables in cosmology are flux measured in a detector, or power spectrum of galaxies measured at some wavenumber k and redshift z.

The relative frequency of the values that a random variable takes are described by its probability distribution. We now discuss the properties of this fundamental concept in statistics. If $P(X)$ is the probability of some event X, then the following fundamental rules hold:

$$P(X) \geq 0 \tag{10.1}$$

$$\int_{-\infty}^{\infty} P(X)dX = 1 \tag{10.2}$$

$$P(X_2) = \int P(X_2|X_1)P(X_1)dX_1. \tag{10.3}$$

Let us comment on the last one of these equations. Here, $P(X_2|X_1)$ means the probability of event X_2 given event X_1. If the events are independent, then $P(X_2|X_1) = P(X_2)$ and Eq. (10.3) becomes a tautology, $P(X_2) = P(X_2)$. Finally, note that one can write the joint probability of X_1 and X_2, $P(X_1, X_2)$, as

$$P(X_1, X_2) = P(X_2|X_1)P(X_1) = P(X_1|X_2)P(X_2). \tag{10.4}$$

Note that the latter equality is essentially the statement of Bayes' theorem, something that we will study in Sec. 10.2.1.

10.1.2 Mean, Median, Mode, and Marginalization: Just Marvelous

The full distribution of random variable X is defined by $P(X)$. An important summary statistic of the distribution of X is its first moment, the **mean** of the distribution, which is defined as

$$\mu \equiv \bar{X} \equiv \langle X \rangle = \int_{-\infty}^{\infty} X P(X) \, dX \quad \text{(mean).} \tag{10.5}$$

Related quantities are the median and the mode:

$$\frac{1}{2} = \int_{-\infty}^{X_{\text{median}}} P(X) \, dX \quad \text{(median)} \tag{10.6}$$

$$\left. \frac{dP}{dX} \right|_{X_{\text{mode}}} = 0 \quad \text{(mode).} \tag{10.7}$$

The median splits the distribution $P(X)$ into two parts with equal areas under the curve, while the mode is simply the location of the peak of $P(X)$. All three of these summary statistics are illustrated in Fig. 10.1.

Finally, it is useful to see how we can **marginalize** (integrate) over some variables

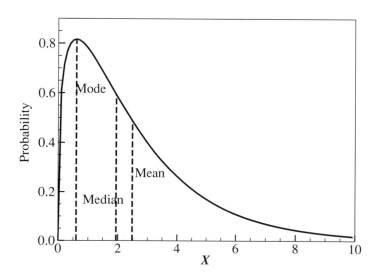

Fig. 10.1   Mean, median, and mode shown for a sample non-Gaussian distribution.

to be left with the marginalized probability distribution in others. Consider two variables X_1 and X_2 with the joint probability function $P(X_1, X_2)$, and say we would like to find out what is the probability in X_1 alone. This is easy to do; plugging Eq. (10.4) into Eq. (10.3), we get

$$P(X_1) = \int P(X_1, X_2) dX_2 \qquad \text{(marginalization).} \qquad (10.8)$$

Marginalization is very common in cosmology because, when we compare theory to data, we are usually interested in the probability of a given parameter taking some values marginalized over the possible values of all other parameters. In an n-dimensional parameter space, a full marginalization over all parameters other than the one we are interested in would look like Eq. (10.8), except with an $(n-1)$-dimensional integral over the remaining parameters $X_2, \ldots, X_n$ on the right-hand side.

10.1.3 Variance and Higher Moments

We have already introduced the lowest (first) moment of a distribution of a variable, its mean, in Eq. (10.5). The second moment of the distribution is called the **variance**, and it is a natural measure of the width (squared) of the distribution of a variable. Variance around the mean of the random variable X that is described by the probability distribution $P(X)$ is given by

$$\text{Var}(X) \equiv \sigma^2 \equiv \langle (X - \mu)^2 \rangle = \int_{-\infty}^{\infty} (X - \mu)^2 P(X) \, dX \quad \text{(variance)}, \qquad (10.9)$$

where μ is the mean, while the angular brackets denote ensemble average. Here we have introduced the standard deviation σ as a square root of the variance.

Higher moments of a distribution are also commonly used. An nth moment is defined by $\langle X^n \rangle \equiv \int X^n P(X) dX$. It is often useful to consider higher moments around the mean (so with $(X - \mu)$ in the integral), and scaled by the appropriate power of the standard deviation σ. For example, **skewness** is a scaled third moment around the mean, defined as

$$S \equiv \left\langle \left(\frac{X - \mu}{\sigma} \right)^3 \right\rangle \qquad \text{(skewness)}, \qquad (10.10)$$

and it measures the asymmetry of the probability distribution of $P(X)$ around the mean. The scaled fourth moment around the mean is called **kurtosis**:

$$K \equiv \left\langle \left(\frac{X - \mu}{\sigma} \right)^4 \right\rangle \qquad \text{(kurtosis)}, \qquad (10.11)$$

and it measures the "peakedness" of $P(X)$. Another useful concept is **excess kurtosis**, defined as $K - 3$, which accounts for the fact that a Gaussian random variable has kurtosis of 3.

We now cover basic properties of probability distributions that are used in data analysis in cosmology.

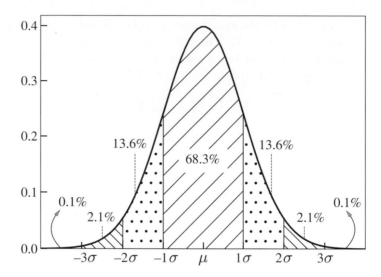

Fig. 10.2 Gaussian distribution with $\pm 1\sigma, \pm 2\sigma$, and $\pm 3\sigma$ regions, along with the percentage of the total area that these regions encompass.

10.1.4 Gaussian (or Normal) Distribution

The **Gaussian distribution** (also known as the **normal distribution**, illustrated in Fig. 10.2) is by far the most common statistical distribution, and is commonly found in phenomena describing physics, astronomy, and other natural and social sciences. The Gaussian distribution is also the simplest distribution to work with.

The probability density function of the Gaussian distribution in one dimension (one parameter X) is

$$P(X) = \frac{1}{\sqrt{2\pi\sigma^2}} \exp\left[-\frac{1}{2}\left(\frac{X-\mu}{\sigma}\right)^2\right]. \tag{10.12}$$

The 1D Gaussian distribution is specified with two parameters: the mean μ and the standard deviation σ. The variance of the Gaussian $P(X)$ is σ^2. In fact, these two inputs, the mean and standard deviation (or variance), are so important that we often write

$$X \sim \mathcal{N}(\mu, \sigma^2), \tag{10.13}$$

which means "X comes from a Gaussian distribution with mean μ and standard deviation σ." Furthermore, one can show that the skewness and all *odd* higher moments (fifth moment, seventh moment, etc.) of the Gaussian distribution are zero while kurtosis, along with all *even* higher moments, can easily be expressed in terms of the variance.[1]

---

[1] The fact that higher-order, even moments (fourth, sixth, etc.) can be written in terms of the second moment is sometimes called Wick's theorem, which we will discuss in Chapter 13, specifically Box 13.3.

For multiple parameters, ordered in a vector $\mathbf{x}$ with mean $\boldsymbol{\mu} \equiv \langle \mathbf{x} \rangle$ and covariance matrix $C \equiv \langle \mathbf{x}\mathbf{x}^T \rangle - \langle \mathbf{x} \rangle \langle \mathbf{x}^T \rangle = \langle (\mathbf{x} - \boldsymbol{\mu})(\mathbf{x} - \boldsymbol{\mu})^T \rangle$, the Gaussian distribution generalizes to

$$P(\mathbf{x}) = \frac{1}{(2\pi)^{n/2}|\det C|^{1/2}} \exp\left[-\frac{1}{2}(\mathbf{x} - \boldsymbol{\mu})^T C^{-1}(\mathbf{x} - \boldsymbol{\mu}) \right]. \tag{10.14}$$

In one dimension, the $\pm 1\sigma$, $\pm 2\sigma$, and $\pm 3\sigma$ intervals around the mean enclose respectively 68.3 percent, 95.4 percent, and 99.7 percent of the total area under the curve; see Fig. 10.2.[2] These particular fractions are particularly popular and used even when the distribution is not Gaussian. For example, when we quote a "one sigma" constraint on some parameter, we usually mean the interval that encloses 68.3 percent of total probability in that parameter, irrespective of whether the probability distribution in that parameter is actually Gaussian or not. Thus, a "one sigma" range in some parameter is not necessarily equal to the interval corresponding to plus or minus one standard deviation around the mean (or mode) of that parameter. In short: if possible, we prefer to calculate 68.3 percent, 95.4 percent, and 99.7 percent of the total area under the curve of our probability distribution regardless of whether it is Gaussian or not.

A huge boost to the importance of the Gaussian distribution is provided by the **Central Limit Theorem** which, loosely speaking, states that a sum of arbitrarily distributed random variables has a Gaussian normal distribution. More specifically, let X_1, X_2, X_3, ..., X_n be a sequence of n independent and identically distributed (i.i.d.) random variables coming from some distribution with expectation μ and variance $\sigma^2 < \infty$. The Central Limit Theorem states that, as the sample size n increases, the distribution of the arithmetic mean of these random variables approaches the Gaussian distribution with mean μ and variance σ^2/n irrespective of the shape of the distribution of the individual terms X_i.

The Central Limit Theorem simplifies the analysis in cosmology and astrophysics. Specifically, it allows us to assume a Gaussian joint probability distribution for a large number of random variables even when their individual distributions are not Gaussian. The Central Limit Theorem justifies why it is often appropriate to assume a Gaussian distribution for the joint probability of observed measurements.

Finally, let us mention the **log-normal** distribution, which is just the normal distribution where X is a logarithm of a random variable (rather than the variable itself). In other words:

$$\log(X) \sim \mathcal{N}(\mu, \sigma^2), \tag{10.15}$$

where μ and σ are the mean and standard deviation of the log-normal distribution.

---

[2] For a generalization of what ranges correspond to a given probability interval in higher dimensions of parameter space, see Problem 10.3.

10.1.5 Chi-Squared Distribution

The chi-squared (χ^2) distribution is also very commonly used in cosmology. This is the distribution that a sum of *squares* of Gaussian variables has. That is, if X_i are Gaussian random variables with mean zero and variance one, then

$$Y = X_1^2 + X_2^2 + \cdots + X_n^2 \tag{10.16}$$

has a chi-squared distribution with n degrees of freedom with the probability distribution

$$P(Y) = \frac{1}{2^{n/2}\Gamma(n/2)} Y^{n/2-1} e^{-Y/2}. \tag{10.17}$$

Most of the time in cosmology, we are not interested in the actual form for $P(Y)$ but rather the following three properties of the chi-squared distribution:

- the mean of $P(Y)$ is equal to the number of degrees of freedom n;

- the variance of $P(Y)$ is equal to $2n$;

- when $n \gg 1$, the chi-squared distribution starts to look like the Gaussian distribution, with mean n and variance $2n$ as mentioned above.

The outsize importance of the chi-squared distribution in cosmology comes from the fact that the CMB anisotropy on the sky, given by the harmonic coefficients $a_{\ell m}$ that we will fully introduce in Chapter 13, is Gaussian to an excellent approximation. [Sec. 13.7 quantifies precisely *how* Gaussian the CMB is.] The same is true for the matter density field $\delta(\mathbf{x}, t)$ which, if evaluated on reasonably large scales ($\gg 10\,\mathrm{Mpc}$), is very nearly Gaussian. The two-point correlation functions of these respective fields – C_ℓ for the CMB anisotropy and $\xi(r)$ for the matter density – are readily measured and used as a meeting point of theory and observations. Because they are the "squares" of the nearly Gaussian quantities, the C_ℓ and $\xi(r)$, as well as ξ's Fourier partner $P(k)$, all become χ^2-distributed. So, for example, at each multipole ℓ, we expect C_ℓ to be χ^2-distributed with $2\ell + 1$ degrees of freedom (again, more details near the beginning of Chapter 13). To be specific: a *realization* of that particular C_ℓ in our universe is drawn from the distribution that has the mean $\overline{C_\ell}$ specified by the ΛCDM (say) theory. And the variance and higher moments around that mean are specified by the properties of the χ^2 distribution with the relevant number of degrees of freedom.

10.1.6 Likelihood

We next introduce an all-important concept in statistics, that of likelihood. Let D be some data described by measurements arranged in a vector $\mathbf{d}$, and M be a model (say, a cosmological model such as ΛCDM) described by parameters arranged in a vector $\mathbf{p}$. Then the likelihood is defined to be the probability of finding the set of measurements $\mathbf{d}$, given the parameter set $\mathbf{p}$ from model M:

$$\mathcal{L} = P(\mathbf{d}|\mathbf{p}, M) \qquad \text{(likelihood)}. \qquad (10.18)$$

Hence:

> Likelihood is the probability of measuring a realization of the data given a model.

Most often we are interested in the probability of physical parameters given some data – say, the probability distribution of values of a dark-energy parameter, or the probability of the sum of neutrino masses, given some measurements. This, however, is *not* given by the likelihood which, as stated in Eq. (10.18), is the probability of data given the model and not vice versa. To go from the likelihood to the desired probability of the model parameters therefore requires additional effort. This leads us to the discussion of Bayesian and frequentist statistics.

10.2　Bayesian vs. Frequentist Statistics

There are two principal approaches to statistics, and their rivalry is as famous as that between the Montagues and Capulets, the Hatfields and McCoys, the Lakers and Celtics, Real Madrid and Barcelona, or Michigan and Ohio State (with Michigan the better team, of course). These are the Bayesian and frequentist approaches.

Frequentist interpretation considers an event's probability as the limit of its relative frequency in a large number of trials. In this picture, if we observe the event unfold many times then, in the limit when that number goes to infinity, the relative frequency of its outcomes becomes its probability.

In contrast, Bayesian reasoning interprets the concept of probability as a measure of a state of knowledge, and not as a frequency. One of the crucial features of the Bayesian view is that a probability can be assigned to a hypothesis, which is not possible under the frequentist view, where a hypothesis can only be rejected or not rejected.

In short, the difference is

> - Frequentist: model is fixed, data are repeatable.
> - Bayesian: data are fixed, model is repeatable.

Historically, Bayesian statistics dominated science in the nineteenth century, the frequentist approach dominated in the twentieth century, and both approaches have been widely used over the first two decades of the twenty-first century. While experimental particle physics has traditionally favored the frequentist approach, cosmology analyses have been largely Bayesian. We now cover the Bayesian approach, which is most immediately applicable to data that are at the heart of the subject matter of this book, but also briefly review the frequentist methodology in Box 10.1.

10.2.1 Bayesian Statistics

Bayesian probability calculus makes use of Bayes' theorem:

$$P(\mathbf{p}|\mathbf{d}, M) = \frac{P(\mathbf{d}|\mathbf{p}, M)\, P(\mathbf{p}|M)}{P(\mathbf{d}|M)}, \tag{10.19}$$

where M represents a model (or hypothesis) made up of the parameter vector $\mathbf{p}$, and $\mathbf{d}$ is data. Here

- $P(\mathbf{p}|M)$ is a **prior** probability of parameters $\mathbf{p}$ (assuming model class M) before the data $\mathbf{d}$ was acquired or seen.
- $P(\mathbf{d}|\mathbf{p}, M)$ is the conditional probability of seeing the data $\mathbf{d}$ given that the parameters $\mathbf{p}$ of model M are true. As we mentioned above, $P(\mathbf{d}|\mathbf{p}, M) \equiv \mathcal{L}(\mathbf{d}|\mathbf{p})$ is the **likelihood** (of the data, given the model).
- $P(\mathbf{p}|\mathbf{d}, M)$ is the **posterior** probability: the probability that the model is true, given the data and the prior state of belief about the models.
- $P(\mathbf{d}|M)$ is the *a priori* probability of witnessing the data $\mathbf{d}$ under all possible realizations of parameters $\mathbf{p}$ of model class M. It is a normalizing constant that only depends on the data, and which in most cases does not need to be computed explicitly. This quantity, however, plays an important role in model selection; it is called the **Bayesian evidence**, and it is given by the likelihood integrated over all model-parameter values:

$$
\begin{aligned}
P(\mathbf{d}|M) &= \int P(\mathbf{d}|\mathbf{p}, M) P(\mathbf{p}|M)\, d\mathbf{p} \qquad \text{(Bayesian evidence)} \\
&\equiv \int \mathcal{L}(\mathbf{d}|\mathbf{p}) P(\mathbf{p})\, d\mathbf{p},
\end{aligned} \tag{10.20}
$$

where in the second line we simplified the notation a little. Note that the evidence may be difficult to numerically evaluate, since it integrates the likelihood (times the prior) over the often huge, multi-dimensional parameter space.

Because each term in Eq. (10.19) is conditioned on the same model class M, we can drop the latter label and simplify Bayes' theorem to

$$P(\mathbf{p}|\mathbf{d}) = \frac{\mathcal{L}(\mathbf{d}|\mathbf{p})\, P(\mathbf{p})}{P(\mathbf{d})} \qquad \text{(Bayes' theorem).} \tag{10.21}$$

The key thing to note is that we are most often interested in the posterior probability of a model parameter's given data, $P(\mathbf{p}|\mathbf{d})$, while what we can typically calculate from the data is the likelihood of the data given a model, $\mathcal{L}(\mathbf{d}|\mathbf{p})$. Bayes' theorem lets us go from the latter to the former. The likelihood and the posterior are very similar if the data are very informative, so that the shape of the prior in the model space doesn't matter much. However, when the data are not very informative, the choice of the prior may play a role. In effect, Bayesian analysis acknowledges the statement that no inference is possible without assumptions, and these assumptions are encoded by the prior.

The Bayesian approach has a number of advantages, and has been widely adopted in cosmology since the data boom in the 1990s. In particular, the Bayesian approach allows easy incorporation of different datasets. For example, we can use one dataset to impose an effective prior on the model space M, and then update this prior probability (to a new posterior) with a new dataset using Bayes' theorem.

10.2.2 Bayesian-Frequentist Example

Say, for example, that we have a measurement of the Hubble constant of $(72 \pm 8)\,\mathrm{km/s/Mpc}$. What would the Bayesian and the frequentist say?

- Bayesian: 68.3 percent of the volume under the posterior probability distribution in H_0 lies between 64 and 80 km/s/Mpc. This posterior can be used as a prior in a new application of Bayes' theorem.
- Frequentist: Performing the same experiment many times will cover the true value of H_0 within the stated limits 68.3 percent of the time. [Of course, "repeating the universe" is a little difficult to do in practice, which is perhaps one reason why the frequentist reasoning in cosmology comes across as less natural than the Bayesian.]

Let us give another example. Say we wish to measure Ω_M and Ω_Λ from type Ia supernova data (which we will discuss in Chapter 12). What should we do in either of the two approaches to statistics?

- Bayesian: Take some prior (say, the uniform prior in both Ω_M and Ω_Λ). Then, for each model parameter set $\mathbf{p} = \{\Omega_M, \Omega_\Lambda\}$, compute the likelihood of the data $\mathcal{L}(\mathbf{d}|\mathbf{p})$ using, for example, a Gaussian likelihood for the data. Then obtain the posterior probability on the two parameters using Bayes' theorem; $P(\mathbf{p}|\mathbf{d}) \propto \mathcal{L}(\mathbf{d}|\mathbf{p})\,P(\mathbf{p})$.
- Frequentist: The general idea is to draw repeated realizations of the data for a given cosmological model, and then repeat the whole procedure for all models (so, e.g., the grid in the $\mathbf{p} = \{\Omega_M, \Omega_\Lambda\}$ space). Particularly nice, though computationally demanding, is the Feldman and Cousins (1998) method. We refer the reader to Box 10.1, and specialized literature on the subject.

Let us further clarify the difference between the probability of the data – the likelihood $\mathcal{L}(\mathbf{d}|\mathbf{p})$ – and the posterior of the cosmological parameters, $\mathcal{P}(\mathbf{p}|\mathbf{d})$ (see Fig. 10.3):

Box 10.1   Frequentist Statistics Tools

We emphasize Bayesian statistics in this chapter, as it is the dominant approach in analyzing data in astrophysical cosmology (LSS and CMB, in particular). However, frequentist approaches are also used in areas of cosmology that use the methodology adopted from particle physics, for example direct detection of dark matter. Hence we briefly review some of the tools used in a frequentist analysis.

Estimators. In statistics, an estimator is a quantity that can be calculated from the data to estimate the desired statistical or physical quantity. For example, we may be interested in estimating, starting from some observational data, the mean and variance of some random variable, the age of the universe, or the ratio of your expected salary if you go work at Wall St. vs. that if you stay in academia. An estimator is usually a direct function of data.

Typically we strive for an estimator that has both small bias (expected deviation from the truth) and small variance (reported error bar). There is often a tradeoff between these two, so that an estimator with a smaller bias may have a larger variance, and vice versa. A useful goal is to find the **best unbiased estimator** for the quantity of our interest – one that has zero bias and the smallest possible variance. Another very popular choice is finding the **maximum-likelihood estimator** (MLE), which maximizes the likelihood but is generally not equivalent to the best unbiased estimator in a given situation.

A simple example is an estimator of the mean of some random variable X, given measurements $x_1, x_2, \ldots, x_N$. The rather obvious – and, as it turns out, best unbiased – estimator of the mean is

$$\hat{\mu} = \frac{\sum_{i=1}^{N} x_i}{N}, \tag{B1}$$

where the hat stands for an "estimate." The variance of this estimator can readily be computed; see Problem 10.1. Another example is finding an estimator for the two-point correlation function, $\xi(r)$, given some galaxy-position data. In that case, the Landy–Szalay estimator, given in Chapter 9 in Eq. (9.72), does a good job.

Estimators are most useful in situations where we cannot explicitly evaluate, or can evaluate but cannot sufficiently quickly numerically compute, the likelihood in our parameter(s) of interest. In those cases, we try to find a suitable estimator, typically derived so that it finds the parameter values that maximize the likelihood. In a frequentist analysis, repeated application of the estimator on simulated data can be used to obtain the confidence intervals in the parameters (which are discussed below).

Monte-Carlo simulations. The Monte-Carlo method (not to be confused with MCMC discussed in this chapter) is one of the first statistical techniques one learns. In our context, Monte Carlo refers to repeated random draws of experimental data based on their measured (or expected) statistical distribution. For example, for a type Ia supernova Hubble diagram represented with the theoretical apparent magnitude $\bar{m}_i \equiv m(z_i, \mathbf{p})$ (where z_i is the SN redshift and $\mathbf{p}$ are the cosmological parameters; see Chapter 12), one could draw the jth Monte-Carlo realization of the data as

Box 10.1   **Frequentist Statistics Tools (continued)**

$$m_i^{(j)} = \bar{m}_i + r\sigma_i, \tag{B2}$$

where σ_i is the standard error on the ith measurement (ignoring any covariance for simplicity), and r is a random number that comes from the $\mathcal{N}(0,1)$ distribution, drawn separately for each i and j.

In frequentist statistics, Monte-Carlo simulations are often used to simulate the likelihood $\mathcal{L}(\mathbf{d}|\mathbf{p})$ for many different realizations of the data. Note that this is fully in the spirit of frequentist statistics – that data $\mathbf{d}$ are variable and the parameters $\mathbf{p}$ are fixed.

Frequentist confidence limits. Let us assume that we have already calculated the best-fit cosmological model using either a direct search of the likelihood or by applying an estimator. Then how do we estimate the likely range that the parameters $\mathbf{p}$ can take – the so-called **confidence intervals**? Note that this is a little tricky since frequentist statistics does not readily report the most probable range of parameters $\mathbf{p}$ as Bayesian statistics does, but rather the range in the true values of the parameters that would be obtained by repeating the experiment many times. This constitutes the subtle difference between the frequentist **confidence** intervals and the Bayesian **credible** intervals.

The Monte-Carlo realizations of the likelihood under repeated draws of data that we introduced above, $\mathcal{L}(\mathbf{d}^{(j)}|\mathbf{p})$, can be used for this purpose. There exist several schemes to obtain confidence intervals in the parameters starting from the likelihood, and discussing them all is beyond the scope of this book. However, we briefly describe one of the most sensible and powerful such approaches, the Feldman and Cousins (1998) algorithm. Consider, for example, all parameter sets (k), where the superscript (k) runs through all possible combinations of parameters $\mathbf{p}$ that are laid out on a grid. First, find the global best-fit (maximum-likelihood) parameter set, $\mathbf{p}_{\max}$. Then, for each parameter set $\mathbf{p}^{(k)}$, generate a suite of likelihoods using Monte-Carlo draws of the data, $\mathcal{L}(\mathbf{d}^{(j)}|\mathbf{p}^{(k)})$. Consider next the chi-squared statistic $\Delta\chi^2(\mathbf{p}^{(k)}) \equiv -2\ln[\mathcal{L}(\mathbf{d}^{(j)}|\mathbf{p}^{(k)})/\mathcal{L}(\mathbf{d}^{(j)}|\mathbf{p}_{\max})]$. For this suite (i.e., at a fixed $\mathbf{p}^{(k)}$), one finds $\Delta\chi_c^2$ such that the desired fraction of experimental realizations, say 68.3 percent, have $\Delta\chi^2(\mathbf{p}^{(k)}) < \Delta\chi_c^2(\mathbf{p}^{(k)})$. This provides $\Delta\chi_c^2$ at each point $\mathbf{p}^{(k)}$ in the parameter space. Now one simply calculates the *actual* $\Delta\chi^2(\mathbf{p})$ at each point, based on the real and not simulated data, and asks whether it is smaller or larger than $\Delta\chi_c^2(\mathbf{p})$. If it is smaller, it is included in the final credible interval, and vice versa if it is larger. As usual in the Bayesian–frequentist comparisons, when the data are informative, then the Feldman–Cousins confidence intervals agree with Bayesian credible intervals.

1. **Data space** (measured quantities): $\mathbf{d} \equiv \{d_i\}$. By the Central Limit Theorem, it is often okay to assume a Gaussian joint distribution in the data, $\mathcal{L}(\mathbf{d}|\mathbf{p})$, even if individual data points are not Gaussian distributed. [There are, of course, examples where a Poisson or a log-normal distribution for the data is more appropriate.]

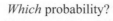

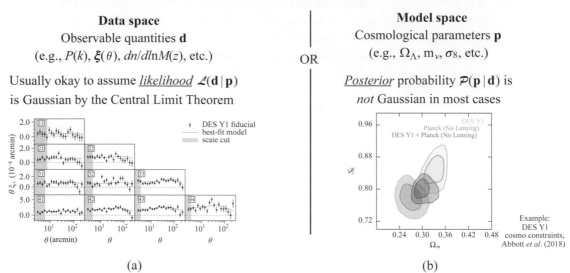

Data space

Observable quantities **d**

(e.g., $P(k)$, $\boldsymbol{\xi}(\theta)$, $dn/d\ln M(z)$, etc.)

OR

Model space

Cosmological parameters **p**

(e.g., Ω_Λ, m_ν, σ_8, etc.)

Usually okay to assume *likelihood* $\mathcal{L}(\mathbf{d}\,|\,\mathbf{p})$
is Gaussian by the Central Limit Theorem

Posterior probability $\mathcal{P}(\mathbf{p}\,|\,\mathbf{d})$ is
not Gaussian in most cases

(a)

(b)

Fig. 10.3   Intuitive description of the two parameter spaces, each with an associated probability, which are often discussed (but sometimes conflated) in cosmology. (a) Data space **d** (e.g., summary statistics such as the power spectrum) with its associated likelihood. (b) Space while on the right is the space of cosmological parameters **p** with its associated posterior. The likelihood $\mathcal{L}(\mathbf{d}|\mathbf{p})$ can often be assumed Gaussian thanks to the Central Limit Theorem, while the posterior $P(\mathbf{p}|\mathbf{d})$ is rarely Gaussian. Both images reprinted by permission from Abbott *et al.* (2018), copyright (2018) by the American Physical Society.

2. **Model space** (theoretical quantities): $\mathbf{p} = \{p_j\}$. It is usually *not* okay to assume that the posterior in the parameters, $\mathcal{P}(\mathbf{p}|\mathbf{d})$, is Gaussian. Nonlinear relations between the true underlying observable quantities and theory parameters often make the posterior quite non-Gaussian, requiring tools like Markov chain Monte Carlo, discussed in Sec. 10.5, to sample the posterior properly.

10.2.3 What Prior to Use?

In general, Bayesian analysis results will depend on the prior. Consider, for example, a flat prior on some model parameter p (equal probability per dp), or a prior flat in the log of p (so equal probability per $d\ln p$). When the data are very informative, they will completely dominate the prior and the prior itself will be irrelevant. Conversely, when the data are weak, the choice of the prior matters. Arguments about what prior to assume for some parameter (e.g., the sum of the neutrino masses m_ν) usually indicate that the data determine the parameter in question only weakly, and that no strong cosmological inferences can be made about the parameter. Finally, it is sometimes both desirable and sensible to apply a *physical* prior on the parameters of interest, as such a prior may reflect "secure" knowledge or a known physical constraint. A simple example is requiring that the sum of the neutrino masses be non-negative, $\sum_i m_{\nu,i} \geq 0$, or similarly for the matter density parameter, $\Omega_M \geq 0$.

10.3 Likelihood Analysis

We now discuss various aspects of constraining the cosmological parameters of interest. Recall that the probability distribution of parameters given some data is called the posterior in a Bayesian analysis, while in a frequentist analysis one ends up with a likelihood. Despite this conceptual and semantic difference, to streamline our exposition we will speak generically about the probability distribution of the model parameters, $P(\mathbf{p})$, and remember that it will, in the commonly used Bayesian analysis, refer to the posterior probability (so likelihood multiplied by the prior).

Our goal is to study how to analyze the probability distribution in cosmological parameters in order to infer the best-fit values of the parameters, and the highest-probability intervals in those parameters, which are called **credible intervals** in Bayesian statistics. [Cosmologists often call these intervals *confidence* intervals even when doing a Bayesian analysis, but that nomenclature should technically be used in a frequentist statistics only; see Box 10.1.]

10.3.1 Credible Intervals – an Amateur's Attempt

Once the best-fit parameters are obtained, how can one calculate the credible interval (or confidence region) around the best-fit parameters? A reasonable choice is to find a region in the n-dimensional parameter space that contains a given percentage of the probability distribution. That region would also contain the best-fit model.

For a purely Gaussian distribution in the parameters and assuming that the data covariance does not depend on the parameters $\mathbf{p}$, $P \propto \exp(-\chi^2/2)$. Consider also the best-fit model that has the maximum probability $P_{\max} \propto \exp(-\chi^2_{\min}/2)$. Then the natural choice is given by regions of constant χ^2 boundaries:[3]

$$\chi^2 - \chi^2_{\min} = -2\ln\left[\frac{P(\mathbf{p})}{P_{\max}}\right]. \tag{10.22}$$

---

[3] In identifying probability and chi-squared we have neglected the term $[(2\pi)^{n/2}|\det C|^{1/2}]^{-1}$. If the covariance does not depend on the model or model parameters, this is just a normalization factor which drops out in the probability ratio. However, in cosmology the covariance often depends on the model: this happens, for example, if our random variable X is the overdensity $\delta \equiv \delta\rho/\rho$, then its mean is zero (and clearly doesn't depend on the cosmological parameters), while the covariance of δ is the two-point correlation function $\xi(r)$ in real space (or power spectrum $P(k)$ if we are talking about $\delta = \delta_{\mathbf{k}}$ in Fourier space), and the latter quantity depends on the cosmological parameters.

For a Gaussian probability, all is well and good: for example, $\chi^2 < 1$ (< 4) specifies the 68 percent (95.4 percent) region in one dimension (see Problem 10.3 for generalizations to two and three dimensions). However, it is less obvious what to do when the posterior probability is non-Gaussian, or even multi-modal, with multiple peaks in the parameter space. In that case, the χ^2 boundaries just described correspond to different fractions of the total volume enclosed under the probability distribution. We now address this.

10.3.2 Credible Intervals – a Professional's Recipe

Our goal is to specify how to define the commonly used 68.3 percent, 95.4 percent, and 99.7 percent credible intervals in some parameter. [Exactly the same procedure holds for the credible interval for two or more parameters jointly – it is just applied to the probability in two dimensions, three dimensions, etc. – but for simplicity we keep the discussion to one parameter.] Recall that these limits are habitually called 1σ, 2σ, and 3σ ranges, even though they are really 68.3 percent, 95.4 percent, and 99.7 percent highest-posterior regions, respectively. So just remember that quoting "sigmas" is just a shorthand way of quoting these fixed percentages.

The general prescription to compute a credible interval is illustrated in Fig. 10.4 and goes as follows:

1. Find the parameter values that maximize the probability (or, loosely speaking, the maximum-likelihood parameter values). Call the associated maximum probability P_{max}.
2. Going to values smaller and larger than this value, go "down the probability" until you enclose 68.3 percent of the total – that is, find a and b such that

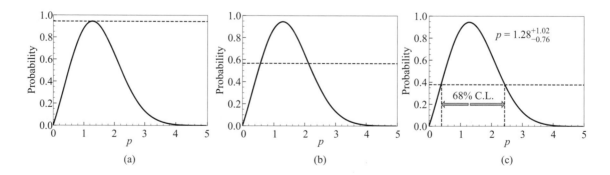

(a)   (b)   (c)

Fig. 10.4 Algorithm for how to find the 68.3 percent credible interval, illustrated on a 1D probability distribution. One starts from the maximum of the probability (a), works one's way down the probability distribution (b) until the encompassed area – and, in a higher-dimensional space, volume – is 0.683 of the total (c). This procedure guarantees that (1) higher-probability models are always accepted before those with a lower probability and (2) two models with the same probability are either both accepted to or both rejected from the desired credible interval.

$P(a) = P(b)$ and

$$\int_a^b P(p)dp = 0.683 \int_{-\infty}^{\infty} P(p)dp, \qquad (10.23)$$

where the full range of the integral on the right-hand side may be different in a particular situation (e.g., if p is the mass of a particle, it starts at zero value).

3. If desired, repeat for another credible interval (say, 95.4 percent), and find c and d such that $P(c) = P(d)$ and

$$\int_c^d P(p)dp = 0.954 \int_{-\infty}^{\infty} P(p)dp. \qquad (10.24)$$

Then we would say, for example, that the "95.4 percent credible interval for p is $[c, d]$," or else the "two-sigma credible interval for p is $[c, d]$." Note that this procedure is the same as that described in Eq. (10.22), where one finds the lowest χ^2 and considers models around the best-fitting one; the key improvement here is that the procedure described in Eqs. (10.23) and (10.24) tells us when to stop – that is, how to determine the quantitative boundaries – for an *arbitrary* probability distribution $P(\mathbf{p})$.

The procedure described above guarantees that two models with equal probability are either both accepted (to a desired credible-interval range), or else both rejected.

This easily generalizes in several directions. For multiple parameters, the credible interval will be a multi-dimensional contour that encloses 68.3 percent, 95.4 percent, etc. of the total integrated probability. And if our probability does not go to zero sufficiently fast at one end, for example the low end of p, then we will only have the upper limit on p, so the "95.4 percent credible limit on the sum of the neutrino masses is $\sum m_\nu < 0.12\,\text{eV}$" (from Planck plus external data). However, even in the one-sided limit case, the procedure is precisely the same as outlined in the steps above.

10.4 Goodness-of-Fit and Model Selection

Apart from finding the parameter values (and, e.g., their 68.3 percent ranges) of a model given some data, there is a basic question of whether the model itself is a good fit to the data. In fact, when we are fitting some model to the data, we ought to perform two separate calculations:

1. Find the best-fit values of the parameters, as well as their credible intervals.
2. Answer yes or no: is the model, evaluated at the best-fit parameter values, a good fit to the data?

If we are considering multiple theoretical models (say ΛCDM and wCDM), we also need the following:

3. Answer yes or no: for two models A and B, do the data favor model A relative to model B?

We now discuss the latter two of these questions.

10.4.1 Goodness-of-Fit with Chi-Squared

If the model is a good fit to the data, we expect the data to scatter around the model symmetrically, and that roughly 68 percent of the data points encompass the model value within their errors, and 32 percent do not. In other words, we expect that the data $\mathbf{d}$ are, "on average," 1σ away from their theoretical expectation based on the model, $\mathbf{d}^{\mathrm{model}}(\mathbf{p})$ (where σ is the error on each data point). More generally, this is codified by the simple requirement that

$$\chi^2_{k_{\mathrm{dof}}} \simeq k_{\mathrm{dof}} \qquad \text{(when model is a good fit to data)}, \qquad (10.25)$$

with $\chi^2_{k_{\mathrm{dof}}} = [\mathbf{d} - \mathbf{d}^{\mathrm{model}}(\mathbf{p})]^T \mathbf{C}^{-1} [\mathbf{d} - \mathbf{d}^{\mathrm{model}}(\mathbf{p})]$, while $\mathbf{C}$ is the covariance of the data and k_{dof} is the number of degrees of freedom:

$$k_{\mathrm{dof}} \equiv N_{\mathrm{data\ points}} - N_{\mathrm{fitted\ parameters}}. \qquad (10.26)$$

We can use the properties of the chi-squared distribution – particularly, the fact that $\mathrm{Var}(\chi^2_{k_{\mathrm{dof}}}) = 2k_{\mathrm{dof}}$ – to quantify when the fit is not good. For example, the DES Y1 key paper (Abbott et al., 2018) contains 457 measurements and 26 free parameters (6 cosmological parameters and 20 nuisance ones), leading to $k_{\mathrm{dof}} = 457 - 26 = 431$. Therefore, we expect chi-squared in the range of $\chi^2 \simeq 431 \pm \sqrt{2 \times 431} \simeq 431 \pm 30$. So $\chi^2 \gg 461$ would imply that either the model is not a good fit, or that we have residual systematic errors, or else that we underestimated the data error bars.[4]

It often happens that χ^2 is higher than that expected due to any of the aforementioned reasons. Somewhat rarely, it also happens that χ^2 is *lower* than that expected, and this typically implies that we *overestimated* the data error bars.

10.4.2 Model Selection with Chi-Squared

If we have two models with potentially different number of parameters, which one fits the data better? This question is most properly answered using Bayesian techniques that we discuss just below in Sec. 10.4.3. However, a simpler (and certainly more widely used) estimate of the answer can be obtained using chi-squared.

Let A and B be the two models. We can calculate the difference in their fit to the data as

$$\Delta\chi^2 = \chi^2_B - \chi^2_A, \qquad (10.27)$$

---

[4] Or... that something else went wrong.

where $\chi_A^2 = -2 \ln \mathcal{L}_{\mathrm{max},A}$ is directly related to the maximum likelihood of the fit of model A, and the same for model B. The qualitative sense of how the $\Delta\chi^2$ model comparison should work is quite clear. For example, if the two models have the same number of degrees of freedom (or free parameters), $k_{\mathrm{dof},A} = k_{\mathrm{dof},B}$, then we expect $\Delta\chi^2 \simeq 0$ provided the two models are about an equally good fit to the data. Conversely, if model B is a better fit to the data than model A, for example, then $\chi_B^2 < \chi_A^2$. However, one is often in a situation where one of the models has more free parameters than the other; for example, the model with a few more parameters may correspond to some extension of ΛCDM, whose proponents hope that it will be a really good fit to the data. In this case, $\Delta\chi^2$ had better be *substantially* smaller for the model with more parameters in order for it to be proclaimed as favored by the data. The challenge is to quantify this, and find the scale on which to evaluate $\Delta\chi^2$ while taking into account the difference in the number of parameters in the two models.

Things are the simplest if the two models are **nested**, that is, if one of them fully contains the other as its subspace. Let us take A to be the simpler model, and B to be the more complicated one that contains A as its subset. Then model B is guaranteed to fit the data at least as well as model A (because it contains A as a special case), so that $\Delta\chi^2 \equiv \chi_B^2 - \chi_A^2 \leq 0$. In this nested-model scenario, we can make use of **Wilks' theorem**, which says that $|\Delta\chi^2|$ between the two models can be evaluated on the chi-squared scale corresponding to the number of degrees of freedom *difference* between the two models. Schematically:

$$|\Delta\chi^2| \xrightarrow{\text{can be quantified as}} \chi^2(k_{\mathrm{dof},B} - k_{\mathrm{dof},A}) \qquad \text{(Wilks' theorem).} \qquad (10.28)$$

For example, the Planck 2018 analysis finds some intriguing evidence for nonzero curvature; their non-flat ΛCDM model (the one with Ω_k allowed to vary) has a better fit than the flat ΛCDM by $\Delta\chi^2 = -11$. Adopting the Wilks' theorem, the significance of this can be evaluated on the chi-squared scale with one degree of freedom (the one extra degree in the more complex model being Ω_k), and is equal to $\sqrt{11} \simeq 3.3$ "sigmas." [When other data are added to Planck CMB, this apparent preference for nonzero curvature disappears.]

What is to be done when the models A and B are *not* nested? The preferred procedure is again to carry out a Bayesian evidence calculation as described below. However, a quick-and-dirty estimate with $\Delta\chi^2$ is still possible. There exist several criteria that can enable this. For model comparison, perhaps the most relevant one is the **Bayesian information criterion (BIC)**, which defines the eponymous quantity as

$$\mathrm{BIC} = -2 \ln \mathcal{L}_{\mathrm{max}} + k \ln N, \qquad (10.29)$$

where k is the number of parameters to be estimated from a model, and N is the number of data points. Then the difference between the BIC of two models A and B is equal to

$$\Delta\mathrm{BIC} \equiv \mathrm{BIC}_B - \mathrm{BIC}_A = \Delta\chi^2 + (k_B - k_A) \ln N, \qquad (10.30)$$

where $\Delta\chi^2$ is defined as in Eq. (10.27). If $\Delta\text{BIC} > 0$ then model A is preferred, and vice versa if $\Delta\text{BIC} < 0$. The main feature of the BIC is that it tells us how to penalize the model with more parameters. [A closely related quantity is the **Akaike information criterion (AIC)**, defined as $\text{AIC} = -2\ln\mathcal{L}_{\max} + 2k$, without the dependence on the number of measurements N.] Both the BIC and the AIC are easily evaluated, but do not easily translate to a reliable scale on which to quantify the preference for one model in favor of the other.

We conclude that evaluating $\Delta\chi^2$ between two cosmological models is straightforward and can lead to quick insights into the relative goodness-of-fit of the two models, but is challenging to translate to a reliable scale on which to quantify which model is favored. For the latter, one should resort to the more complicated-to-implement but also more robust Bayesian model comparison, which we now describe.

10.4.3 Bayesian Model Comparison

Bayesian statistics enables hypothesis testing, and in particular comparisons of different models. For example, it can answer questions like:

- Given some data, what is the preference for models with dark energy ($\Omega_{\text{DE}} > 0$) compared to those without ($\Omega_{\text{DE}} = 0$)?
- Given some data, is my favorite two-parameter modified-gravity extension of ΛCDM actually favored compared to ΛCDM?

Let us consider two models, M_1 and M_2, that we would like to compare. We are typically interested in the ratio of the posterior probabilities, or **posterior odds**, given by

$$\frac{P(M_2|D)}{P(M_1|D)} = \frac{P(D|M_2)}{P(D|M_1)}\frac{P(M_2)}{P(M_1)} \equiv B_{21}\frac{P(M_2)}{P(M_1)}, \tag{10.31}$$

where we used Bayes' theorem to write the posterior $P(M_1|D)$ as proportional to the likelihood $P(D|M_1)$ and prior $P(M_1)$, and the same for M_2. Here the **Bayes factor** B_{21} is the ratio

$$B_{21} \equiv \frac{P(D|M_2)}{P(D|M_1)} \equiv \frac{\int P(D|\mathbf{p}_2, M_2)P(\mathbf{p}_2|M_2)\,d\mathbf{p}_2}{\int P(D|\mathbf{p}_1, M_1)P(\mathbf{p}_1|M_1)\,d\mathbf{p}_1} \quad \text{(Bayes factor)}. \tag{10.32}$$

Here, $\mathbf{p}_1$ and $\mathbf{p}_2$ are the parameters making up the models M_1 and M_2, respectively, and $P(\mathbf{p}_1|M_1)$ and $P(\mathbf{p}_2|M_2)$ are *their* prior distributions. Note also that the priors are properly normalized, so that

$$\int P(\mathbf{p}|M)\,d\mathbf{p} = 1, \tag{10.33}$$

which is important to implement when the parameter spaces $\mathbf{p}_1$ and $\mathbf{p}_2$ have different dimensionalities.

Table 10.1 Jeffreys' scale – an empirically calibrated scale for evaluating the odds in favor of one model vs. another. Here B_{21} is the Bayes factor. See text for details.

$|\ln B_{21}|$	Odds	Probability	Strength of evidence
< 1.0	$\lesssim 3:1$	< 0.750	Inconclusive
1.0	$\sim 3:1$	0.750	Weak evidence
2.5	$\sim 12:1$	0.923	Moderate evidence
5.0	$\sim 150:1$	0.993	Strong evidence

The Bayes factor B_{21} quantifies how the relative odds between the two models changed after the arrival of the data. A value $B_{21} > 1$ indicates an increase of the support in favor of model M_2 versus model M_1, and vice versa for $B_{21} < 1$. The Bayes factor is independent of the relative odds of the models *before* the data (which is given by the ratio of the models' prior probabilities). This makes the Bayes factor the relevant quantity to consider when evaluating the belief in two competing models.

How do we interpret a specific value of the Bayes factor B_{21} (beyond it being greater or smaller than one)? This is usually done using an empirically calibrated scale – the so-called Jeffreys' scale; see Table 10.1, adopted from Trotta (2008). Here we assume that one of the two models is definitely correct, so $P(M_1) + P(M_2) = 1$, and that the *prior* probabilities for the two models are equal. The Jeffreys' scale's thresholds are set at values of the odds of $3:1$, $12:1$, and $150:1$, representing weak, moderate, and strong evidence, respectively. We can think of these numbers as odds we would apply in a common-sense everyday application; for example, we are "moderately confident" to place trust in an event whose probability has odds $12:1$ compared to some alternative. Note, from Table 10.1, that the relevant quantity in using Jeffreys' scale is the *logarithm* of the Bayes factor, so that evidence only accumulates slowly, and strongly favoring a model requires a many times higher B_{21} than only moderately favoring it.

The Bayesian model comparison is typically carried out after the parameter constraints were completed, and is somewhat optional; see Fig. 10.5. For example, once we have constrained the parameters of some model (say wCDM, which allows a constant equation-of-state ratio of dark energy, w), we may want to see if there is evidence for this one-parameter extension relative to the fiducial ΛCDM model which fixes $w = -1$. To do that, we compute the Bayes factor for the two models, then make use of the Jeffreys' scale. Note that we do not need to worry about the fact that the two models being compared have a different number of parameters,

Bayesian analysis	**Frequentist analysis**
Collect data **d** and estimate their covariance **C**	Collect data **d** and estimate their covariance **C**
↓	↓
Write down likelihood $\mathscr{L}(\mathbf{d} \mid \mathbf{p})$	Write down likelihood $\mathscr{L}(\mathbf{d} \mid \mathbf{p})$
↓	↓
Choose priors and use Bayes' theorem to write down posterior $\mathcal{P}(\mathbf{d} \mid \mathbf{p})$	Find best-fit parameters (maximize $\mathscr{L}$ or use an estimator)
↓	↓
Map out posterior (e.g., using MCMC)	Test goodness-of-fit of best-fit model to data
↓	↓
Test goodness-of-fit of best-fit model to data	Set confidence intervals (e.g., using Monte-Carlo or Feldman–Cousins)
↓	↓
Report credible intervals in parameters **p** of interest	Report confidence intervals in parameters **p** of interest
↓	
[Run model selection to compare one whole model class to another]	

Fig. 10.5 Typical steps in Bayesian and frequentist analysis, from data collection to constraints on cosmological parameters and tests of the theory model.

as the definition of the Bayes factor, which incorporates the Bayesian evidence for the two models, automatically takes care of that.

This discussion concludes the recipes on how to obtain credible intervals of parameters in a model, as well as compare models. Now we come back to arguably the most challenging aspect of statistical data analysis: sampling the posterior in a multi-dimensional parameter space.

10.5 Markov Chain Monte Carlo

Say we want to constrain n cosmological parameters; let us take $n = 10$, which is typical for applications in cosmology. Say, for simplicity, that we want to allow each parameter to take v discrete values; let us take $v = 10$, which is the bare minimum we would want (since it would lead to posterior probability plots with just 10 values in each parameter, surely a rather jagged plot). Then the total number of models to explore (and calculate observables for) is $v^n = 10^{10}$, which is very large – this might be doable for a simpler dataset, but if we consider running `CAMB` to calculate the CMB power spectrum (see Chapter 13), even though it only takes seconds per model, this is about 100 years. And if we want to allow the still-modest $v = 20$ values per parameter, then likelihood calculations would take 100,000 years, which means that an early Neanderthal starting the numerical calculation on a good-quality Intel processor would make it just in time for his/her paper to be published this year.[5]

MCMC methods are an extremely powerful tool to overcome these problems.[6] MCMC are a class of algorithms for *sampling* from probability distributions based on constructing a Markov chain that has the desired distribution as its equilibrium distribution. The state of the chain after a large number of steps is then used as a sample from the desired distribution. The quality of the sample improves as a function of the number of steps. Instead of scaling exponentially with the number of parameters, the MCMC calculation scales approximately linearly with n.

It is not difficult to construct a Markov chain with the desired properties. The more difficult problem is to ensure that the results have converged to the stationary distribution within an acceptable error. We now first describe the most basic (yet ingenious) MCMC algorithm, then talk about what other ingredients are needed to ensure rapid "mixing" of MCMC, so that the desired posterior is mapped out as efficiently as possible.

10.5.1 MCMC: The Science

We will only consider the **Metropolis–Hastings** algorithm here, which is the simplest and most famous variant of MCMC. Those of you interested in parameter analyses are advised to take a look at the Gibbs sampler, Hamiltonian Monte-Carlo, Polychord, and other MCMC (or similarly inspired) algorithms.

The Metropolis–Hastings algorithm draws samples from the likelihood (or probability distribution) $\mathcal{L}(\mathbf{p})$. How does it do that? The algorithm generates a Markov chain where the state $\mathbf{p}_{t+1}$ at the next step $t+1$ depends only on the parameter values $\mathbf{p}_t$ at the current step t. The algorithm uses a proposal density $Q(\mathbf{p}'|\mathbf{p}_t)$, which depends on the current state $\mathbf{p}_t$, to generate a new proposed sample $\mathbf{p}'$. This proposal is either accepted as the next value (so, $\mathbf{p}_{t+1} = \mathbf{p}'$) or rejected (so, $\mathbf{p}_{t+1} = \mathbf{p}_t$) according to the following rule:

- Calculate the ratio of the posterior probability evaluated at the proposed parameter values to that at the current point in parameter space:

$$r \equiv \frac{P(\mathbf{p}')}{P(\mathbf{p}_t)}. \tag{10.34}$$

- If $r > 1$ (i.e., if the proposed point's probability is greater than the current), then ACCEPT the move to the new point: $\mathbf{p}_{t+1} = \mathbf{p}'$.
- If $r < 1$ (i.e., if the proposed point's probability is smaller than the current), then draw a random number α from a uniform distribution $U[0, 1]$.
  - If $\alpha < r$, then ACCEPT the move to the proposed point: $\mathbf{p}_{t+1} = \mathbf{p}'$.

---

[5] Although if the Neanderthal has resources to perform parallel computing, then he/she can start the calculation at a later time, for example during overlap with Cro-Magnons.

[6] A **Markov process** (or a **Markov chain**) is the probability of the (first subsequent) future state only depending on the present state, but not on the past states.

– If $\alpha > r$, then REJECT the move and do not move to the proposed point: $\mathbf{p}_{t+1} = \mathbf{p}_t$.

One can mathematically prove that an algorithm based on this rule converges to the desired true posterior distribution $P(\mathbf{p}|\mathbf{d})$. Figure 10.6 schematically shows the Metropolis–Hastings algorithm, as well as a visual illustration of rejected/accepted steps for a 1D probability. Figure 10.7 shows a 2D example, along with several attempted steps and their respective rejections or acceptances.

While the MCMC algorithm is running, we compile the *weight* at each point in parameter space – the number of times our proposal was rejected and we did not move to a new model. For example, if we arrived at a new point in the parameter space and it was followed by a reject–reject–accept, then we spend a total of three steps there so weight = 3. As mentioned above, the weight is proportional to the true posterior probability that we would like to recover. Typically, the number of steps we need for convergence scales linearly with n_{params}, and evaluates to something like 10,000 or 100,000. This sounds large but is very modest compared to the exponential number (scaling as $e^{n_{\mathrm{params}}}$) that we would need with a brute-force grid exploration of the parameter space.

Let us finally try to clear up a potential confusion here: why do we need MCMC when we *can* readily evaluate the likelihood (and the posterior, given that we have the prior) at any given parameter point $\mathbf{p}$? The answer: it is all about the efficiency of sampling the huge parameter space. In fact:

▸ choose proposal function Q(p' |p$_t$)
▸ at step t, at some parameters p$_t$
▸ propose move to p' =p$_t$+$\varDelta$p$_t$ (draw $\varDelta$p$_t$ from Q)
▸ evaluate r = P(p')/P(p$_t$)
▸ MH step:
  ▸ if r > 1 accept move
  ▸ if r < 1 generate random number $\alpha \in [0, 1]$
    ▸ if α < r, accept move
    ▸ if α > r, reject move
▸ t=t+1
▸ end when MCMC has converged

(a)

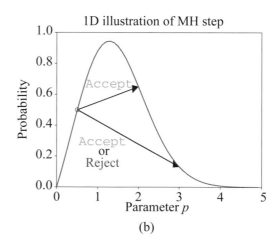

(b)

Fig. 10.6 Pseudo-code for the Metropolis–Hastings MCMC algorithm (a), and illustration of how the model always takes the proposed step in parameter space if the new probability is higher, but only sometimes if it is lower (b).

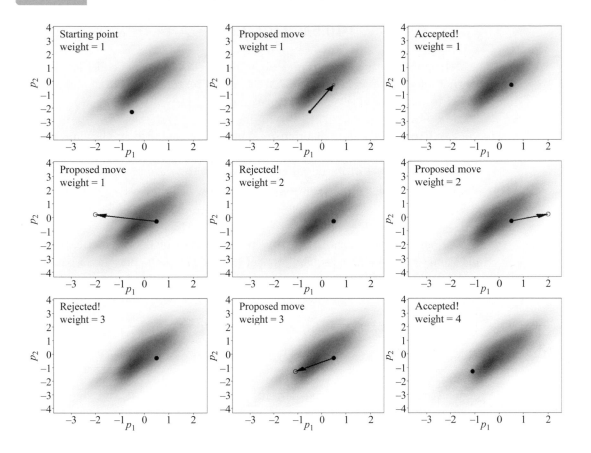

Fig. 10.7 Illustration of a series of MCMC steps in the Metropolis–Hastings algorithm. The sequence of proposed steps proceeds as illustrated in the panels going from left to right (and top to bottom across three rows). In this example, the first proposed step is accepted, the next two are rejected, and the last one is again accepted. At any given point in parameter space, the number of rejections for proposed steps is reflected in the assigned weight (equal to the number of rejections plus one). The weight of points in parameter space is, at the end of the MCMC walk, identified with the posterior probability of the model corresponding to those parameters.

> MCMC enables us to cleverly sample, or "walk through," the parameter space.

In this way, we effectively map out the shape of the posterior much faster than we could ever do by some other way (say gridding, or else random sampling) of the posterior. This efficiency becomes all the more critical given that we wish to find not just the best-fitting model but also *uncertainties* in our n_{params}-dimensional space, so that we really need to map out the full shape of the posterior.

10.5.2 MCMC: The Art

The Metropolis–Hastings algorithm is about ten lines of computer code; see Fig. 10.6. So is that all we need to implement?? Well, not quite: to make a successful MCMC, we need to take care of several additional things.

1. We need to assure that the **burn-in** stage is not included in the final results. Burn-in refers to some initial number of steps where MCMC is in the process of "losing memory" of its initial conditions (notably, *where* on the likelihood it started). Addressing this typically means running the MCMC for a number of steps (say, 10,000), discarding these burn-in results, and then doing a "production run" (with, say, 100,000 steps).

2. The parameter excursions, $\Delta \mathbf{p}_t \equiv \mathbf{p}' - \mathbf{p}_t$, should not be deterministic, but equal to estimated 1σ errors (or square roots of eigenvalues of the parameter covariance matrix) *times* the Gaussian normal variable of mean zero and variance one. That is, the excursion in the ith parameter at step t is

$$\Delta(p_i)_t = \sigma_i^{(\text{est})} \times \mathcal{N}(0, 1), \tag{10.35}$$

where $\mathcal{N}(0, 1)$ is a Gaussian (normal) variate with zero mean and unit variance.

3. The **acceptance rate** of the chain should not be too high or too low. An overly high acceptance rate (say, 90 percent) implies that the MCMC steps are too small so that the chain is exploring a relatively flat part of the likelihood – in other words, the samples are too highly correlated. An overly low acceptance rate (say 2 percent) implies that the steps are too large, and the chain is trying to jump to parameter values where the likelihood is overwhelmingly lower than at the present point in parameter space; these steps are thus being overwhelmingly rejected. One can show that the ideal acceptance rate is roughly $1/3$.

4. We need to ensure that the MCMC is **efficient** – ideally, it will move from $\mathbf{p}_t$ to the proposal value $\mathbf{p}'$ about one-third of the time, as mentioned just above. Inefficiencies, however, lurk everywhere. Consider, for example, the scenario with two highly degenerate parameters – say, Ω_M and h in CMB measurements where only the combination $\Omega_M h^2$ is well determined. Say we use an otherwise reasonable proposal distribution which is a multivariate Gaussian with standard deviation equal to the forecasted projected error in each of the two parameters, but without correlation between them. Then mutually uncorrelated steps in Ω_M and h will lead to a rejection of the proposed steps the vast majority of the time, as the steps will overwhelmingly lead to parameter points outside of the skinny banana-shaped region of high likelihood in the $\Omega_M - h$ plane. However, if we are clever and reparameterize the problem so that we have $\Omega_M h^2$ as one parameter, and Ω_M as the other, then the acceptance will be much higher, and the asymptotic distribution will be reached sooner. This is because the latter choice will generate two parameters that are less correlated, and thus the MCMC steps will not be missing the high-likelihood part of their parameter space as often. Summariz-

ing, it is often beneficial to make the proposal function include any correlation between different parameters, with:

- directions specified by eigenvalues of the covariance matrix in the chosen parameters (which can be pre-computed with, for example, a short MCMC run); and

- parameter excursions equal to approximately 1σ steps along the eigenvectors – that is, steps are equal to (square roots of) the eigenvalues of the covariance matrix.

5. Finally, we need to ensure the **mixing** of our chain. To do so, we can **thin** the chain, writing out every 100th (for example) parameter value $\mathbf{p}_t$, so that we decrease the (otherwise very high) correlation between the steps. [In practice, one typically first calculates the correlation length of a finished chain, then thins.] Likewise, we should run several (say, four) chains, and test convergence using one of the criteria (say, the **Gelman–Rubin criterion**) that typically compare variance within a chain with variance between different chains.

Phew, that was a lot of things to consider for what at first looked like a completely straightforward task of implementing the MCMC. Satisfying the conditions above requires tweaking the code, and adjusting it to the specificities of our problem at hand. This is why it is sometimes facetiously said that MCMC is as much art as science.

10.5.3 MCMC: Enjoying the Fruits of Labor

MCMC is a fantastic tool, enabling efficient exploration and mapping of the posterior probability in a multi-dimensional parameter space – something that would be entirely unfeasible using a naive gridding of the parameter space.

To analyze the results from the MCMC, what we need to do is write out the chains, together with the weight (the number of times the chain is "stuck" at that value after the proposed move was rejected) for each step; see Fig. 10.8. Then, for any parameter set of choice – a single parameter (e.g., Ω_M), or joint contour of two parameters (e.g., (Ω_M, w)) – we rank-order their weights, and add them until we get 68.3 percent or 95.4 percent, or any other desired fraction of the total weight. Remember that the weight is proportional to the posterior probability of the corresponding parameter(s).

Moreover, computing constraints on any *derived* quantities of interest, once we have run our chains, is trivial as it can be done with post-processing of the MCMC output. Say, for example, we would like to constrain the age of the universe, which in a flat ΛCDM model is a function of two parameters, $t_0 = t_0(\Omega_M, h)$. In that case, we simply add another column to the MCMC output shown in Fig. 10.8 with the age t_0 calculated for the parameter values in that row. We then obtain the constraints on t_0 in exactly the same way as we did for any of the other cosmological parameters in the MCMC output.

Weight	p_1	p_2	p_3	.....	p_N
5	0.2	−0.3	0.15	.....	2.8
1	−0.7	0.4	0.13	.....	3.5
12	0.7	0.1	0.19	.....	1.7
.....	.....	.....	.....	.....	.....

(∼ million rows)

Fig. 10.8 Illustration of an MCMC output for a Metropolis–Hastings sampler. Each row lists a single model that was accepted at least once in the MCMC walk. The weight entry shows the number of attempted steps that the sampler waited at that model. For example if, after arriving at the model, the sampler rejected moving to a new model three times before moving on, then weight $= 4$. The weight is proportional to the posterior of each model. See text for other details.

Finally, let us say that, after this hard work, we decide we would like to combine our constraints with some other data. That is easy – we just use constraints from our chain, and use them as a *prior* to be combined with new data to produce the new posterior.

Since the beginning of the twenty-first century, MCMC has become an indispensable, dominant tool in the cosmological-parameter analyses, and a number of specialized software packages have been developed in this regard. In Problem 10.7, I ask you to write your own MCMC from scratch. Do it, it's fun!

10.6 Fisher Information Matrix: Forecasting the Errors

We have just seen that, given some data **d** and their relation to theory, one can obtain the constraints on the theory parameters **p** using MCMC. While in principle relatively straightforward, building the whole machinery including MCMC is still a serious undertaking, and even using publicly available MCMC software packages requires implementing your functions, checking for convergence, and (often) using substantial computing resources. Is there a simpler approach if we are only interested in an approximate *forecast* of the constraints on theory parameters given some data, and do not particularly care about the central values of **p** but rather their errors (and parameter error covariance), given data **d** and their covariance?

Yes there is, and it is called the **Fisher matrix**! Consider again the picture in Fig. 10.3, with the data space **d** and parameter space **p**. The question is: given data $\mathbf{d}$ of some quality (i.e., consistent with some underlying model, and with some data covariance matrix $\mathbf{C}$), what are the errors on the cosmological parameters $\mathbf{p}$? An MCMC-type computation could in principle give us an answer, but it requires effort and time to set it up. However, if we are only interested in an approximate answer, then the Fisher matrix calculation provides an essentially instantaneous result with no stochastic noise. We now explain how it works.

10.6.1 Fisher Matrix Definition

Let us start from the data likelihood $\mathcal{L}(\mathbf{d}|\mathbf{p})$. The Fisher matrix is formally defined as the curvature of the likelihood in model parameters $\mathbf{p}$ – that is, the matrix of second derivatives of the log likelihood around the peak:

$$
\ln \mathcal{L} = \ln \mathcal{L}|_{\max} + \left.\frac{\partial \ln \mathcal{L}}{\partial p_i}\right|_{\max} (p_i - \bar{p}_i) + \frac{1}{2} \left.\frac{\partial^2 \ln \mathcal{L}}{\partial p_i \partial p_j}\right|_{\max} (p_i - \bar{p}_i)(p_j - \bar{p}_j) + \cdots
$$

$$
= \ln \mathcal{L}|_{\max} + \frac{1}{2} \left.\frac{\partial^2 \ln \mathcal{L}}{\partial p_i \partial p_j}\right|_{\max} (p_i - \bar{p}_i)(p_j - \bar{p}_j) + \cdots , \tag{10.36}
$$

where the summations over repeated indices are implied. The linear term vanishes, since the derivative at the maximum likelihood is zero. The Fisher matrix is defined as the negative of the second derivative term (i.e., the **Hessian**) of the log likelihood:

$$
F_{ij} = \left\langle -\frac{\partial^2 \ln \mathcal{L}}{\partial p_i \partial p_j} \right\rangle , \tag{10.37}
$$

where $\mathbf{p} \equiv \{p_i\}$ is the set of cosmological parameters as usual. In other words, the Fisher matrix is defined as a (negative) curvature around the peak of the likelihood. The higher (by absolute value) the curvature, the better the parameters are determined, and the more information ("Fisher information") is available in the data regarding cosmological parameters. Moreover:

> The Fisher matrix F is an approximation for the inverse covariance matrix of parameters $\mathbf{p}$:

$$
F_{ij} \simeq [\text{Cov}(p_i, p_j)]^{-1} . \tag{10.38}
$$

Therefore, as long as we can calculate F, we have the errors and covariances of the cosmological parameters of interest. We now turn to the practical evaluation of the Fisher matrix.

In the Fisher matrix formalism, we always assume that the probability distribution of both the data $\mathbf{d}$ and of the cosmological parameters $\mathbf{p}$ are each distributed as a multivariate Gaussian. With the latter assumption, it follows that the mean $\boldsymbol{\mu}$ and covariance $\mathbf{C}$ of the data contain all information:

$$\mathcal{L} = \frac{1}{(2\pi)^{n/2}|\det \mathbf{C}|^{1/2}} \exp\left[-\frac{1}{2}(\boldsymbol{d} - \boldsymbol{\mu})^T \mathbf{C}^{-1}(\boldsymbol{d} - \boldsymbol{\mu})\right], \tag{10.39}$$

where $\mathbf{d}$ are the data and $\boldsymbol{\mu}$ are the theoretical observable quantities evaluated at the parameter values $\mathbf{p}$ at which we assume the Fisher contour is centered (e.g., $\mathbf{p} = \{\Omega_M = 0.3, \Omega_\Lambda = 0.7\}$). One can show that the Fisher matrix explicitly evaluates to

$$F_{ij} = \boldsymbol{\mu}_{,i}^T \mathbf{C}^{-1} \boldsymbol{\mu}_{,j} + \frac{1}{2}\mathrm{Tr}[\mathbf{C}^{-1}\mathbf{C}_{,i}\mathbf{C}^{-1}\mathbf{C}_{,j}], \tag{10.40}$$

where, i is partial derivative with respect to p_i.

Equation (10.40) is the master expression for the Fisher matrix. Note that both $\boldsymbol{\mu}$ and $\mathbf{C}$ are the theoretically calculated quantities; we need them, and their respective partial derivatives (usually calculated numerically) with respect to the parameters $\mathbf{p}$. If the mean $\boldsymbol{\mu}$ of the data depends on the parameters $\mathbf{p}$, then the first term in Eq. (10.40) is nonzero. This is the simplest case and occurs, for example, when the mean depends on the theory but the errors (or elements of the full covariance matrix, $\mathbf{C}$) take some values that potentially depend on the data or experimental conditions, but not the underlying theoretical model. If, on the other hand, the data covariance $\mathbf{C}$ depends on the parameters $\mathbf{p}$, then the second term in Eq. (10.40) contributes; this happens, for example, in galaxy-clustering measurements when the mean is $\langle\delta\rangle = 0$ (so the first term is zero), but its covariance – the correlation function – depends on the cosmological parameters.

In summary: if we (1) assume the distribution in the cosmological parameters $\mathbf{p}$ is multivariate Gaussian, (2) know how the data is related to model parameters, and (3) know errors in the data (or their full covariance, if applicable), then we can forecast the expected errors in the cosmological parameters and *their* full covariance. The Fisher matrix also requires selecting the fiducial model (i.e., central values of the parameters $\mathbf{p}$); errors on the cosmological parameters typically depend only weakly on the precise choice of the fiducial model.

10.6.2 Fisher Estimates of Parameter Errors

Most of the time, Fisher matrix users rely on the **Cramer–Rao bound**, which says that an error in a cosmological parameter p_i will be greater than or equal to the corresponding Fisher matrix element:

$$\sigma(p_i) \geq \begin{cases} \sqrt{(F^{-1})_{ii}} & \text{(marginalized)} \\ 1/\sqrt{F_{ii}} & \text{(unmarginalized)}, \end{cases} \tag{10.41}$$

where "marginalized" is the uncertainty integrated (marginalized) over all other $n - 1$ parameters, while the "unmarginalized" case is when we ignore the other parameters, assuming them effectively fixed and known. The marginalized case has inverse of F, which lets the parameters "inform each other" about mutual degeneracies. Most often in cosmology we are interested in the marginalized errors; the unmarginalized ones are often much smaller and correspond to an unrealistic case when we somehow independently know the values of all the other parameters.

While the Cramer–Rao bound just tells us about the *best possible* error, we usually assume that it forecasts the *actual* error on the cosmological parameters of interest given data of some quality.

10.6.3 Fisher Matrix Examples

Let us give some examples of the expressions for probe-specific Fisher matrices.

For type Ia supernova observations, the covariance matrix of SNe doesn't depend on cosmological parameters, and can further be assumed to be constant. To provide a simple pedagogical example, let us further assume that the covariance is diagonal, $C_{ij} \to \sigma_m^2 \delta_{ij}$, with $\sigma_m \sim 0.15$ mag. Then Eq. (10.40) evaluates to

$$F_{ij}^{\mathrm{SNe}} = \sum_{n=1}^{N_{\mathrm{SNe}}} \frac{1}{\sigma_m^2} \frac{\partial m(z_n)}{\partial p_i} \frac{\partial m(z_n)}{\partial p_j}, \tag{10.42}$$

where $m(z) = m(z, \Omega_M, \Omega_\Lambda, \mathcal{M}...)$ is the theoretically expected apparent magnitude; see Chapter 12. Notice that, if we had a full off-diagonal covariance matrix C_{ij} which is still independent of the cosmological parameters, the equation above would generalize trivially to $F_{ij} = (\partial \mathbf{m}/\partial p_i)^T \mathbf{Cov}^{-1}(\partial \mathbf{m}/\partial p_j)$, with matrix multiplication implied.

Let us next consider the example of counting the clusters of galaxies in mass and redshift, and comparing it to theory. The cluster-count Fisher matrix can be calculated as follows. Let N_k be the number of clusters in the kth bin (e.g., in flux) and O_k be an observable (say, X-ray flux) in that same bin. Then

$$F_{ij}^{\mathrm{clus}} = \sum_{k=1}^{k_{\max}} \frac{N_k}{\sigma_{O,k}^2} \frac{\partial O_k}{\partial p_i} \frac{\partial O_k}{\partial p_j}, \tag{10.43}$$

where $\sigma_{O,k}$ is the error in the kth measurement of the observable O.

Finally consider the case of measurements of the CMB angular power spectrum C_ℓ that we will discuss at length in Chapter 13. The mean temperature of the CMB is zero and anyway doesn't depend on cosmology, but the covariance of temperature, encoded in harmonic space with coefficients C_ℓ where ℓ are the multipoles, carries all cosmological information. Then the Fisher matrix simplifies to

$$F_{ij}^{\mathrm{CMB}} = \sum_\ell \frac{\partial C_\ell}{\partial p_i} \mathbf{Cov}^{-1} \frac{\partial C_\ell}{\partial p_j}, \tag{10.44}$$

where $\mathbf{Cov}^{-1}$ is the inverse of the covariance matrix between the observed power spectra whose elements are given by Wick's theorem (see Box 13.3 in Chapter 13):

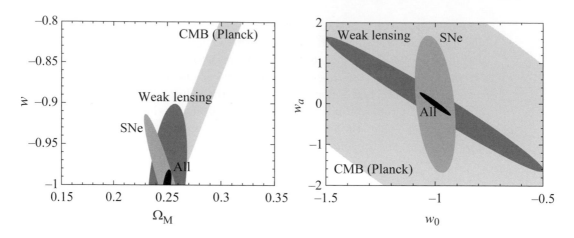

Fig. 10.9 Illustration of forecast constraints on dark-energy parameters, adopted from the Frieman *et al.* (2008) review of dark energy. All contours have been computed using the Fisher matrix formalism.

$$\mathrm{Cov}\left[C_\ell, C_\ell'\right] = \frac{2C_\ell^2}{(2\ell + 1)f_{\mathrm{sky}}\Delta\ell}\,\delta_{\ell\ell'}, \qquad (10.45)$$

where $\mathrm{Cov}\left[C_\ell, C_\ell'\right]$ is the covariance of the convergence power spectra at multipoles ℓ and ℓ' (clearly being uncorrelated if $\ell \neq \ell'$), $\Delta\ell$ is the width of the multipole bin, and f_{sky} is the fraction of the sky covered by observations.

In Fig. 10.9, we show Fisher matrix forecasts (made a while ago, before Planck and some other recent surveys) for CMB, weak lensing, and type Ia supernova measurements. Note that the contours are ellipses by construction, implied by the multivariate Gaussian distribution in the cosmological parameters that the Fisher matrix assumes. We now discuss how to calculate and plot these contours in practice.

10.6.4 Marginalization over Parameters

If we have n cosmological parameters, how do we marginalize over $n - m$ of them to be left with a desired joint constraint on m parameters? In the Fisher formalism, this is easy:

- Calculate the full $n \times n$ matrix F.
- Invert it to get F^{-1}.
- Take the desired $m \times m$ subset of F^{-1}, and call it G^{-1}.
- Invert G^{-1} to get G.

And *voilà* – the matrix G is the Fisher matrix projected on the m-dimensional space. Notice that the step of inverting F assures that the parameters "talk to each other," effectively accounting for the marginalization (and increasing the error bars significantly compared to the unmarginalized error).

Most often, the total parameter space is larger than the 2D space in which we are plotting the constraints. In that case, it is the projected Fisher matrix G that is to be used when plotting the confidence contours, and when quoting the figure of merit, both of which we talk about next.

10.6.5 Fisher Ellipses

How do we plot the Fisher matrix contour such as that in Fig. 10.9? To plot a 2D parameter-error ellipse, we first want to project down to that space, and be left with a marginalized 2×2 Fisher matrix G. Let $\bar{p}_1$ and $\bar{p}_2$ be the mean (fiducial) values of each parameter. Then the equation for the 2D ellipse is

$$G_{11}(p_1 - \bar{p}_1)^2 + 2G_{12}(p_1 - \bar{p}_1)(p_2 - \bar{p}_2) + G_{22}(p_2 - \bar{p}_2)^2 = \Delta\chi^2_{2-\text{dof}}, \qquad (10.46)$$

where, for two parameters (i.e., for a 2D ellipse), $\Delta\chi^2_{2-\text{dof}} = 2.3$ for a 68.3 percent credible-interval ellipse, and $\Delta\chi^2_{2-\text{dof}} = 6.2$ for a 95.4 percent ellipse (see Problem 10.3). More generally, the equation of the n-dimensional ellipsoid (i.e., with n degrees of freedom) would be

$$(\mathbf{p} - \bar{\mathbf{p}})^T F (\mathbf{p} - \bar{\mathbf{p}}) = \Delta\chi^2_{n-\text{dof}} \quad \text{(equation of Fisher ellipsoid)}, \qquad (10.47)$$

where F is naturally to be replaced by the corresponding projected matrix G if we are plotting the ellipsoid in a lower-dimensional subspace.

An example of the 68.3 percent uncertainty Fisher ellipse in two dimensions, along with the approximate marginalized errors in each parameter, is shown in Fig. 10.10. Note the rule of thumb that the projected 1D errors are slightly smaller than the size of the projection of the 2D ellipse on each axis; this is simply due to the mathematics of the 2D → 1D projection for a Gaussian distribution.

Another useful quantity is the volume of the Fisher ellipsoid, which quantifies how much information the data contain on the parameters of interest. The smaller the volume, the more information there is. More generally, the volume of an n-dimensional ellipsoid is related to the determinant of the Fisher matrix:

$$\text{Volume} \propto (\det F)^{-1/2} \quad \text{(volume of Fisher ellipsoid)}. \qquad (10.48)$$

Roughly speaking, the larger the Fisher *information* matrix is, the smaller the volume. [The off-diagonals of F, corresponding to parameter correlations, generally work toward making $(\det F)$ smaller, and hence to less information.] Specifically, in two dimensions, the area of the 68.3 percent ellipse is given by $A = 2.3\pi/(\det G)^{1/2}$, where G is the Fisher matrix projected on this subspace. For the 95.4 percent ellipse, the relation is the same except the coefficient changes from 2.3 to 6.2, corresponding to the "2 sigma" chi-squared in two dimensions (see Problem 10.3).

The inverse area of the Fisher contour is sometimes used as a figure of merit, a number that quantifies how accurately we are measuring two parameters of interest (e.g., the dark-energy equation-of-state parameters w_0 and w_a, defined in

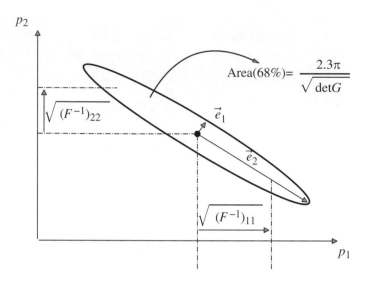

Fig. 10.10 Cosmological constraints on two parameters, p_1 and p_2, in the Fisher matrix calculation (i.e., assuming a Gaussian parameter likelihood). The "1 sigma" (68.3 percent) contour is obtained using G, the 2×2 Fisher matrix projected on this subspace (see text). We also show the marginalized errors in the two parameters, $(F^{-1})_{ii}^{1/2}$ for $i = 1, 2$; note that these are somewhat smaller than the extent of the 2D ellipse due to projection. The directions $\mathbf{e}_1$ and $\mathbf{e}_2$, calculated as the eigenvectors of the projected Fisher matrix G, represent respectively the well-determined and poorly determined linear combination of p_1 and p_2. The area of the Fisher contour is inversely proportional to the square root of the determinant of the projected Fisher matrix G.

Chapter 12). The smaller the area, the larger the figure of merit. The figure of merit (FoM) is then given in terms of the projected 2×2 Fisher matrix G as

$$\text{FoM} \propto (\det G)^{1/2}. \tag{10.49}$$

The constant of proportionality is unimportant, as one is usually interested in the ratios of the FoMs between different experiments, different experimental configurations, or different cosmological probes. This figure of merit easily generalizes to larger parameter spaces simply by using the projected matrix G corresponding to the desired parameter space.

10.6.6 Fisher Bias

Another neat application of the Fisher matrix is to calculate the bias in parameters p_i given biases in the observables.

As an example, consider type Ia supernova measurements of apparent magnitudes $m(z)$, whose theoretical prediction depends on the cosmological parameters $\{p_j\}$. Assuming constant covariance with diagonal errors σ_m, the bias in the parameters takes the form

$$\delta p_i \approx F_{ij}^{-1} \sum_n \frac{1}{\sigma_m^2} \left[m(z_n) - \bar{m}(z_n) \right] \frac{\partial \bar{m}(z_n)}{\partial p_j}, \tag{10.50}$$

where $[m(z_n) - \bar{m}(z_n)]$ is the bias in the observed apparent magnitudes, and the sum over the repeated index j is implied. [See Problem 10.4 for the derivation of the Fisher bias formula in a similar scenario.]

Next consider weak gravitational lensing, which we will cover in detail in Chapter 14. Assuming measurements of the convergence power spectrum $C_\alpha^\kappa(\ell)$ in tomographic redshift bins labeled by α and at multipole ℓ, the bias in the cosmological parameters is

$$\delta p_i \approx F_{ij}^{-1} \sum_\ell \left[C_\alpha^\kappa(\ell) - \bar{C}_\alpha^\kappa(\ell) \right] \text{Cov}^{-1} \left[\bar{C}_\alpha^\kappa(\ell), \bar{C}_\beta^\kappa(\ell) \right] \frac{\partial \bar{C}_\beta^\kappa(\ell)}{\partial p_j}, \tag{10.51}$$

where $[C_\alpha^\kappa(\ell) - \bar{C}_\alpha^\kappa(\ell)]$ is the bias in the observed (or theoretically modeled) shear covariance, and the sum over j is implied.

The Fisher bias formula is extremely useful to gauge what effect on the cosmological parameter errors an arbitrary shift in the observed *or theoretical* quantities makes. Given arbitrary biases in the observed quantities, for example $[C_\alpha^\kappa(\ell) - \bar{C}_\alpha^\kappa(\ell)]$ in Eq. (10.51), we can find biases in the cosmological parameters δp_i. For example, we can ask what the effect on the derived parameters $\mathbf{p}$ is if we assume an additional systematic uncertainty that shifts the observed quantities by a small amount. Or, we can ask the same question if we change the underlying *theory* model, as that too will lead to a difference in the (observed minus theoretical) term. Because the Fisher bias formula contains no stochastic noise (i.e., the calculation is fully "theoretical"), we can obtain accurate answers in the limit when the observable shift, and hence the parameter shifts, are very small, something that would be challenging and much more time-consuming to achieve using the standard method of running Monte-Carlo simulations or MCMCs.

Bibliographical Notes

There are lots of readings about the application of statistics to science, physics, and astrophysics, but they tend to be either very basic or very technical. There is no one source that contains everything that you need to carry out basic research in cosmology, though a few of the references listed below come close.

Lupton (1993) provides a range of useful results and simple worked examples. For Bayesian statistics, Roberto Trotta's "Bayes in the sky" (Trotta, 2008) review article is an essential read, as is the very useful article by Padilla *et al.* (2021). For frequentist statistics, the original article by Feldman and Cousins (1998) that introduced their eponymous frequentist method (see Box 10.1) is very readable. To clarify the difference between the Bayesian and frequentist approaches, I have also found the article by Efstathiou (2003), where he performs a simple CMB analysis using both the frequentist and the Bayesian methodology, very helpful. For the

Fisher matrix, there is no single, comprehensive read, though an influential early paper by Tegmark *et al.* (1997) is a good starting point. Finally, the *Numerical Recipes* book (Press *et al.*, 2007) provides a multitude of computational techniques, and a good amount of statistics along the way; it is highly recommended.

Problems

10.1 Variance of the mean. Assume N random variables X_i ($i = 1, 2, \ldots, N$) that are uncorrelated and where each has the same variance of σ^2. Consider the (best unbiased, as discussed in the text) estimator of their mean

$$\hat{\mu} = \frac{\sum_{i=1}^{N} X_i}{N}.$$

Find the variance of $\hat{\mu}$.

10.2 Two correlated random variables. Consider two random variables with zero means ($\mu_1 = \mu_2 = 0$) and covariance matrix given by

$$C = \begin{pmatrix} 1 & \epsilon \\ \epsilon & 1 \end{pmatrix},$$

so that the variances in the two parameters are 1, while $\mathrm{Cov}(X_1, X_2) = \epsilon$. In this problem do *not* assume that $|\epsilon| \ll 1$. Assume that the joint likelihood in X_1 and X_2 is multivariate Gaussian.

(a) Write down the full expression for the joint likelihood $P(X_1, X_2)$ including the correct normalization. Be sure to get an expression which is a function of X_1, X_2, and ϵ.

(b) Sketch the probability $P(X_1, X_2)$ in the X_1–X_2 plane; imagine you are doing it for some arbitrary credible interval. Point out the most obvious features: what is the tilt of the probability contour relative to the X_1-axis (and why); what determines its elongation; and how does the contour look in the limit $\epsilon \to 1$ or $\epsilon \to -1$.

(c) What is the standard deviation in X_1, marginalized over X_2? That is, what is the standard deviation of $P(X_1)$? *Note:* You can either explicitly prove this, or simply evaluate it by inspection of what is given in the problem. Or both.

(d) Find analytically the two *uncorrelated* linear (parameter) combinations of X_1 and X_2; make sure they are properly normalized. What are the standard deviations of these two (eigen-)parameters? Does your result make intuitive sense, and why? [You may assume in this part that $\epsilon > 0$ for simplicity.]

10.3 [Computational] $\Delta\chi^2$ **values for "famous" credible intervals.** Calculate numerically, and round to one decimal place and report, the values of $\Delta\chi^2$ for the 68.27 percent, 95.45 percent, and 99.73 percent credible intervals for the cases of $n = 1$, $n = 2$, and $n = 3$ parameters. Recall that the above-mentioned percentage values are chosen so that they represent the fractional area under a one-parameter Gaussian distribution spanned by $\pm 1\sigma$, $\pm 2\sigma$, and $\pm 3\sigma$ ranges around the mean, respectively.

To help you think about this, I supply a table, with three "easy" entries already filled in. Whatever approach you use to fill out the rest of the table, explain in detail how you arrived at your answer.

Probability	$\Delta\chi^2$ values		
	1-par	2-par	3-par
68.27%	1.0	?	?
95.45%	4.0	?	?
99.73%	9.0	?	?

10.4 Bias from the Fisher matrix formalism. In the Fisher matrix formalism (i.e., assuming the parameter likelihood is a multivariate Gaussian), one can estimate the bias in the parameters, given bias in the observable quantity and the Fisher matrix. This is extremely useful. For example, you can very easily answer common questions like "if my observations/measurements are off by such-and-such, how much bias in the inferred values of the cosmological parameters does that introduce?"

Derive a *linearized* formula (i.e., valid for small biases in a sense that they are much smaller than the statistical error) for the bias in the ith cosmological parameter, δp_i, given the bias in the observed quantities, δm_j. [For notational simplicity, I am specializing in SN observations, where apparent magnitudes $m_j \equiv m(z_j)$ are the observable.] Your formula should look something like this:

$$\delta p_i = \sum_j (\text{some terms})\, \delta m_j,$$

which you can alternatively write in a vector form too. *Hints:* Remember that the (minus two log) likelihood in the parameters $\mathbf{p}$ is $-2\ln\mathcal{L} = (\mathbf{p} - \bar{\mathbf{p}})^T F(\mathbf{p} - \bar{\mathbf{p}})$. Moreover, the likelihood in the observables $\mathbf{m}$ is $-2\ln\mathcal{L} = (\mathbf{m} - \bar{\mathbf{m}})^T C^{-1}(\mathbf{m} - \bar{\mathbf{m}})$. Perturb both expressions to first order in $\mathbf{p}$ and $\mathbf{m}$ and equate them. You will need another small trick.

10.5 [Computational] Frequentist and Bayesian analysis. In this problem, you will practice solving a rather simple data-fitting problem using several different approaches. Assume the following dataset (x_i, y_i):

$$y(x = 1) = 2.0 \pm 1.0$$
$$y(x = 2) = 4.0 \pm 2.0$$
$$y(x = 3) = 6.0 \pm 3.0$$
$$y(x = 4) = 8.0 \pm 4.0,$$

where you can assume that these four measurements are independent and identically distributed, and that each comes from a Gaussian distribution. You may also assume that all sources of uncertainty are captured in the statistical errors on the measurements quoted above. Then consider a deterministic model with a single free parameter m, defined as

$$y = mx.$$

Your task is to find the best-fit value of m, as well as its 68.3 percent confidence/credible interval in a frequentist/Bayesian analysis.

(a) Write down the (Gaussian) likelihood for m. Then find the expression for the maximum-likelihood estimate (MLE), $\widehat{m}$, in terms of x- and y-values of the data and errors on the latter, x_i, y_i, and σ_i. Evaluate this to find $\widehat{m}$ given the data above. *Hint:* Maximizing $\mathcal{L}$ is equivalent to minimizing χ^2, and the latter is easier.
Optionally, find also the MLE for the *variance* of m, and evaluate it.

(b) **Frequentist analysis**. To evaluate the error in the MLE value, perform a simple Monte-Carlo simulation. Simulate your data: draw values of y from Gaussian normal distributions given the means and errors as given by the data, that is

$$y_i^{\mathrm{sim}} \in \mathcal{N}(y_i, \sigma_i^2),$$

where $i \in \{1, 2, 3, 4\}$ for the four data points. Repeat this some large number of times (say, 10,000), and in each instance calculate the MLE estimate of the parameter, $\widehat{m}^{\mathrm{sim}}$. Then evaluate the mean of this distribution, $\widehat{m}^{\mathrm{freq}}$, and its 68 percent range. A proper way to do this would be to perform the algorithm described in Fig. 10.4, but I recommend that you use a (possibly slightly inaccurate) shortcut of calculating the p-values of $p = 0.16$ and $p = 0.84$ fractions of the volume under the distribution.

(c) **Bayesian analysis**. Now perform a Bayesian analysis. Assume a flat prior on the parameter of interest, say $m \in [0, 3]$. Set up a simple grid over this space, and map out the posterior. Find the best-fit model, and the 68 percent credible interval in m. Try varying the prior on m, and comment how your results change, and whether that is to be expected or not.

If everything goes well, you should be getting a very comparable error from the frequentist and Bayesian analyses, both in agreement with the direct estimator in part (a).

10.6 [**Computational**] **Wilks' theorem and BIC.** Given two models M_1 and M_2, it is simplest to gauge their goodness-of-fit using the difference in chi-squared, $\Delta\chi^2 \equiv \chi_2^2 - \chi_1^2$. As Sec. 10.4.1 describes, there are several ways in which to interpret $\Delta\chi^2$ (though none of them is overly accurate). Specifically, when the models are nested – say, model M_1 is completely subsumed within model M_2 – then one can use the Wilks' theorem; see Eq. (10.28). If the models are not nested, one can still use the BIC. *Note:* Both Wilks' theorem and the BIC only strictly apply in the limit when the number of data is much greater than the number of parameters, which is clearly not satisfied in the parts that follow. In the interest of pedagogical clarity, we ignore this issue in the problem.

(a) Adopt the $y(x)$ dataset from Problem 10.5. Compare two nested models – a constant and a line:

$$M_1 : y = b$$

$$M_2 : y = mx + b.$$

Numerically find the best-fit parameters for both models and their corresponding $\chi^2 = -2 \ln \mathcal{L}$ (note, for model M_2 this is trivial). Then use Wilks' theorem to quantify any evidence for the more complex model.

(b) For the same data and analysis from part (a), use the BIC to compare the two models. [Simply find ΔBIC between the two models, comment on whether it is positive or negative, and do not worry about the scale at which it is to be evaluated.]

10.7 [**Computational**] **MCMC.** Setting up a basic Markov chain Monte Carlo is quite straightforward, as you will now demonstrate. Use the same dataset as in Problem 10.5, with four measurements with uncorrelated error bars, and consider the same model as there:

$$y = mx \qquad \text{(model 1)}.$$

(a) Set up the MCMC algorithm in the single parameter m and run it for some reasonably large number of steps (say, 10,000). [If you wish, you can look up the Gelman–Rubin convergence criterion and track it as well.]

(b) Report the acceptance rate of your chain. If the value is either far too large or too small (relative to the optimal ballpark rate of 0.25), then change the stepsize accordingly, repeat the MCMC run, and see if the acceptance rate improves.

(c) Plot the posterior in m. Find the best-fit value of this parameter, and its 68 percent credible interval.

(d) Calculate the Bayesian evidence of your model:

$$P(D|\text{model}) = \int \mathcal{L}(D|m)P(m)\, dm.$$

Note that you should use the properly normalized prior so that

$$\int P(m)dm = 1.$$

Adopt the range $m \in [0, 5]$, and a uniform prior so that $P(m) = \text{const.}$ *Note:* It might be easiest to not use your MCMC calculation, but rather calculate the evidence directly by summing on a grid in m.

(e) Now consider a more complicated model:

$$y = mx + b \qquad \text{(model 2)},$$

with two free parameters, m and b. Calculate the Bayesian evidence of this model as well. You don't need to run MCMC, just evaluate the evidence by summing on a grid of (m, b). Adopt the same m range as in part (d), and the range $b = [-2, 2]$. And a uniform prior as before, normalized as $\int P(m, b)\,dm\,db = 1$.

Then evaluate the Bayes factor

$$B_{21} = \frac{P(D|\text{model 2})}{P(D|\text{model 1})} = \frac{\int \mathcal{L}(D|m, b)P(m, b)\,dm\,db}{\int \mathcal{L}(D|m)P(m)\,dm},$$

where the numerator and denominator are Bayesian evidences for the two models, respectively. With your result for B_{21}, use the Jeffreys' scale (Table 10.1) to assess if the data prefer model 2 over model 1.

10.8 [**Computational**] **Fisher matrix for type Ia supernova data.** Forecast the errors on the cosmological parameters in the flat wCDM model (dark energy with a constant equation-of-state parameter w) using current type Ia supernova data. We are only covering type Ia supernovae as a cosmological probe in Chapter 12, but the analysis is pretty simple and revolves around using Eq. (12.7). The wCDM model has three parameters: matter density Ω_M, dark-energy equation of state w, and the offset in the Hubble diagram $\mathcal{M}$ (see again Eq. (12.7)).

Download the data from `https://github.com/dscolnic/Pantheon`; the file you want is `lcparam_full_long_zhel.txt`. You will only need the redshift (`zcmb`) and the magnitude error (`dmb`) columns. Assume the errors are uncorrelated, and ignore any covariance in the data. You will *not* need the apparent magnitudes, as you should calculate the theoretical ones using Eq. (12.7). This way, you can take the derivatives of the magnitude with respect to the cosmological parameters. Take Ω_M and w to have fiducial values from Table 3.1, while you can take any value for the nuisance parameter $\mathcal{M}$ – say, zero; this is because none of the derivatives depends on the value of $\mathcal{M}$.

(a) Calculate the marginalized and unmarginalized errors on the cosmological parameters $(\Omega_M, w, \mathcal{M})$.

(b) Plot the 68 percent credible-interval contour in the $\Omega_M - w$ plane. Note that you will have to marginalize over $\mathcal{M}$ as discussed in Sec. 10.6.4. *Note:* In Problem 12.7 you will be asked to do actual analysis of real SN Ia data, which you can compare to your forecasted contours here.

(c) Find the best-constrained linear combination in the $\Omega_M - w$ space. How well is this linear combination measured? *Hint:* You will want to diagonalize the 2×2 Fisher matrix projected to this space.

11 Dark Matter

Dark matter is one of the most familiar concepts in cosmology and astrophysics, as perhaps evidenced by the fact that there is a Hollywood movie (featuring Meryl Streep), a rock band, and a TV series with that name. The byline for a dark-matter-themed event could be that there is a lot of stuff in the universe that does not emit or reflect light, and we do not know what it is. Roughly speaking, this simple premise is correct. In this chapter, we briefly review the evidence for dark matter, and outline key directions in theoretical and experimental research on the topic.

11.1 Historical Evidence for Dark Matter

A brief history of evidence for dark matter is given in Box 11.1. We now go on to talk about it in more detail. We will first review the historical evidence for the presence of dark matter in order to explain what motivated generations of astronomers – and it was mostly astronomers and not physicists in the early years – to grapple with the possibility that there is unseen mass in the universe. Rather than being chronological, our presentation here will go from smaller to larger astronomical objects.

11.1.1 Mass-to-Light (M/L) Ratio

A traditional way to speak about dark matter is to consider the **mass-to-light ratio** of various cosmological objects. Consider first the mass of the Sun and its luminosity in the B-band (the B-band is sensitive to light between 400 and 490 nm):

$$M_\odot = 1.989 \times 10^{30} \, \text{kg} \tag{11.1}$$

$$L_{\odot,B} \simeq 5.7 \times 10^{25} \text{W}. \tag{11.2}$$

The ratio between the two is the mass-to-light ratio of the Sun:

$$\frac{M_\odot}{L_\odot} \approx 40,000 \, \text{kg/W}. \tag{11.3}$$

It is conventional in the astronomy community to quote the mass-to-light ratio of an object (or class of objects) in units of $M_\odot/L_\odot$.

Box 11.1	**Dark Matter: A Brief History**

Dark matter has a long, almost 100-year history, featuring discoveries and developments in both astronomy and physics. Highlights of this history are as follows.

- 1933: Fritz Zwicky, a Swiss-American astronomer working at Caltech, noticed that the total mass of the Coma cluster of galaxies must be larger than that corresponding to the luminous matter in the cluster.
- 1970s: Vera Rubin and others confirm dark matter by establishing "flat rotation curves" of galaxies.
- 1970s: Early numerical simulations of galaxy dynamics indicate that galaxies need a halo of dark matter to stabilize their rotation.
- 1980s: Dark matter is gradually confirmed by a variety of dynamical measurements in galaxies and galaxy clusters.
- 1970s–1990s: Big Bang nucleosynthesis measurements indicate that the total amount of *baryons* is small, about 5 percent of critical ($\Omega_B \approx 0.05$), and thus insufficient for baryons to be all matter in the universe.
- 1990s–present: Gravitational lensing helps map out the distribution of dark matter in the universe, while numerical simulations present a detailed picture of its expected distribution and evolution.
- Late 1990s–present: Increasingly accurate measurements of the dark-matter abundance using galaxy clustering from Sloan Digital Sky Survey and other experiments, along with the discovery of dark *energy* using type Ia supernovae and mapping out the microwave background anisotropy, help confirm the mathematical relation $0.30 + 0.70 = 1.0$. That is, dark and baryonic matter together comprise 30 percent of critical energy density, dark energy 70 percent, and the universe is flat so that the total energy density is equal to the critical.

11.1.2 M/L **Ratio of Stars**

Masses of relatively nearby stars in our galaxy can be obtained for objects that reside in binary stellar systems. In those systems, one can apply Kepler's laws to the observed orbits. Such estimates indicate that the stellar mass-to-light ratio is on average several times that of the Sun:

$$\left\langle \frac{M}{L_B} \right\rangle_{\text{stars}} \approx 4 \, \frac{M_\odot}{L_\odot}. \tag{11.4}$$

While not immediately enlightening, this number will become interesting once we see the much bigger M/L numbers for galaxies and clusters of galaxies.

Meanwhile it is interesting to calculate the total energy density in the universe of stars. For that, we make use of the measured luminosity density of stars in a broad region around galaxies (hundreds of megaparsecs away), which is $j_{B,\text{stars}} \approx 1.2 \times 10^8 L_{\odot,B} \, \text{Mpc}^{-3}$. Using this as a proxy for the average luminosity density of the universe, we get

$$\Omega_{\text{stars}} = \frac{\rho_{\text{stars}}}{\rho_{\text{crit}}} = \frac{1.2 \times 10^8 L_{\odot,B} \,\text{Mpc}^{-3} \times 4 M_{\odot}/L_{\odot,B}}{10^{11} M_{\odot} \,\text{Mpc}^{-3}} \approx 0.004. \qquad (11.5)$$

The most notable feature of the night sky – stars – contribute only about 0.4 percent of the critical (and total, because $\Omega_{\text{TOT}} \simeq 1$) energy density in the universe.

11.1.3 Dark Matter in Galaxies

Numerous ways exist to determine the total mass of a galaxy. Typically, one applies Newton's second law to the observed motion of stars, or else neutral hydrogen regions for example, to infer the galaxy's mass. Other methods include observations of X-ray gas in galaxy clusters, where application of hydrostatic equilibrium leads to inference of the mass.

Here we review historically probably the most famous piece of evidence for dark matter, which comes from the so-called **rotation curves** of galaxies. Consider a star of mass m moving with velocity v which is perpendicular to the radius vector **r** between the center of the galaxy and the star. Assume, for simplicity, that the galaxy is spherical and its material is isotropically distributed from the center. Newton's second law gives

$$\frac{mv^2}{r} = G\frac{mM(r)}{r^2}, \qquad (11.6)$$

where $M(r)$ is the enclosed mass of the galaxy within radius r. The velocity is thus given by

$$v = \sqrt{\frac{GM(r)}{r}}. \qquad (11.7)$$

If we observe a star located beyond where the light of the galaxy stops – where (naively) there is no more matter – then the enclosed mass should be $M(r) \approx M_{\text{total}} = \text{const}$. For such a star, we therefore expect

$$v(r) \propto r^{-1/2} \qquad \text{(expectation from luminous-only matter)}. \qquad (11.8)$$

However, when Rubin, Ford, and others measured the rotation curves (of stars in M31 galaxy, in this case) in the 1970s, they found that the velocities do not decrease as $r^{-1/2}$. Instead, the velocities remain roughly constant as a function of distance from the center of the galaxy, even at distances several times larger than the radius out to which the galaxy light is observed.

This development means that there has to be significant **dark matter** beyond the typical light radius of $\sim 10\,\text{kpc}$. In fact, it turns out that the dark-matter distribution picks up roughly where the (luminous) disk stops so that the total mass increases with radius, $M(<r) \propto r$, and the velocity is constant as per Eq. (11.7) (see Figure 11.1). Traditionally, cosmologists refer to a dark-matter **halo** – a spherical distribution of dark matter centered at each galaxy (see Fig. 11.2). The halo's mass does not increase indefinitely with radius; the halo rather extends out to some

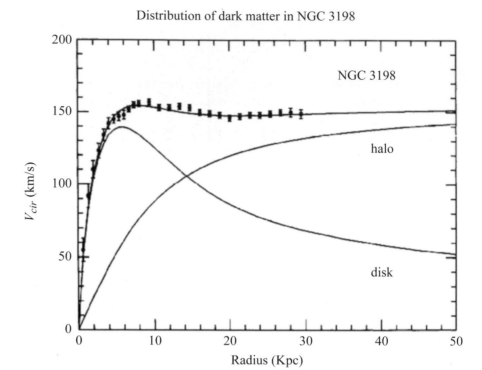

Distribution of dark matter in NGC 3198

The measured flat rotation curve in galaxy NGC 3198. Note that the flat curve needs a (dark-matter) halo in addition to the usual contribution of the luminous disk. Reproduced with permission from van Albada *et al.* (1985), © AAS.

finite distance that is much larger than the distribution of light (roughly 100 kpc for galaxies like the Milky Way, for example).

Taking into account the dark-matter halo, one can estimate the total mass of a galaxy:

$$\left\langle \frac{M}{L_B} \right\rangle_{\text{galaxy}} \approx (50 - 150) \, \frac{M_\odot}{L_\odot} \left(\frac{R_{\text{halo}}}{100 \, \text{kpc}} \right), \tag{11.9}$$

where the range encompasses the values found for spiral galaxies (where the prefactor is closer to 50) and ellipticals (closer to 150). While there are fairly significant systematic uncertainties in the rotation-curve method, and galaxies display heterogeneity making the above M/L not universal, one thing is clear: galaxies' mass-to-light ratio is roughly 25 times larger than that of stars (for which, recall, $M/L \simeq 4 M_\odot/L_\odot$). This is a significant piece of evidence for the presence of unseen mass.

Galaxies therefore mostly consist of dark matter, an unseen component that Zwicky first identified in 1933. As a rule of thumb, dark matter starts to dominate luminous matter roughly at the half-light radius (the radius that encompasses

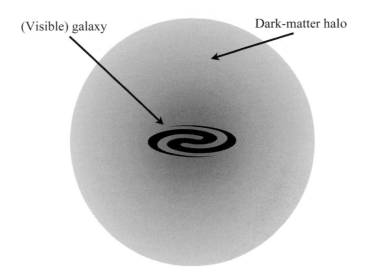

(Visible) galaxy   Dark-matter halo

Fig. 11.2   Illustration of a typical galactic dark-matter halo (not necessarily to scale). The halo is roughly spherical, and its density falls off from the center according to the NFW profile; see Eq. (11.20). At the center of the halo is a galaxy which is largely made up of baryons and emits light at various wavelengths.

half of the total light emitted by the galaxy), and extends far beyond. Moreover, different types of galaxies are affected by dark matter differently. For example, the so-called dwarf spheroidal galaxies apparently consist almost entirely of dark matter.

Another, ingenious argument for the existence of dark matter in galaxies was put forth by Ostriker and Peebles in 1973. They carried out a purely numerical study of the evolution of a galaxy, with only several hundred particles in their pioneering N-body simulation. They found that the galaxy is unstable to formation of a bar – a long linear structure that crosses the galaxy disk – *unless* the galaxy contains "a substantial unobserved mass" (in their own words). This is one of the earliest applications of numerical simulations to directly inform our knowledge about the universe.

11.1.4 Dark Matter in Clusters

Clusters of galaxies are the largest collapsed objects in the universe. They each contain up to 1000 galaxies, and are a few megaparsecs in size.

Dark matter in clusters is most easily discussed with the help of the **virial theorem**. Consider the velocities v_i of galaxies with masses m_i and positions $\mathbf{r}_i$ in a cluster. The total kinetic and potential energy of the system are given via

$$K = \frac{1}{2} \sum_i m_i v_i^2 \equiv \frac{1}{2} M \langle v^2 \rangle \tag{11.10}$$

$$V = -\frac{G}{2} \sum_{i \neq j} \frac{m_i m_j}{|\mathbf{r}_j - \mathbf{r}_i|} \equiv -\frac{GM^2}{R_c}, \tag{11.11}$$

where the factor of $1/2$ in the potential energy accounts for double-counting of pairs. Here M is the total mass of the cluster, while R_c is the characteristic radius for which the last expression holds; it has been found empirically that R_c is approximately twice the half-light radius of the cluster, of order $3\,\mathrm{Mpc}$ for rich clusters.

In Problem 11.2 you will be asked to prove the virial theorem. The result is

$$\ddot{I} = 2V + 4K \qquad \text{(virial theorem)}, \tag{11.12}$$

where I is the moment of inertia of the cluster, $I = \sum m_i |\mathbf{r}_i|^2$. Useful progress can be made by applying the virial theorem to *relaxed* clusters – those that show no observable signs of recent mergers. Then the moment of inertia is stationary, so that $\ddot{I} = 0$ and

$$V \simeq -2K \qquad \text{(virial theorem for relaxed clusters).} \tag{11.13}$$

Using the relations above for K and V, we have

$$-\frac{GM^2}{R_c} \simeq -M \langle v^2 \rangle. \tag{11.14}$$

For the Coma cluster of galaxies, the velocity dispersion along the line of sight is about $v_{\mathrm{los}}^2 = (900\mathrm{km/s})^2$ and, assuming isotropy, the full 3D mean velocity squared is about $\langle v^2 \rangle = 3\,v_{\mathrm{los}}^2 \approx (1500\mathrm{km/s})^2$. Coma's characteristic radius is $R_c \simeq 2R_{\mathrm{half\ light}} \simeq 3\,\mathrm{Mpc}$. This finally gives the mass of Coma:

$$M_{\mathrm{Coma}} \simeq \frac{R_c \langle v^2 \rangle}{G} \simeq 1.5 \times 10^{15} M_\odot. \tag{11.15}$$

Adopting the measured B-band luminosity of Coma, $L_{B,\mathrm{Coma}} \simeq 4 \times 10^{12} L_\odot$, we finally get its mass-to-light ratio

$$\left\langle \frac{M}{L_B} \right\rangle_{\mathrm{Coma}} \approx 400 \,\frac{M_\odot}{L_\odot}. \tag{11.16}$$

There is a lot of dark mass in Coma – roughly 100 times more than stellar mass!

11.1.5 Why not Baryons?

As we saw above, the consensus of the cosmology community toward the existence of dark mass in galaxies and clusters had started to shift in the 1970s, and by the 1980s this became part of the standard picture. Nevertheless, it was still, at the time, considered possible – maybe even likely – that the unseen mass was made up of dark baryons (so normal matter, like protons and neutrons).

One particularly simple yet powerful argument against this notion was put forth by Peebles in 1982. With baryons only, and initial density fluctuation $\delta_{\text{init}} \simeq 10^{-5}$, which was expected though not yet experimentally measured at the time, one would expect a present-day overdensity of

$$\delta_0 \simeq \frac{a_0}{a_*}\delta_{\text{init}} \simeq 10^{-2} \qquad \text{(with baryonic dark matter).} \qquad (11.17)$$

This expression assumes a matter-dominated universe with $\delta \propto a$, and the fact that the baryonic perturbations can grow only once baryons are decoupled from photons, which is at recombination when $a_* \simeq 10^{-3}$. The small present-day overdensity in Eq. (11.17) would have a hard time explaining the highly nonlinear structure in the universe today, $\delta_0 \gtrsim 1$. Peebles suggested that a massive, non-relativistic particle would have largely resolved this problem, as it would have decoupled from the thermal bath well before recombination. The universe whose matter content is dominated by this particle – the dark-matter particle – would allow an earlier start of perturbation growth, and hence a much larger overdensity today.

11.2 Modern Evidence for Dark Matter

Much modern evidence for dark matter comes from gravitational lensing. As we will discuss at some length in Chapter 14, gravitational pull from unseen dark matter is consistently observed in a variety of lensing measurements, most famously in the Bullet cluster, which shows clear evidence of the dark component being spatially offset from the X-ray luminous mass.

Evidence for a shortfall of baryons to explain the total mass of clusters of galaxies is also well established. Hot gas is readily observed in X-rays, whose characteristic energy is $E \sim$ keV, and temperature $T \sim 10^8$ K. A variety of measurements indicate that the mass in the X-ray gas dominates the stellar mass by a factor of 6–7, so that $\Omega_{\text{gas}} \sim 0.025$. This still falls far short of the total amount of matter measured. In fact, it cannot even account for all of the easily observed *baryonic* matter ($\Omega_B \simeq 0.05$); most of these "missing baryons" probably reside in a colder (and less X-ray-emitting) intergalactic medium.

However, the strongest statistical evidence for the presence of dark matter comes from the angular power spectrum of the cosmic microwave background radiation. The CMB is sensitive to both the baryon physical density $\Omega_B h^2$ and the cold dark matter (CDM) physical density $\Omega_{\text{CDM}} h^2$ (where "cold" roughly means non-relativistic, as will be explained in Sec. 11.3 just below). We already discussed the CMB and BBN constraints on the baryon density in Eq. (7.44), namely

$$\Omega_B h^2 = 0.0224 \pm 0.0002 \quad \text{(CMB)}$$
$$\Omega_B h^2 = 0.0223 \pm 0.0004 \quad \text{(BBN).} \qquad (11.18)$$

The dark-matter density, on the other hand, modulates the overall morphology of the peaks in the CMB power spectrum. For $\Omega_{\mathrm{CDM}} = 0$, for example, the CMB spectrum's wiggles would be strongly suppressed relative to large-scale plateau, in contradiction to measurements. The constraint from Planck and external data is

$$\Omega_{\mathrm{CDM}} h^2 = 0.1197 \pm 0.0014 \qquad (\mathrm{CMB}). \tag{11.19}$$

Take a moment to appreciate the significance of this Ω_{CDM} constraint: it shows a nearly 100-sigma (!) evidence – from the CMB alone – for the presence of a component that looks, walks, and talks like dark matter; that is, for $\Omega_{\mathrm{CDM}} > 0$. Moreover, Eqs. (11.18) and (11.19) indicate that $\Omega_B \ll \Omega_{\mathrm{CDM}}$, and thus that the dark matter is non-baryonic – again at a very high statistical significance.

Therefore:[1]

> The CMB angular power spectrum provides the strongest single piece of evidence for the existence of dark matter. It, along with the BBN constraints, also indicates that dark matter is probably not baryonic.

This leads us to the discussion of the physical nature of dark matter.

11.3 Particle Dark Matter

There is near consensus in the cosmology community that dark matter is made up of particles – most likely elementary particles, but perhaps instead macroscopic-scale objects which we will put in the same category. Near the end of this chapter, we will discuss another possibility – that of modified gravity – but for the moment let us discuss the dark-matter-particle candidates and their properties.

Two principal stories of structure formation, vis-à-vis dark matter, have already been covered in Chapter 9. These are hot and cold dark matter. Then there is also warm dark matter. We now briefly review these.

- **Cold dark matter** (CDM): These are particles that are non-relativistic at decoupling. Their mass is typically in the giga-electronvolt range, though it could be as low as mega-electronvolts and, for a non-thermal relic, as high as 10^{15} GeV (these are the so-called WIMPs, to be discussed in Sec. 11.3.1). It is a fairly standard assumption that dark matter is cold, principally because the numerical (N-body) simulation of structure in the universe with CDM agrees remarkably well with the observed cosmological structure and various galaxy-clustering and gravitational-lensing measurements.

    The foremost feature of cold dark matter is freezeout, which we discussed at length in Chapter 6. Recall that, when the self-annihilation rate of the particle

---

[1] It is still *possible* that dark-matter particles *are* baryonic because they are somehow made before Big Bang nucleosynthesis, thus evading the BBN and CMB constraints (on $\Omega_B h^2$). Primordial black holes, discussed next, are such baryonic dark-matter candidates.

Γ falls below the expansion rate H, the particle freezes out – its comoving abundance stops diluting and asymptotes to a constant. Without freezeout, the CDM abundance today would be negligible.

Formation of CDM halos has been modeled with computer N-body simulations for the past 50 years, and has become particularly sophisticated since the late 1980s. CDM is relatively easy to model; it simply involves gravitational interactions between dark-matter particles.[2] Simulations have also indicated a **universal density profile** of the CDM halos: it is universal since the same form applies to all dark-matter halos regardless of their mass. This is called the **Navarro–Frenk–White (NFW) profile**:

$$\rho_{\text{NFW}}(r) = \frac{\rho_0}{\left(\dfrac{r}{r_s}\right)\left(1 + \dfrac{r}{r_s}\right)^2}, \tag{11.20}$$

where ρ_0 and r_s are the parameters which vary from halo to halo. Here r_s is the scale radius – the characteristic radius at which the profile goes from $\rho(r) \propto r^{-1}$ to $\rho(r) \propto r^{-3}$. See Problem 11.6 for more on NFW.

- **Warm dark matter** (WDM): These particles, of typical mass $m_{\text{DM}} \simeq$ keV, are mildly relativistic – hence "warm" – at decoupling. The principal feature of WDM is free streaming: because WDM particles are fast, they do not readily form structure at small scales. It can be shown that the free-streaming wavelength and the corresponding wavenumber are

$$\lambda_{\text{FS}} \simeq 300 \left(\frac{1\,\text{eV}}{m_{\text{DM}}}\right)\,\text{Mpc}; \qquad k_{\text{FS}} \equiv \frac{2\pi}{\lambda_{\text{FS}}} \simeq 0.02 \left(\frac{m_{\text{DM}}}{1\,\text{eV}}\right)\,\text{Mpc}^{-1}, \tag{11.21}$$

which you can work out in Problem 11.5. Therefore, WDM suppresses power on scales below a certain scale λ_{FS} that depends on m_{DM}; see Fig. 11.3. This makes WDM also appealing as a mechanism to suppress the amount of structure on small scales, and help address some of the claimed challenges to CDM discussed in Sec. 11.3.3 below. Note that the numerical prefactors in Eq. (11.21) are very approximate, as the onset of free streaming is gradual, as illustrated for neutrinos in Fig. 11.4.

- **Hot dark matter** (HDM): This is the dark-matter candidate that is totally relativistic (hence "hot") at decoupling. A prototypical example is a massive neutrino, with mass $m_{\text{DM}} \sim$ eV. The free-streaming bound from Eq. (11.21) very much applies here, and implies that all structure below scales of tens of megaparsecs is suppressed (see Fig. 11.3). This is in strong conflict with observations; for an early evidence of this, see Fig. 9.10 in Chapter 9. Therefore, HDM cannot correspond to all or even most of the dark matter in the universe.

It is possible, however, that HDM makes up a small fraction of dark matter – in fact this is guaranteed for neutrinos, as the neutrino oscillation experiments indicate that these particles have mass, and thus $\Omega_\nu h^2 = \sum_i m_{\nu,i}/(94\,\text{eV}) > 0$.

---

[2] We are referring to pure-dark-matter N-body simulations here. One can also add baryonic matter in (considerably more challenging to implement) *hydrodynamical* simulations.

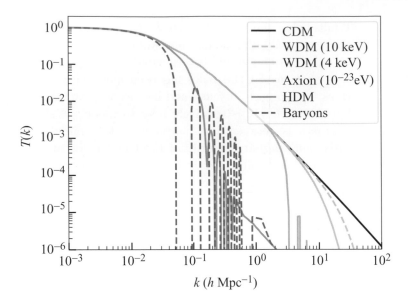

Fig. 11.3 Transfer functions for cold dark matter (CDM), warm dark matter (WDM) with two particle masses, hot dark matter (HDM; massive neutrinos), ultra-light axions, and a baryon-only case. In all cases we assume that all of the dark-matter content, $\Omega_{\mathrm{dm}}h^2 \simeq 0.112$, is made up of the particle in question (in addition to the usual $\Omega_B h^2 \simeq 0.022$). Warm and hot dark matter $T(k)$ are suppressed because of the free streaming of the dark-matter particle. Axion $T(k)$ is also suppressed, but for a different reason: because of the quantum pressure of these extremely light, long-wavelength particles. The baryonic $T(k)$ oscillates because baryons are coupled to radiation in the early universe. The plot was produced using CLASS (Blas *et al.*, 2011) and axionCAMB (Hlozek *et al.*, 2015). A black and white version of this figure will appear in some formats. For the color version, please refer to the plate section.

Currently, CMB and large-scale structure constrain the sum of the neutrino masses to be $\sum_i m_{\nu,i} \lesssim 0.12\,\mathrm{eV}$; see Eq. (5.33). Below the free-streaming scale, the density fluctuations are suppressed – proportionally to the energy density contribution (and thus mass) of the neutrinos. A rough approximation for the relative suppression of the matter power spectrum at $k \gg k_{\mathrm{FS}}$ is

$$\frac{\Delta P(k)}{P(k)} \simeq -8\frac{\Omega_\nu}{\Omega_M}. \qquad (11.22)$$

For example, for a 0.1 eV mass neutrino, $\Omega_\nu h^2 \simeq 0.001$, and given that $\Omega_M h^2 \simeq 0.1$, we have $\Delta P/P \simeq -0.1$ – a 10 percent suppression, potentially observable with the current and next generation of large-scale structure surveys. This kind of accuracy in measuring the sum of the neutrino masses will enable tests of the neutrino mass hierarchy discussed in Chapter 6.

We now discuss specific particle-physics scenarios for dark matter.

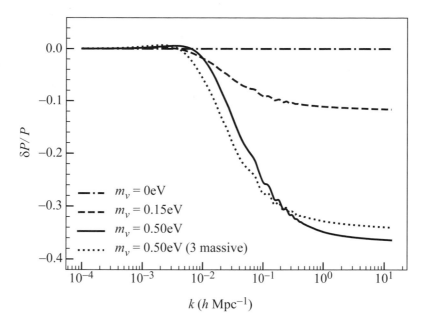

Fig. 11.4 Effects of neutrino mass on the matter power spectrum. We show the relative difference of power spectra with neutrinos of mass $0.15\,\mathrm{eV}$ or $0.50\,\mathrm{eV}$ and the fiducial case with zero neutrino mass. When varying the neutrino mass, we adjust the CDM mass fraction so that the total mass, $\Omega_M = \Omega_{\mathrm{cdm}} + \Omega_B + \Omega_\nu$, is held fixed. We assume the linear-theory power spectrum evaluated at redshift zero. We further assume that all of the mass is in a single massive neutrino species, except for the dotted curve where we assume that the mass is equally split between three degenerate species. The plot was produced using CAMB computer code (Lewis *et al.*, 2000).

11.3.1 Dark-Matter Candidates

The number of proposed dark-matter candidates is large, and their properties are extremely diverse. Let us review some major candidates discussed over the past 50 years.

- **WIMPs** (weakly interacting massive particles) are probably the most popular dark-matter candidate. Typically, one has in mind non-baryonic particles in the mass range between a few giga-electronvolts and a few hundred giga-electronvolts or more (the so-called "WIMPZILLAs," whose mass is as high as $10^{15}\,\mathrm{GeV}$). WIMPs are weakly interacting, as we haven't yet seen them scatter off protons, neutrons, or electrons, despite decades of experimental searches. One particularly interesting candidate for WIMPs is the lightest supersymmetric particle (LSP). The "lightest" in its name comes from the requirement that it does not decay into another light particle (that would already have been detected). Current searches at the Large Hadron Collider at CERN are also hoping to see the LSP. WIMPs are a prototypical CDM candidate. Most past and ongoing direct searches for dark matter (discussed further below) specifically target WIMPs.

- **Axions and ultra-light dark matter**. Axions are particles originally postulated to exist in order to solve a technical problem in quantum chromodynamics (QCD; theory of strong interactions). Paradoxically, even though they are expected to be very light ($m \sim 10^{-5}\,\text{eV}$), axions are a CDM candidate since they were never in thermal equilibrium with other species.[3] The principal way to find axions is to search for their interactions with magnetic fields, where they are expected to (occasionally!) produce a photon. They might also be observed via their emission, and subsequent conversion to photons, in atmospheres of compact objects such as neutron stars.

  More broadly, there is interest in considering **ultra-light** axions; they are principal representations of a more general class of **fuzzy dark matter**. These ultra-light dark-matter candidates have extremely small mass and huge de Broglie wavelength that is a fraction of the size of a galactic halo (see Problem 11.4):

$$m_{\text{ultra-light}} \sim 10^{-22}\,\text{eV}; \qquad \lambda_{\text{ultra-light}} \simeq (m_{\text{ultra-light}}v)^{-1} \lesssim R_{\text{halo}}. \qquad (11.23)$$

  Ultra-light dark matter also suppresses small-scale power, but not because of free streaming (as WDM and HDM), rather because of quantum pressure – essentially, a lack of localization of the wavepacket to a region much smaller than the de Broglie wavelength. This is illustrated in Fig. 11.3. Overall, the rich phenomenology of axions and their arguably strong theoretical motivation prompted one prominent cosmologist to declare that axions are the "thinking man's dark-matter candidate."

- **Primordial black holes and MACHOs**. Primordial black holes (PBHs), proposed in the 1970s by Bernard Carr, Stephen Hawking, and others, are a purely baryonic candidate. They are compact objects with masses anywhere from that of a small planet to that of a large star. The PBHs could have formed in the early universe, with mass comparable to the horizon mass at their creation, or

$$m_{\text{PBH}} \sim \frac{t}{G} \simeq M_\odot \left(\frac{t}{10^{-5}\,\text{s}} \right), \qquad (11.24)$$

where t is their formation time. PBHs formed prior to Big Bang nucleosynthesis lock up their baryons and thus effectively hide them; hence the PBHs evade the BBN (and CMB) constraint on baryon abundance. The appealing feature of PBHs is that no exotic physics or a non-baryonic scenario need be postulated for their creation. The challenge for creating PBHs in the early universe is that one requires a substantial primordial power at scales characteristic of PBH (the horizon size at PBH creation), far greater than that predicted by scale-invariant

---

[3] Axions, if they exist, would have been produced at the QCD phase transition ($T \sim 200\,\text{MeV}$; see Chapter 4), rather than as a by-product of reheating after inflation as other particles of the Standard Model of particle physics. This is why they were never in equilibrium with other species.

fluctuations with $n_s \simeq 1$. Therefore, one requires a substantial blue tilt $(n_s < 1)$ at scales smaller than observed by the CMB, or else a "bump" in power at PBH scales.

PBHs would show up as massive compact halo objects (MACHOs), a more generic term coined in the 1980s that describes planet- or star-mass objects forming the galactic dark-matter halos. An ingenious idea for detecting MACHOs was proposed by Bohdan Paczynski in 1986: to observe thousands of stars and search for the gravitational-lensing signature – amplification of light – when a MACHO crosses the line of sight between the observer (us) and the star (see Chapter 14 for further discussion of this). There was a lot of excitement about this in the early 1990s when apparent evidence for MACHOs was found (see Box 14.1 in Chapter 14), as it looked like dark matter was finally identified. Unfortunately, further searches by several experiments (MACHO, OGLE, EROS) did not find many MACHO candidates, and eventually imposed only upper bounds. MACHOs can, at best, form only a fraction of the dark matter in the universe.

We now discuss dark matter from a more theoretical angle, and outline some general constraints on the mass of the dark-matter particle.

11.3.2 General Conditions on the Properties of Dark Matter

We have already mentioned the de Broglie limit on the mass in the context of ultra-light dark matter and axions, $m_{\text{ultra-light}} \gtrsim 10^{-23}\,\text{eV}$ (give or take a few orders of magnitude; see also Problem 11.4). This very general bound requires no assumptions about the nature of dark matter.

Thermal relics – the particles that were in thermal equilibrium until freezeout – are particularly popular among dark-matter model builders because such *weakly interacting* relics benefit from the **"WIMP miracle"** feature, discussed in Box 11.2. Thermal relics generally satisfy some general constraints, which we now review.

Earlier in Chapter 5 we calculated the energy density in massive neutrinos; Eq. (5.32) says that $\Omega_\nu h^2 = \sum m_\nu/(94\,\text{eV})$. This generalizes to any hot-dark-matter (relativistic at decoupling) candidate. Requiring that the energy density not exceed that of dark matter today, $\Omega_{\text{HDM}} h^2 \lesssim 0.1$, we get the upper limit

$$m \lesssim 10\,\text{eV} \qquad \text{(HDM upper mass limit; assumes thermal)}. \qquad (11.25)$$

Note that this limit only assumes that the energy density in HDM not be too large; detailed structure formation modeling actually implies that HDM cannot be the dominant dark-matter component, and the corresponding upper limit on the mass is therefore much stronger (currently in the ballpark of $0.12\,\text{eV}$; see Eq. (5.33)).

For much more massive (order GeV) cold-dark-matter candidates, there is a *lower* limit on the mass, the so-called Lee–Weinberg bound.[4] Assuming electroweak

---

[4] In addition to Lee and Weinberg, several other groups simultaneously made essentially the same argument in 1977.

Box 11.2   **WIMP Miracle?**

It turns out that a thermal relic particle with a cross-section characteristic of weak interactions – the "WIMP" – would freeze out with the density roughly equal to the observed dark-matter density today. We now review this calculation.

We would like to evaluate the (WIMP) dark-matter energy density relative to critical today. Because WIMPs are non-relativistic, $\rho_{DM,0} = mn_{DM,0}$, where m is the WIMP mass. Let us first evaluate the WIMP number density today: following Eq. (6.13)

$$n_{DM,0} = Y_\infty s_0 = \frac{x_f}{\lambda} s_0 = x_f \left[\left(\frac{2\pi^2}{45}(g_{*S})_{T=m} \right) \frac{m^3\langle\sigma v\rangle}{H(T=m)} \right]^{-1} \frac{2\pi^2}{45} g_{*S} T_0^3$$

$$= x_f T_0^3 \frac{(g_{*S})_0}{(g_{*S})_{T=m}} \frac{H(T=m)}{m^3\langle\sigma v\rangle}, \tag{B1}$$

where we used the expression for entropy density from Eq. (4.32). Next, we make use of Eq. (4.36):

$$H(T=m) \simeq 1.66 g_*^{1/2} \frac{T^2}{m_{\rm Pl}} \simeq 1.66 g_*^{1/2} \frac{m^2}{m_{\rm Pl}}, \tag{B2}$$

where m is the WIMP mass. Now it is easy to evaluate omega in WIMPs:

$$\Omega_{\rm WIMP} \equiv \frac{\rho_{DM,0}}{\rho_{\rm crit,0}} = \frac{mn_{DM,0}}{\rho_{\rm crit,0}} = \frac{1.66}{\rho_{\rm crit,0}} \frac{x_f T_0^3}{m_{\rm Pl}\langle\sigma v\rangle} g_*^{1/2} \frac{(g_{*S})_0}{(g_{*S})_{T=m}}. \tag{B3}$$

Note that the WIMP mass m canceled out in this expression, though the cross-section itself can depend on m. Note also that a smaller annihilation cross-section $\langle\sigma v\rangle$ allows more particles to survive, and thus yields a larger relic density $\Omega_{\rm WIMP}$. Let us first evaluate the ratio of the g_{*S} parameters. Recall from Chapter 4 that $(g_{*S})_0 = 3.94$. Because the WIMP mass is typically well above a giga-electronvolt, the freezeout temperature is at $T \simeq x_f m \simeq 10m$; at that time, $(g_{*S})_{T=m} \simeq 10 - 100$, depending on m (see Fig. 4.1). Let us adopt $(g_{*S})_{T=m} = 10$; then $(g_{*S})_0/(g_{*S})_{T=m} \simeq 1/3$.

Plugging this and the other constants into Eq. (B3), we get

$$\Omega_{\rm WIMP} h^2 \simeq 0.1 \left(\frac{x_f}{10} \right) \left(\frac{3 \times 10^{-9} \,{\rm GeV}^{-2}}{\langle\sigma v\rangle} \right). \tag{B4}$$

For $\Omega_{\rm WIMP} h^2$ to come out "correct" (recall, $\Omega_{\rm WIMP} h^2 \equiv \Omega_{\rm CDM} h^2 = (\Omega_M - \Omega_B) h^2 \simeq 0.11$; see Table 3.1), and given that $x_f \simeq 10$, we require

$$\langle\sigma v\rangle \simeq 3 \times 10^{-9} \,{\rm GeV}^{-2} \simeq 3 \times 10^{-26} \,{\rm cm}^3/{\rm s} \simeq G_F \times \frac{v_{\rm WIMP}}{c}, \tag{B5}$$

where the last expression contains the Fermi constant ("G Fermi"), $G_F = 1.16 \times 10^{-5} \,{\rm GeV}^{-2}$, and we assumed $v_{\rm WIMP} \simeq 10^{-3}c$, roughly expected for the velocity of non-relativistic particles. In particle physics, the Fermi constant gives *roughly* the expected cross-section of electroweak interactions. Hence: **dark-matter (WIMP) freezeout with typical electroweak cross-section leads roughly to observed present-day dark-matter density.** This is, perhaps somewhat hyperbolically, called the "WIMP miracle" (hallelujah!). Note, however, that a fixed cross-section does not give us the *rate* of interactions, as the latter also depends on the *density* of dark matter: the more massive the dark-matter particle is, the less abundant it is (for a fixed $\Omega_{\rm CDM}$), and the lower the rate is.

interactions, the annihilation cross-section typically depends on the WIMP mass squared, $\langle \sigma_{\mathrm{EW}} v \rangle \simeq G_F^2 m^2$. Plugging this into Eq. (B4) in Box 11.2 we get (setting $x_f = 10$, and with $G_F \simeq 1.16 \times 10^{-6} \, \mathrm{GeV}^{-2}$)

$$\Omega_{\mathrm{DM}} h^2 \simeq 0.1 \frac{3 \times 10^{-9} \, \mathrm{GeV}^{-2}}{G_F^2 m^2} \simeq 0.1 \left(\frac{5 \, \mathrm{GeV}}{m} \right)^2, \qquad (11.26)$$

indicating the lower limit on the cold-dark-matter candidate whose interaction scale is electroweak

$$m \gtrsim 5 \, \mathrm{GeV} \qquad \text{(CDM lower mass limit; assumes thermal).} \qquad (11.27)$$

This number is again approximate; more detailed analysis of weak interactions of the WIMP are required to get a more precise answer.

Finally, there is a very general upper limit from unitarity – roughly, the requirement that "probabilities must add to no more than unity": the cross-section must be less than some mass-dependent value; $\sigma \lesssim 4\pi/m^2$. Plugging this into Eq. (B4) and assuming $v \simeq 10^{-3}$, we get

$$m \lesssim 100 \, \mathrm{TeV} \qquad \text{(limit from unitarity; assumes thermal).} \qquad (11.28)$$

All of these constraints are still fairly model-independent. Once the early-universe physics is fully taken into account, the constraints (for specific scenarios) can be much stronger. For example, *fermion* dark matter must satisfy the so-called Tremaine–Gunn limit; see Problem 11.3.

11.3.3 "Challenges" to the Standard CDM Picture

Cold dark matter is the observationally favored dark-matter paradigm, and therefore a target for tests with a wide variety of new data. Occasionally, some discrepancies between theory (largely based on N-body numerical simulations with CDM) and observations have been claimed. Here we purposely put quotation marks around the word "challenges," since it has been shown that some or all of these challenges can be explained by better modeling of baryons and (in some cases) improved observations. Meanwhile, new challenges are appearing in the literature fairly regularly. At any rate, there are at least three such claimed tensions between the simulated and observed phenomenology of dark-matter halos:

1. **Missing satellite problem**. N-body simulations appear to show a much larger number of satellites – smaller structures within galaxies (sometimes also called subhalos) – than seen in observations of the Milky Way and M31.

2. **Core/cusp problem**. At the center of dark-matter halos, numerical simulations predict density profiles that are "cuspy" or steep, going roughly as $\rho(r \to 0) \propto r^{-1}$, as given by the NFW profile. Observations, on the other hand, seem to indicate a "core," where the density at the center levels off, $\rho(r \to 0) \propto r^0$.

3. **Overcooling problem**. Cooling is necessary for the formation of galaxies and stars. Adopting standard calculations for various channels of dissipating energy,

cooling is too efficient at early times, leading to too much star formation at high redshift (and therefore not as much as needed today) relative to what we observe.

Much has been written about these problems. A consensus has emerged in recent years that more careful modeling of baryons, combined with better observations, helps explain these tensions. For example, the missing satellite problem is alleviated by incorporating additional physical ingredients (supernova feedback, tidal stripping, and photoionization) into simulations, which suppress the formation of satellites, as well as by finding many more satellites in the Sloan Digital Sky Survey and Dark Energy Survey than were previously known to exist. Similarly, the NFW cusps start looking more core-ish once baryonic effects are added to simulations. Similar solutions have been proposed for the overcooling problem. At the present time, none of these challenges is in danger of ringing a death knell for CDM. However, these tensions present a major motivation for tests of the CDM physics with simulations and data, which is a very active area of cosmology.

11.4 Detection of Dark Matter

There is an impressive variety of ways to search for dark matter. By this we mean finding evidence for a dark-matter particle, rather than the gravitational effect of dark matter on the dynamics and lensing signal of galaxies and clusters which are already very well established. The three principal detection methods are direct, indirect, and accelerator searches; for an overview of what they basically measure, see Fig. 11.5. We now briefly discuss them in turn.

11.4.1 Direct Detection

The most secure way to definitively confirm the presence of dark matter and learn about its microphysical properties is to detect its interaction with ordinary matter in a controlled environment. Specifically, the dark-matter particle will, rarely but predictably, scatter off a nucleus of ordinary matter. If $m \gtrsim 1\,\mathrm{GeV}$, this scatter imparts sufficient recoil energy to the nucleus that the impact can be detected via either a lattice vibration or ionization of the material. Lower ($\sim$MeV) dark-matter particle masses can be detected by studying their interaction with electrons, rather than nuclei, leaving ionization, scintillation (i.e., light), or molecular-dissociation signatures in the detector.

Schematically, direct-detection methods search for interactions of the form

$$\mathrm{DM} + \mathrm{SM} \longrightarrow \mathrm{DM} + \mathrm{SM}, \qquad (11.29)$$

where DM stands for the dark-matter particle, and SM for a standard model particle. Popular SM choices are metals like germanium or silicon, or else noble gases such as argon or xenon.

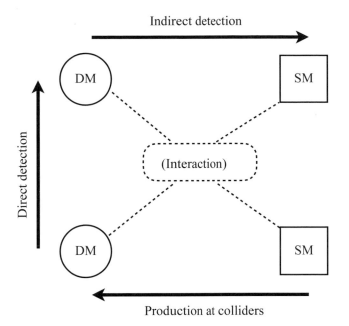

Fig. 11.5   Schematic representation of the possible ways to find dark-matter (DM) particles. Going clockwise, starting on the bottom right: in particle colliders, two Standard-Model (SM) particles may interact to produce DM particles. In direct detection, a DM and an SM particle interact and may produce both DM and SM particles, along with a thermal, chemical, or acoustic signal that can be detected. In indirect detection, two DM particles interact to produce SM particles (say photons) that can be detected.

Dark matter interacts with ordinary matter exceedingly feebly, as evidenced by extremely low upper limits on the dark-matter cross-section; see Fig. 11.6. The expected rate of interactions R (number of interactions per second) is given by

$$\frac{dR}{dE} = N_T \frac{\rho_\odot}{m} \int_{v_{\min}}^{v_{\rm esc}} v f(\mathbf{v} + \mathbf{v}_E) \frac{d\sigma(E_R, v)}{dE_R} d^3 v, \qquad (11.30)$$

where N_T is the total number of nuclei in the detector, $f(\mathbf{v})$ is the phase-space velocity density of dark-matter particles,[5] $v_{\min}$ is the minimal velocity of dark matter to scatter a nucleus at recoil energy E_R, $v_{\rm esc}$ is the escape velocity of the particle from our galaxy, and $\mathbf{v}_E$ is the instantaneous velocity of Earth (through the galaxy). Here, we also introduce the *local* density of dark matter

$$\rho_\odot \simeq 0.3 \frac{\rm GeV}{\rm cm^3}. \qquad (11.31)$$

---

[5] The quantity $f(\mathbf{v})$ is equal to the full phase-space density $f(\mathbf{x}, \mathbf{p}, t)$ discussed in Chapter 6, integrated over $d^3 x$ (but not over momentum).

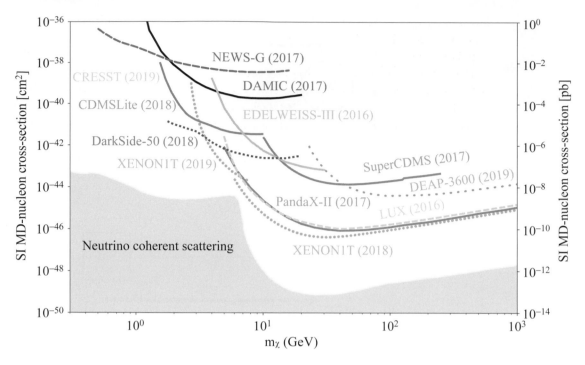

Fig. 11.6 Current limits on the dark-matter-nucleon cross-section as a function of dark-matter mass. The plot shows recent constraints; decades of improvements to the limits are not shown. Adopted from the Particle Data Group (Zyla *et al.*, 2020). A black and white version of this figure will appear in some formats. For the color version, please refer to the plate section.

Clearly, the rate R depends on factors that are not under our control – the aforementioned local dark-matter density, and also the velocity distribution of dark-matter particles. What *is* under our control is detector design. The larger the detector, the larger N_T and hence the rate R, and the better chance of detecting the dark-matter particle. A key experimental challenge is protecting the detector against backgrounds – unwanted scatters that may masquerade as the dark-matter signal. The backgrounds are principally due to radioactivity of the detector material itself, as well as the scattering of cosmic-ray particles (e.g., muons) off target nuclei. Much experimental effort in modern dark-matter detection is focused on minimizing these backgrounds.

Decades of direct dark-matter searches have produced impressive progress in the limits on the dark-matter scattering cross-section, as well as the development of a variety of new ideas and techniques. But they have produced no confirmed dark-matter detections. Figure 11.6 shows that, after one to two more order-of-magnitude improvement expected in the current and near-future generations of experiments, we will hit the so-called neutrino floor, where the guaranteed signal of neutrinos interacting with the nuclei in the detector will likely swamp the dark-matter signal, and thus make direct detection very much more challenging.

It is finally worth noting the implications of Earth's velocity $\mathbf{v}_E$ in Eq. (11.30). Earth moves with speed ~30 km/s around the Sun, which itself moves with velocity ~220 km/s through the Milky Way galaxy. Hence there is an ~10 percent modulation of the expected rate of observed dark-matter interactions – a potential tell-tale signal that we are indeed moving through the galactic halo made up mostly of dark matter. The DAMA experiment saw such a signal in 1998, and later found increasing statistical significance for it, but other direct-detection experiments have not confirmed their findings thus far, finding instead results in apparent conflict with DAMA.

11.4.2 Indirect Detection

Another very general method to detect dark matter is the so-called indirect detection, where we look for products of dark matter annihilating in our galaxy, or perhaps in other nearby galaxies, and producing normal (Standard-Model) particles that we can easily observe. Schematically, the reaction is

$$\text{DM} + \text{DM} \longrightarrow \text{SM} + \text{SM}, \qquad (11.32)$$

where the Standard-Model particles that are expected are chiefly photons (which are easiest to detect), or else positrons, anti-protons, or even neutrinos. Typically, large gamma-ray telescopes are tuned to look toward the galactic center or else toward dwarf galaxies or other sources, and search for an excess signal. The rate that one expects, observed at an angle θ away from the center of a galaxy, and per solid angle Ω, is

$$\frac{dR(\theta)}{d\Omega} = N_T \frac{1}{2} \frac{\langle \sigma_{\text{ann}} v \rangle}{m_{\text{DM}}^2} \frac{1}{4\pi} \int \rho^2 \left(\sqrt{r^2 + d^2 - 2dr\cos\theta} \right) dr, \qquad (11.33)$$

where N_T is again the total number of nuclei in the detector, $\langle \sigma_{\text{ann}} v \rangle$ is the velocity-averaged annihilation cross-section, and the factor of $1/2$ accounts for double-counting of dark-matter particle–antiparticle pairs. Here, ρ is the dark-matter density, and its argument (the square root) is the distance from the center of the halo, with d the distance to its center, and r our distance to an arbitrary point within the halo. Note the factor of density *squared* due to the fact that two DM particles interact, rather than a DM and an SM particle in direct detection with only a single power of density.

The principal challenge for indirect detection is separating the dark-matter signal from known astrophysical sources which also emit gamma-rays (and positrons and other particles). For example, pulsars and active galactic nuclei produce such background signals. Sophisticated experimental and analysis tools have been applied to indirect detection, and extended beyond nearby galaxies to objects such as neutron stars. No solid evidence for dark matter has been found to date, although a number of potential dark-matter signals were reported, then keenly followed up and investigated. The near consensus in the community is that a confirmed indirect detection of dark matter will nevertheless have to wait for a further confirmation by direct detection in order to be seen as a final proof of the arrest of the dark-matter outlaw.

11.4.3 Laboratory Searches

There is also hope that dark matter could rear its head in a particle accelerator – notably the Large Hadron Collider. Here, dark matter could show as missing energy in a particle reaction, or else as a "bump" in the distribution of detected Standard-Model particles. Schematically:

$$\text{SM} + \text{SM} \longrightarrow \text{DM} + \text{DM}. \tag{11.34}$$

So far no evidence for dark matter has emerged from laboratory searches. Other detectors that involve resonant cavities in a strong magnetic field, or other specialized designs, have been built to search for axions.

11.5 MOND – an Alternative to Dark Matter?

There does exist one somewhat controversial alternative to particle dark matter. MOdified Newtonian Dynamics (MOND) was proposed by cosmologist Mordehai Milgrom in 1983 and has had a turbulent life since. The idea is to fundamentally change the physical law in order to explain the phenomena usually ascribed to dark matter. A single, fixed parameter is claimed to be sufficient to explain the astrophysical observations of dynamics in and around objects of a huge range of masses, from $\sim 10^5 M_\odot$ dwarfs, to $\sim 10^{10} M_\odot$ galaxies, and arguably all the way to $\sim 10^{15} M_\odot$ clusters.

Milgrom proposed that *acceleration* around an astrophysical object of mass M, $a = GM/r^2$, be modified to take the form

$$\mu\left(\frac{a}{a_0}\right) a = \frac{GM}{r^2}, \tag{11.35}$$

where the function μ takes the following asymptotic limits:

$$\mu(x) = \begin{cases} 1 & x \gg 1 \\ x & x \ll 1 \end{cases} \tag{11.36}$$

with a single free parameter that has effectively been fit to data

$$a_0 \simeq 10^{-10} \frac{\text{m}}{\text{s}^2} \qquad \text{(MOND parameter)}. \tag{11.37}$$

Therefore, things would be unchanged near the centers of objects – gravity is Newtonian. In the regime of *low accelerations* (note the subtle difference between the similar regimes of low acceleration and great distance from the center of some object), Eq. (11.35) gives $a^2/a_0 = GM/r^2$, or

$$a = \frac{\sqrt{GMa_0}}{r}, \tag{11.38}$$

which, when combined with the response of velocity to acceleration $a \equiv v^2/r$, leads to

$$v = (GMa_0)^{1/4} = \text{const.} \tag{11.39}$$

MOND therefore explains flat rotation curves by fiat. In fact, on scales corresponding to the size of a galaxy or smaller, the MOND empirical law does surprisingly well, owing chiefly to the fact that the acceleration scale a_0 is basically constant across a wide range of mass/size scales.

The big challenge for MOND is creating a "father" relativistic theory from which the Milgrom rule, Eq. (11.35), can be derived. Various attempts – the most notable of them called TeVeS (Tensor–Vector–Scalar theory) – have managed to produce deflection of light and other effects that general relativity predicts given dark matter, but they still have problems. Notably, it appears that additional dark-matter components, say massive neutrinos, are required in galaxy clusters in addition to whatever effects are produced by MOND-type relativistic theories. Moreover, even the relativistic TeVeS-type theories typically have a hard time fitting the very large-scale cosmological observations – CMB angular power spectrum, galaxy clustering, and cosmic shear. On these large scales, observations clearly favor the cold-dark-matter scenario over MOND.

Bibliographical Notes

Dark matter is by now a grandfatherly 90-year-old subject, and there are many good reviews and books about it. The exposition in this chapter was influenced by an excellent basic introduction to the topic by Ryden (2016), and fine review articles by Strigari (2013), Lisanti (2017), Slatyer (2018), and Ferreira (2021). Some of the material for our discussion of dark-matter fraction in astrophysical objects loosely followed Rood (1981), Lang (1999), and Raine and Thomas (2001). I also highly recommend the review of the *history* of our understanding of dark matter by Bertone and Hooper (2018).

Problems

11.1 Local density of dark matter. Compare the local density of dark matter in Eq. (11.31) to the global average dark-matter density in the universe today. By what factor, roughly, is the former larger than the latter?

11.2 Virial theorem. Prove the virial theorem, Eq. (11.12). Assume N particles of mass m_i and position $\mathbf{r}_i$, and start with Newton's second law for the jth particle:

$$m_j \frac{d^2 \mathbf{r}_j}{dt^2} = \sum_{i \neq j} \frac{G m_i m_j}{|\mathbf{r}_i - \mathbf{r}_j|^3} (\mathbf{r}_i - \mathbf{r}_j),$$

then

- Multiply both sides with $\mathbf{r}_j$, and sum them over j.
- On the left-hand side, use the identity $(d^2 \mathbf{r}/dt^2)\mathbf{r} = d/dt(d\mathbf{r}/dt \cdot \mathbf{r}) - (d\mathbf{r}/dt)(d\mathbf{r}/dt)$, then write this as a second derivative of the moment of inertia plus a remainder.
- On the right-hand side, write $\mathbf{r}_j = 1/2[(\mathbf{r}_i + \mathbf{r}_j) - (\mathbf{r}_i - \mathbf{r}_j)]$ and recognize that one of the resulting terms is zero.

... and you should have a proof of the virial theorem.

11.3 Tremaine–Gunn bound. Tremaine and Gunn (1979) determined a lower limit on the mass of any fermion dark-matter particle by applying basic quantum-mechanical reasoning to dark-matter halos. The Pauli exclusion principle limits the number of fermionic particles in phase space, so if the mass of the particle is below a certain threshold, the particles' number density (for a fixed total mass) will be too high, and will violate the Pauli bound.

The total phase-space volume is

$$\mathcal{V} = \int f(\mathbf{x}, \mathbf{p}) d^3 x d^3 p \simeq R_{\text{halo}}^3 (m_{\text{ferm}} \sigma_v)^3,$$

where m_{ferm} is the mass of the fermionic dark-matter particle, R_{halo} is the size of the galaxy (dark-matter) halo, and σ_v is the velocity dispersion of dark matter. The maximum number of fermions that can fit in the phase-space element $\hbar^3$ is $N = \mathcal{V}/\hbar^3$. Hence the maximum total halo mass is

$$M_{\text{tot}} = N m_{\text{ferm}} = \frac{R_{\text{halo}}^3 m_{\text{ferm}}^4 \sigma_v^3}{\hbar^3}.$$

By requiring that this mass be equal to or greater than the virial mass (i.e., the mass obtained by applying the virial theorem), we can obtain the lower limit on the fermion mass; write it as

$$m_{\text{ferm}} \gtrsim A \left(\frac{R_{\text{halo}}}{10\,\text{kpc}} \right)^B \left(\frac{\sigma_v}{200\text{km/s}} \right)^C \text{eV},$$

where A, B, and C are constants that you should determine (you can round A to one or two decimal places).

11.4 **Minimum mass of ultra-light particles.** Particles with extremely small masses – the so-called ultra-light particles – are dark-matter candidates with a very large de Broglie wavelength

$$\lambda_{\text{dB}} \simeq \frac{\hbar}{p}.$$

On scales below λ_{dB}, clustering is suppressed due to "quantum pressure," a kind of quantum repulsion that resists gravity. Find the minimum mass of those particles by requiring that λ_{dB} not be larger than the typical size of a galactic dark-matter halo, $R_{\text{halo}} \simeq 100\,\text{kpc}$. Assume the ultra-light dark-matter particle velocity to be comparable to that of CDM particles.

11.5 **Free streaming.** Neutrinos and other hot- and warm-dark-matter candidates move so fast prior to matter–radiation equality that they erase structure below some characteristic comoving scale λ_{FS}.

Derive this scale. Start by evaluating the comoving distance

$$\lambda_{\text{FS}}(t) \equiv \frac{\lambda_{\text{FS}}^{\text{prop}}(t)}{a(t)} = \int \frac{v(t')dt'}{a(t')},$$

where $v(t')$ is the speed of relativistic particles, given by

$$v = \begin{cases} 1 & t < t_{\text{NR}} \\ \left(\dfrac{a}{a_{\text{NR}}} \right)^{-1} & t_{\text{NR}} < t < t_{\text{eq}}, \end{cases}$$

where t_{NR} is the time at which the neutrinos become non-relativistic. [The contribution to free-streaming length at $t > t_{\text{eq}}$ is negligible.] The reason that $v \propto a^{-1}$ for a non-relativistic particle in an expanding universe is that the momentum of a free particle just stretches with the expansion; of course $p = mv$, so v does the same.

Assume that the particle becomes non-relativistic when $T \simeq m_{\text{DM}}/3$. You may find it useful to use Eq. (4.37) to relate time t and temperature T. Assume further that g_* is equal to its value today, $g_* = g_{*,0}$, at the relatively recent times when the dark-matter particle becomes non-relativistic. Finally, your final answer should look like Eq. (11.21), but there should also be an additional logarithmic term in m_{DM} which contributes more massive particles and when $a_{\text{NR}} < a_{\text{eq}}$ (and which we ignored in Eq. (11.21)).

11.6 Hail to the NFW. The Navarro–Frenk–White density profile (Navarro *et al.*, 1997) describes the radial density profile, $\rho(r)$, of halos in the universe. Because luminous objects are dynamically dominated by dark matter, NFW is also a realistic density profile of galaxies (although, near the galaxy centers, baryons play an important role and typically steepen the profile). The NFW profile is defined as

$$\rho_{\text{NFW}}(r) = \frac{\rho_0}{\left(\dfrac{r}{r_s}\right)\left(1+\dfrac{r}{r_s}\right)^2},$$

where the density parameter ρ_0 can also be expressed in terms of the critical density times the **characteristic overdensity** for collapse δ_{char}:

$$\rho_0 = \delta_{\text{char}}\rho_{\text{crit}}.$$

The choice of δ_{char} is a bit arbitrary and depends on where we want to "cut off" the halo from its surroundings. One common choice is $\delta_{\text{char}} = 200$; in this definition, the halo is defined as the region in which the enclosed, averaged density of the halo is 200 times bigger than the critical density of the universe.

(a) Find the characteristic mass, $M_{200} \equiv M(r \leq r_{200})$, in terms of the central density ρ_0, scale radius r_s, and concentration parameter $c = r_{200}/r_s$. Here r_{200} is the radius in which the enclosed density of the object is 200 times bigger than the critical density (i.e., the "average" density of the universe). M_{200} is basically one way to define the mass of a halo.

(b) The NFW profile is determined solely by the halo's mass, M_{200}, and its concentration parameter, c. Find, for example, the characteristic overdensity δ_{char} in terms of concentration c (and nothing else!).

12 Dark Energy

The discovery that the universe's expansion is accelerating, made near the end of the twentieth century, sent shock waves through the field of cosmology. What is the origin of dark energy, the component that causes this accelerated expansion? In this chapter, we review this exciting discovery, as well as the known facts about dark energy and the mysteries that it presents. We then address the fascinating possible links between dark energy and our fundamental understanding of particles and forces in the universe.

12.1 The Discovery of Dark Energy

The discovery of dark energy is one of the great detective stories in cosmology, one that we now spend some time describing.

12.1.1 Early Evidence of Something Amiss

Inflationary theory – the most important idea in theoretical cosmology in the past 50 years – explains how tiny quantum-mechanical fluctuations in the early universe formed and grew to become structures that we see on the sky today. One of the factors that motivated inflation is that it also explains the observations that indicate that the universe is very close to being spatially flat. In fact, inflation convinced many theorists that the universe must be *precisely* flat.[1]

Around the same time that inflation was proposed, in the 1980s, a variety of dynamical probes of the large-scale structure in the universe were starting to indicate that the *matter* energy density is much lower than the value needed to make the universe flat. Perhaps the most specific case was made by the measurements of the clustering of galaxies, which are sensitive to the parameter combination $\Gamma \equiv \Omega_M h$ (which, as you may see in Eq. (9.64), controls the wavenumber $k_{\rm eq}$ at which the matter power spectrum turns over). The measured value at the time was $\Gamma \simeq 0.25$, with rather large errors however. One way to preserve a flat, matter-only universe was to postulate that the Hubble constant itself was much lower than the measurements indicated ($0.5 \lesssim h \lesssim 1$ at the time), so that $\Omega_M = 1$ but $h \sim 0.3$! This

---

[1] More precisely, the prediction from standard inflation is $\Omega_{\rm TOT} = 1 \pm 10^{-5}$, where the plus-or-minus theory rms variation comes from the fact that horizon-scale modes of amplitude $\delta \sim 10^{-5}$ induce tiny departures from flatness of the same magnitude.

proposal would have explained the large-scale cosmological observations, but had the obvious problem of being in conflict with direct measurements of the Hubble constant. Another way out was to postulate the presence of Einstein's cosmological constant, as we discussed in Chapter 3. The cosmological-constant term in Einstein's equations has $P = -\rho$, and contributes to the total energy density while generally behaving very differently than matter (in particular, it slows down the growth of density fluctuations, as explained in Chapter 9). The cosmological constant was suggested as far back as 1984 as a possible missing ingredient that could alleviate the tension between data and matter-only theoretical predictions; specifically, it would allow a flat universe with a low value of the matter density. It turns out that this second possibility – that of a new component with $P \simeq -\rho$ – was correct.

The revolutionary discovery of the accelerating universe took place in the late 1990s, but to understand its astrophysical origins, we have to step a few decades back in time.

12.1.2 Standard Candles

It is very difficult to measure *distances* in astronomy. Contrast this with the relatively straightforward measurements of angular locations of objects; or else the in-principle easy access to objects, redshifts from the measurements of their spectra. Accurate distance measurements are a hard nut to crack, however. Getting them typically involves empirical – and uncertain – astronomical methods: parallax, period–luminosity relation (Leavitt law) of Cepheids, main-sequence fitting, surface brightness fluctuations, etc. Typically, astronomers construct an unwieldy "distance ladder" to measure distance to a galaxy: they use one of these relations (say, parallaxes – apparent shifts due to Earth's motion around the Sun) to calibrate distances to nearby objects (e.g., variable stars Cepheids), then go from those objects to more distant ones using another relation that works better in that distance regime. In this process the systematic errors add up, making the distance ladder flimsy.

"**Standard candles**" are hypothetical objects that have a nearly fixed luminosity, that is, fixed total power that they radiate. Having standard candles would be useful since then we could infer distances to objects just by using the flux–luminosity inverse square law

$$f = \frac{L}{4\pi d_L^2}, \tag{12.1}$$

where d_L is the luminosity distance which can therefore be inferred by comparing the measured flux to the known (fixed) luminosity of the standard candle. Note that the distance and redshift, both measured, are related by the theory formula that depends on the energy density of components in the universe (see Chapter 3). Therefore, by measuring d_L and z, we can constrain what the universe is made up of. In fact, we don't even need to know the luminosity of the standard candle; all we require is measurements of the *relative* luminosities of, and hence distances to, objects in order to perform this test.

In astronomy, flux is expressed in terms of **apparent magnitude** m, which is a logarithmic measure of flux. Luminosity, on the other hand, is related to the **absolute magnitude** M of the object. The apparent and absolute magnitude, m and M, are respectively defined as

$$m \equiv -2.5 \log_{10} \left(\frac{f}{f_x} \right)$$

$$M \equiv -2.5 \log_{10} \left(\frac{L}{L_x} \right), \tag{12.2}$$

where f_x and L_x are some fixed but as yet unspecified values of the flux and luminosity. The difference between m and M, known as the **distance modulus**, is then a measure of distance

$$m - M = 2.5 \log_{10} \left(\frac{L}{f} \right) + \text{const.} = 5 \log_{10} \left(\frac{d_L}{10\,\text{pc}} \right), \tag{12.3}$$

where f_x and L_x have been conventionally chosen so that, for an object which is 10 parsecs away, the distance modulus is zero. For a standard candle, the absolute magnitude M (or, equivalently, luminosity L) is known to be approximately the same for each object. Therefore, measurements of the apparent magnitude to each object provide information about the luminosity distance, and thus the makeup of the universe. The plot of $m(z)$ vs. z of all objects in a survey – the Hubble diagram – is then sensitive to the cosmological parameters that enter the luminosity distance.

12.1.3 Type Ia Supernovae

Having access to standard-candle objects sounds fantastic. But where could we find them? It turns out that well-known astrophysical events called type Ia supernovae fit the bill.

Type Ia supernovae (SN Ia) are explosions seen to distances of many gigaparsecs, and are thought to be the cases where a rotating carbon–oxygen white dwarf accretes matter from a companion star, approaches the Chandrasekhar limit, starts thermonuclear burning, and then explodes. The Ia nomenclature refers to spectra of SN Ia, which have no hydrogen, but show a prominent silicon (Si II) line at $6150\,\text{Å}$. Light from type Ia supernovae brightens and fades over a period of about a month (in the SN Ia rest frame). At its peak flux, an SN Ia can be a sizable fraction of the luminosity of the entire galaxy in which it resides. SN Ia had been studied extensively by Fritz Zwicky who also gave them their name, and by Walter Baade who noted that SN Ia have very uniform luminosities.

The fact that SN Ia can potentially be used as standard candles has been realized long ago, at least as far back as the late 1960s. However, scheduling telescopes to detect and follow up (observe multiple times as the SN Ia luminosity increases then decreases) SN Ia that are discovered presented a real challenge. If you simply point a telescope at a galaxy and wait for the SN Ia to go off, you will wait an average of ∼100 years. There had been a program in the 1980s to find supernovae Ia but, partly due to the inadequate technology and equipment available at the time, it discovered only one SN Ia, and after the peak of its light curve.

Four crucial developments contributed to the robustness of using SN Ia as cosmological tools, which in turn led to the discovery of dark energy. These developments are:

1. The first major breakthrough came in the 1990s when two teams of SN researchers – Supernova Cosmology Project (SCP; led by Saul Perlmutter and organized in the late 1980s) and High-z Supernova Search Team (High-z; organized in the mid 1990s and led by Brian Schmidt) – developed an efficient approach to use the world's most powerful telescopes working in concert to discover and follow up high-redshift SN, and thus complement the existing efforts at lower redshift led by the Calán/Tololo collaboration. These teams had been able to essentially guarantee that they would find batches of SN in each run.

2. The second breakthrough came in 1993 by Mark Phillips, an astronomer working in Chile. Based upon earlier similar findings by Pskovskii and Rust, he noticed that the SN Ia luminosity – or absolute magnitude — is correlated with the decay time of the SN light curve. Phillips considered the quantity Δm_{15}, the attenuation of the (log of) flux of the SN between the light maximum and 15 days past the maximum. He found that Δm_{15} is strongly correlated with the absolute magnitude – so, the log luminosity – of SN Ia; see the left panel of Fig. 12.1. The "Phillips relation" roughly goes as

> Broader is brighter.

[This mnemonic, while easy to remember, is slightly inaccurate in that the broader light curves have a larger luminosity, or *intrinsic* brightness.]

The Phillips relation therefore says that supernovae with broader light curves have a larger intrinsic luminosity. One way to quantify this relation is to use a "stretch" factor which is a (calibration) parameter that measures the width of a light curve; see Fig. 12.1(b). By applying the correction based upon the Phillips relation, astronomers found that the intrinsic dispersion of SN, which is of order ∼ 0.5 magnitudes, can be brought down to $\delta m \sim 0.2$ magnitudes once we correct each SN Ia luminosity using its measured stretch factor. The final dispersion in magnitudes corresponds to the error in distance of ballpark 10 percent:

$$\frac{\delta d_L}{d_L} = \frac{\ln 10}{5} \delta m \simeq 0.5 \, \delta m \simeq 0.1. \tag{12.4}$$

The Phillips relation was the second key ingredient that enabled SN Ia to achieve the precision needed to probe the contents of the universe accurately.

3. The third key invention was the development of techniques to correct SN Ia magnitudes for dimming by dust, or "extinction," out of multi-color observations

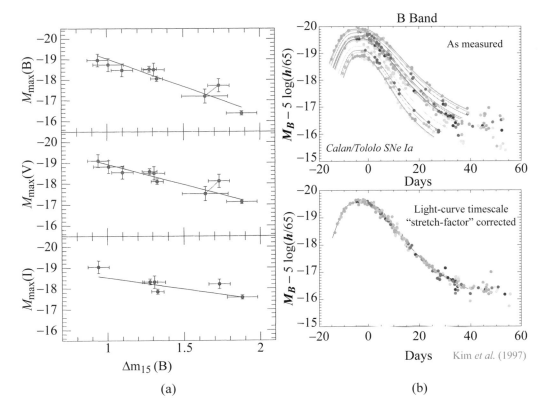

Fig. 12.1 (a) Phillips relation, adopted from his paper and reproduced by permission from Phillips (1993), © AAS. The absolute magnitude of type Ia supernovae, M, is correlated with Δm_{15}, the observed decay of the light-curve light 15 days after the maximum. (b) Light curves of a sample of SNe Ia before correction for the Phillips relation (top) and after (bottom). Figure courtesy of Alex Kim. A black and white version of this figure will appear in some formats. For the color version, please refer to the plate section.

of supernova light. Such corrections are an important part of SN cosmology to this day.

4. The fourth and perhaps the most important ingredient for the discovery of dark energy was the development and application of charge-coupled devices (CCDs) in observational astronomy. Both teams of SN hunters used the CCDs, which had been installed at telescopes at Kitt Peak and Cerro Tololo.

Some of the early results came out in 1997 and paradoxically indicated that the universe is matter dominated and consistent with being flat, but those results included only a handful of SN Ia and therefore large error bars. The definitive results from both the High-z (Riess *et al.*, 1998) and SCP team (Perlmutter *et al.*, 1999) came out shortly thereafter, and indicated that the universe is dominated by a component with negative pressure. This component was thereafter named "dark energy" by cosmologist Michael Turner (Huterer and Turner, 1999).

12.1.4 Cosmology with SN Ia

Starting with Eq. (12.3), one can apply a few simple algebraic operations to get

$$m = M + 5\log_{10}\left(\frac{d_L}{1\,\mathrm{Mpc}}\right) + 25$$

$$\equiv 5\log_{10}(H_0 d_L) + \mathcal{M},$$

(12.5)

where the "script-M" factor is defined as

$$\mathcal{M} \equiv M - 5\log_{10}\left(H_0 \times 1\,\mathrm{Mpc}\right) + 25.$$

(12.6)

To summarize, the relation between the observed apparent magnitude and the inferred luminosity distance is

$$m = 5\log_{10}(H_0 d_L) + \mathcal{M}.$$

(12.7)

Thus, SN Ia measure *relative* distances (see Problem 12.2).

Note that $\mathcal{M}$ is a nuisance parameter that captures *two* uncertain quantities: the absolute magnitude (i.e., intrinsic luminosity) of a supernova, M, and the Hubble constant H_0. Because we do not know M, even accurate independent measurements of H_0 would not help us independently pin down $\mathcal{M}$. Conversely, and for the same reason, SN Ia alone are completely insensitive to the value of H_0. Thus we conclude with one of the main (and sometimes misunderstood) features of SN Ia cosmological analysis:

> SN Ia data measure relative distances, as well as a single nuisance parameter that combines H_0 and the absolute magnitude of SN Ia.

There *is* actually a way to independently get at $\mathcal{M}$, and therefore the Hubble constant, which is to use the distance ladder. Some nearby galaxies hosting SN Ia are also hosts of pulsating stars, Cepheids, which also serve as absolute distance indicators (Cepheids themselves are calibrated by systems in which absolute distance can be measured from the parallax). Because they provide absolute distance measurements, Cepheids could be used to "anchor" the Hubble diagram – that is, to determine absolute distances to nearby SN Ia and, by extension, provide the absolute scale for the whole Hubble diagram. In our language from Eq. (12.7), Cepheids are used to determine the absolute magnitude M, while more distant SN Ia measure $\mathcal{M}$; the two jointly[2] then determine the Hubble constant H_0.

---

[2] The step-by-step explanation here is pedagogical; in practice, one typically carries out a joint analysis of Cepheids and all SN Ia (both the nearby SN Ia with Cepheids in the same galaxies

To summarize the use of SN Ia in cosmology: astronomers measure m (commonly taken to be the apparent magnitude at the peak of the light curve). They then measure the redshift of the host galaxy of those same SN Ia. With a bunch of SN Ia, they can marginalize over the parameter $\mathcal{M}$ and be left with constraints on the cosmological parameters that enter the luminosity distance, namely the various energy densities relative to critical, Ω_i, as well as the dark-energy equation of state parameter w.

The definitive results of the High-z and SCP teams that came out in 1998–99 were a shot heard around the world. They indicated that SN Ia fluxes are dimmer, and thus their distances are greater, than would be expected in a matter-only universe with *any* amount of curvature. Figure 12.2 illustrates this.

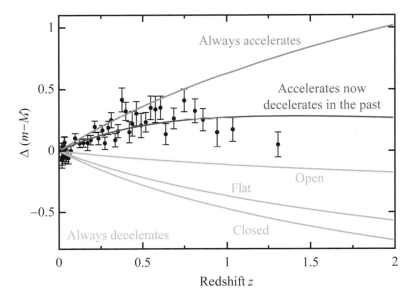

Fig. 12.2 Hubble diagram for the current SN Ia "Pantheon" dataset of 1048 SN Ia (Scolnic *et al.*, 2018). We have binned (grouped) SN Ia measurements in redshift for easier viewing; each datapoint shows the average of about 20 objects, with the error bar suitably scaled. We also show several theory models to guide the eye; the model favored by the data is the "accelerates now, decelerated in the past" model with $\Omega_M \simeq 1 - \Omega_\Lambda \simeq 0.3$. The vertical axis shows the distance modulus relative to some (unimportant) fiducial case. Remember that all theory curves are allowed to slide vertically (corresponding to an *a-priori* unknown $\mathcal{M}$ parameter); it is only the *shape* of the measured Hubble diagram that informs us about the cosmological model. A black and white version of this figure will appear in some formats. For the color version, please refer to the plate section.

and the SN Ia farther away) to simultaneously constrain the Hubble constant along with the cosmological parameters.

The two supernova-hunter teams' results were accepted by the community fairly quickly, given the remarkable agreement between the competing teams' results, as well as the notion that the missing piece of the puzzle – the long-standing tension between $\Omega_M \simeq 0.2 - 0.3$ favored by galaxy surveys and $\Omega_{\rm TOT} = 1$ favored by inflation – has finally been found. In subsequent years, SN Ia results were confirmed with a variety of completely independent observations that indicated a new standard model of cosmology, with about one-third of energy density in matter and two-thirds in the mysterious dark-energy component that causes the acceleration (see Fig. 12.3). The discovery of dark energy was a watershed event in modern cosmology,

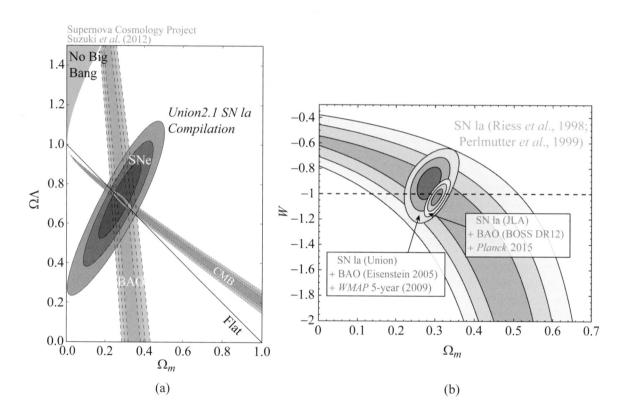

(a)   (b)

Fig. 12.3 (a) Constraints on $\Omega_{\rm M}$ and Ω_Λ in the consensus model using SN, baryon acoustic oscillation (BAO), and CMB measurements. Reprinted with permission from Suzuki *et al.* (2012), © AAS. (b) Constraints on $\Omega_{\rm M}$ and w in the wCDM cosmological model using the combination of SN Ia, BAO, and CMB. The plot shows the evolution of the constraints over the period of 20 years; constraints on panel (a) roughly correspond to data in the red contour on the right, while the blue contour is the more recent data. Reprinted with permission of IOP publishing, from Huterer and Shafer (2018); permission conveyed through Copyright Clearance Center Inc. A black and white version of this figure will appear in some formats. For the color version, please refer to the plate section.

and was rewarded with the 2011 Nobel Prize in Physics to Saul Perlmutter, Adam Riess, and Brian Schmidt.

12.2 Describing and Measuring Dark Energy

We now recapitulate our knowledge about dark energy, then briefly discuss how we describe it and measure it using cosmological probes.

12.2.1 What we Know About Dark Energy

After all this fuss about its discovery, the next logical question is: what *is* dark energy? We simply do not know! The physical nature of dark energy is one of the great mysteries of cosmology and astrophysics, but also theoretical particle physics today. Further below, we will return to this question by examining some attempts to understand the physical cause of the accelerated universe.

Meanwhile, what we do know is that:

- dark energy is spatially smooth;
- it leads to an accelerated expansion of the universe;
- as a corollary, it slows the growth of cosmic structure (discussed in Chapter 9), and leads to an older universe (as discussed in Chapter 3).

All of these facts are in excellent agreement with essentially all measurements in cosmology today. In addition to the accelerated expansion, we have also observed the slowed growth of structure in observations of weak gravitational lensing and galaxy clustering, and the abundance of clusters of galaxies. And while there is no guarantee that dark energy is spatially smooth, there exist fairly stringent constraints on this from combined cosmological measurements.

12.2.2 A Cosmic Coincidence?

Dark matter and energy densities are measured to be comparable today, with $\rho_{DE} : \rho_M \simeq 2 : 1$. Yet they scale very differently with redshift; $\rho_{DE} \propto (1+z)^{(1+w)} \simeq (1+z)^0$, while $\rho_M \propto (1+z)^3$. Hence, they evolve very differently in cosmic time. In the past, dark matter (and, still earlier, radiation) was completely dominant; this is something we know from cosmological measurements, as a significant dark-energy contribution in the past, for example $\Omega_{DE}(z \simeq 1000) \gtrsim 0.1$, would spoil predictions of theory to the CMB and other measurements. And in the future, dark energy will presumably dominate; it already does so today. Hence the question (illustrated in Fig. 12.4): why do we *just happen* to live in an era when dark matter and dark energy are comparable?

Unlike the cosmological-constant problem discussed below, this coincidence problem is not a rigorous physical tension, but rather represents a "guess what happened

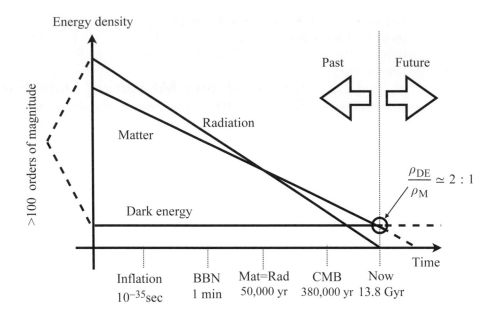

Fig. 12.4 Illustration of the dark-energy coincidence problem. At the present time, the dark energy-to-dark matter ratio is 70 : 30 percent, or about 2:1. In the past, dark matter (and, still earlier, radiation as well) were much greater than dark energy. In the future, dark energy will presumably totally dominate the energy density budget. Why do we just happen to live in an era when dark matter and dark energy are comparable?

to me today?" observation which is specific to the time when the coincidence is observed. Hence, perhaps we should not be taking it too seriously. Nevertheless, it is worth noting that numerous theory explanations for dark energy (e.g., scalar-field models, discussed below) were proposed specifically in order to try to alleviate this apparent coincidence, by making the $\rho_{\mathrm{DE}} \simeq \rho_M$ a less sharp event in the history of the universe.

12.2.3 Describing Dark Energy

The absence of a consensus model for cosmic acceleration presents a challenge in trying to connect theory with observations. The simplest description of the dark energy is obtained by assuming vacuum energy, corresponding to the cosmological-constant term, with $w \equiv P/\rho = -1$. Then a single-parameter energy density relative to critical describes dark energy.

More freedom in the dark-energy sector is provided by allowing the equation-of-state parameter w to vary. Because w is the ratio of pressure to energy density, it is also closely connected to the underlying physics, making this a rather well-motivated extension. Specifically, this form fully describes vacuum energy ($w = -1$) or topological defects ($w = -N/3$, with N an integer dimension of the defect – 0 for monopoles, 1 for strings, 2 for domain walls). Together with $\Omega_{\rm DE}$ and $\Omega_{\rm M}$, w provides a three-parameter description of the dark-energy sector (or two parameters if flatness is assumed). However, it does not describe scalar field or modified gravity models, which generically have a time-varying w. Historically, constraints on (constant) w were very weak, but steadily improved, as seen in Fig. 12.3(b); the constraint from Planck and BAO data quoted in our Table 3.1 is $w = -1.04 \pm 0.06$.

A number of two-parameter descriptions of w have been explored in the literature, for example using the next term in a simple Taylor expansion in redshift, $w(z) = w_0 + w'z$. For low redshift they are all essentially equivalent, but for large z, some lead to unrealistic behavior, for example $w \ll -1$ or $\gg 1$. The parametrization

$$w(a) = w_0 + w_a(1 - a) = w_0 + w_a z/(1 + z) \tag{12.8}$$

avoids this problem and leads to the most commonly used description of dark energy, namely[3] $(\Omega_{\rm DE}, w_0, w_a)$ or, in a non-flat cosmological model, $(\Omega_{\rm DE}, \Omega_{\rm M}, w_0, w_a)$.

More generally, one could consider an arbitrary redshift-dependent equation of state, $w(z)$. From the continuity equation

$$\dot{\rho} + 3H(P + \rho) = 0, \quad \text{or}$$
$$\frac{d\ln\rho}{d\ln a} = -3(1 + w(a)), \tag{12.9}$$

we can calculate the dark-energy density; phrased in terms of redshift, it is

$$\frac{\rho_{\rm DE}(z)}{\rho_{\rm DE}(0)} = \exp\left(3\int_0^z (1 + w(z'))d\ln(1 + z')\right). \tag{12.10}$$

Measuring $w(z)$ without *any* assumptions about its smoothness is impossible, as one would have to constrain infinitely many parameters. Fortunately, schemes exist where suitable effective priors are imposed that do enable measurements that constrain the time variation of the equation of state.

Finally note that measurements of constant w, or the time-varying (w_0, w_a) or any other $w(z)$ description, are very useful regardless of the nature of the true dark-energy model. This is because, whatever the true physical nature of dark energy is, the equation of state can be considered as a purely phenomenological description of the expansion history of the universe, thus encoding important information that effectively compresses the information from the measurements. In particular, one can derive the redshift dependence of $w(z)$ in terms of the first and second derivatives

---

[3] Because $\Omega_M + \Omega_{\rm DE} = 1$ in a flat universe, a more popular but completely equivalent parameterization of this model is (Ω_M, w_0, w_a).

of the comoving distance $r(z)$ (see Problem 12.6):

$$1 + w(z) = \frac{1+z}{3} \frac{3H_0^2\Omega_M(1+z)^2 + 2(d^2r/dz^2)/(dr/dz)^3}{H_0^2\Omega_M(1+z)^3 - (dr/dz)^{-2}}, \qquad (12.11)$$

where we have assumed a flat universe. Therefore, the information content of $w(z)$, along with that of the physical matter density $\Omega_M h^2$, is equivalent to that of the distance-of-z relation, $r(z)$.

To summarize, three popular parameterizations of dark energy, with increasing capability to fit data (but decreasing accuracy in the respective parameters), are

$$\text{dark-energy parameterization} = \{\Omega_\Lambda\} \qquad (\Lambda\text{CDM})$$

$$= \{\Omega_{DE}, w\} \qquad (\text{wCDM}) \qquad (12.12)$$

$$= \{\Omega_{DE}, w_0, w_a\} \qquad (w_0 w_a\text{CDM}),$$

where ΛCDM, wCDM, and $w_0 w_a$CDM are names given to cosmological models with these dark-energy parameterizations.

12.2.4 Probes of Dark Energy

A number of cosmological probes are sensitive to dark energy, as the latter affects both the geometrical quantities (e.g., distances) and the growth of cosmic structure. There are arguably four principal probes of dark energy:

1. Type Ia supernovae (SN Ia).
2. Baryon acoustic oscillations (BAO).
3. Weak gravitational lensing.
4. The abundance of massive clusters of galaxies.

SN Ia were discussed earlier in this chapter, the BAO and cluster abundance in Chapter 9, and weak lensing will be discussed in Chapter 14. In addition, there are several more probes that are important, and here we outline two:

5. The angular power spectrum of CMB fluctuations, which does not probe dark energy by itself, but does provide a very important prior for other measurements by determining a *single* but very precise measurement of the angular diameter distance out to $z \simeq 1000$ (see the discussion below Eq. (13.62) in Chapter 13).
6. Gravitational-wave inspiral events (merging black holes or neutron stars), which serve as "standard sirens" by providing an absolute measurement of the luminosity distance. An independent measurement of the redshift to the host galaxy then enables dark-energy constraints (see Problem 12.4).

Gravitational-wave standard sirens do not yet provide significant constraints on dark energy, but are expected to do so in the near future, when more events out to higher redshifts are detected and followed up by ground and (hopefully) space detectors.

12.3 Theoretical Considerations

The theory behind dark energy is both fascinating and unsettled, as we now explain.

12.3.1 The Cosmological Constant...

In quantum field theory, vacuum is not characterized by a completely empty space, as spontaneous particle–antiparticle creations and annihilations are allowed. While it is very unlikely that you can ever get a basketball and an anti-basketball out of empty space, creation of (say) an electron–positron pair is not unlikely. The more massive the two particles are, the less time they are likely to stick around before annihilating; this is governed by the Heisenberg uncertainty condition $\Delta E \Delta t \simeq \hbar$. This process of creations and annihilations therefore gives vacuum a certain energy. It turns out that the vacuum energy can be described by a single constant term in Einstein's (and Friedmann's) equations, Λ.

General covariance of Einstein's equations dictates that, if we add an energy density term Λ to the stress–energy tensor, then the term should be proportional to the metric $g_{\mu\nu}$ which, in flat space, is proportional to the diagonal form $\mathrm{diag}(-1,1,1,1)$. The term $\Lambda g_{\mu\nu}$ then contributes to $8\pi G T_{\mu\nu}$, where $T_{\mu\nu}$ is the stress–energy tensor which, for components without anisotropic stress, looks like $\mathrm{diag}(\rho, P, P, P)$. If we link the vacuum energy density to the Λ term,

$$\rho_{\mathrm{vac}} = \frac{\Lambda}{8\pi G}, \tag{12.13}$$

then the discussion above implies that

$$P_{\mathrm{vac}} = -\rho_{\mathrm{vac}}. \tag{12.14}$$

Hence, vacuum pressure is equal to minus its energy density (or, restoring the MKS units, $P_{\mathrm{vac}} = -\rho_{\mathrm{vac}}c^2$). In yet other words, $w = -1$ exactly for vacuum energy. With that, vacuum energy can be incorporated in the usual way into equations for the Hubble parameter and the distances, as we have already done in Chapter 3.

In 1917 Einstein introduced the cosmological-constant term, Λ, to the field equations of general relativity in order to produce a static, finite cosmological model. With the subsequent discovery of the expansion of the universe, the motivation for this term disappeared and Einstein allegedly referred to it as his "greatest blunder" (see Sec. 3.2.3 where we discussed this). Nevertheless, the motivation for introducing such a term was reconsidered many decades later, especially since the discovery of the accelerating universe. The cosmological-constant term is a leading – phenomenological, at least (as we discuss next) – explanation for the accelerating universe.

12.3.2 ... and the Problem with it

Dark energy *could* physically be the energy of vacuum, given that $w = -1$ is an excellent fit to present-day cosmological data (see Fig. 12.3(b)). At this point we have an existing, proven physical mechanism that could be the dark energy that powers the accelerated expansion. So do we have a clear winner in the dark-energy theory competition? No, because there is a problem.

The energy density required to explain the accelerated expansion is about three-quarters of the critical density, or about

$$\rho_{\text{DE},0} \simeq \rho_{\text{crit},0} = \frac{3H_0^2}{8\pi G} \sim 10^{-26} \frac{\text{kg}}{\text{m}^3} \sim 4 \times 10^{-47} \,\text{GeV}^4 \simeq (3 \times 10^{-3}\text{eV})^4. \quad (12.15)$$

This is tiny compared to energy scales in particle physics (with the exception of neutrino mass differences). Even more seriously, it is incredibly small relative to what we would theoretically estimate for vacuum energy, which we can see as follows. We can model the energies associated with quantum fluctuations in the vacuum with harmonic oscillators. For each mode of a quantum field there is a zero-point energy $\hbar\omega/2$ or, in natural units, $(1/2)\sqrt{k^2 + m^2}$, where m is the mass of the particle and k is its momentum. The total energy density of vacuum quanta is then obtained by summing over all momenta:[4]

$$\rho_{\text{vac}} = \frac{1}{2} \sum_{\text{fields}} g_i \int_0^{k_{\text{max}}} \sqrt{k^2 + m^2} \, \frac{d^3k}{(2\pi)^3} \simeq \sum_{\text{fields}} \frac{g_i k_{\text{max}}^4}{16\pi^2}, \quad (12.16)$$

where g_i accounts for the degrees of freedom of the field (the sign of g_i is plus for bosons and minus for fermions), and the sum runs over all quantum fields (quarks, leptons, gauge fields, etc.). Here, k_{max} is an imposed momentum cutoff, because the sum diverges quartically in the ultraviolet – that is, at high wavenumber or energy.

To illustrate the magnitude of the problem, let us assume just one field, and require that the energy density it contributes does not exceed the critical density. Then the cutoff k_{max} must be $<0.01\,\text{eV}$ – *well* below a typical energy scale beyond which we could reasonably argue for new physics (which is something like a SUSY or GUT or Planck scale, many tens of orders of magnitude higher). Wolfgang Pauli apparently carried out this calculation in the 1930s, using the electron mass scale for k_{max} and finding that the size of the universe, that is, H^{-1}, "could not even reach to the Moon" (in his words) – because such high energy density would dramatically overclose the universe. Taking the cutoff to be the Planck scale ($\approx 10^{19}\,\text{GeV}$), where one expects quantum field theory in a classical spacetime metric to break down, the zero-point energy density would be

$$\rho_{\text{vac}} \sim k_{\text{Pl}}^4 = m_{\text{Pl}}^4 \sim (10^{19}\,\text{GeV})^4 = 10^{76}\,\text{GeV}^4, \quad (12.17)$$

---

[4] There is a subtlety in this calculation as set up here, in that it doesn't preserve Lorentz invariance. The correct calculation gives essentially the same result but is longer, so for brevity and clarity, we keep this widely used albeit imperfect argument.

thus exceeding the critical density by some 120 orders of magnitude! This very large discrepancy is known as the **cosmological-constant problem**.

Supersymmetry, the hypothetical symmetry between bosons and fermions, appears to provide only partial help. In a supersymmetric (SUSY) world, every fermion in the Standard Model of particle physics has an equal-mass SUSY bosonic partner and vice versa, so that fermionic and bosonic zero-point contributions to ρ_{vac} would exactly cancel. However, SUSY is not a manifest symmetry in nature; none of the SUSY particles has yet been observed in collider experiments, so they must be substantially heavier than their Standard-Model partners. If SUSY is spontaneously broken at a mass scale m_{SUSY}, one expects the imperfect cancellations to generate a finite vacuum energy density. For the currently favored value $m_{\text{SUSY}} \gtrsim 10$ TeV, this leads to $\rho_{\text{vac}} \simeq m_{\text{SUSY}}^4 \gtrsim 10^{12}\,\text{GeV}^4$, implying a discrepancy with observations of >50 (as opposed to 120) orders of magnitude; see Eq. (12.15). We conclude that

> The cosmological-constant problem is one of the top unsolved problems in physics. It may be resolved by as yet unknown physics beyond that encompassed in the Standard Model of particle physics.

Note that, if dark energy is *not* due to vacuum energy (or Lambda), then the cosmological-constant problem remains, but in a very slightly milder form: why is the vacuum energy (presumably) precisely zero, as opposed to the huge value predicted by quantum field theory?

Ongoing and future cosmological measurements will sharply test the cosmological-constant scenario for dark energy by measuring the equation of state w, as well as performing consistency tests of the ΛCDM model such as those described further below.

12.3.3 Scalar Fields

Vacuum energy does not vary with space or time and is not dynamical. However, by introducing a new degree of freedom, a scalar field ϕ, one can make vacuum energy effectively dynamical. For a free (i.e., with no interactions) scalar field ϕ, the stress energy takes the form of a perfect fluid, with

$$\rho = \dot{\phi}^2/2 + V(\phi), \quad P = \dot{\phi}^2/2 - V(\phi), \tag{12.18}$$

where ϕ is assumed to be spatially homogeneous, that is $\phi(\mathbf{x}, t) = \phi(t)$, $\dot{\phi}^2/2$ is the kinetic energy, and $V(\phi)$ is the potential energy. Note that we already encountered these expressions in the context of inflation – the physics is essentially the same, albeit at an energy scale (as we will see below) some 60 orders of magnitude lower than during the inflationary epoch. The evolution of the field is governed by the Klein–Gordon equation

$$\ddot{\phi} + 3H\dot{\phi} + V'(\phi) = 0, \tag{12.19}$$

where a prime denotes differentiation with respect to ϕ. Scalar-field dark energy can be described by the equation-of-state parameter

$$w = \frac{\dot{\phi}^2/2 - V(\phi)}{\dot{\phi}^2/2 + V(\phi)}. \tag{12.20}$$

If the scalar field evolves slowly, $\dot{\phi}^2/2V \ll 1$, then $w \approx -1$, and the scalar field behaves like a slowly varying vacuum energy, with $\rho_{\text{vac}}(t) \simeq V[\phi(t)]$. Equation (12.20) shows that w can take on any value between -1 (rolling very slowly) and $+1$ (evolving very rapidly), and it typically varies with time.

Scalar-field models raise new questions: is cosmic acceleration related to inflation; and is dark energy related to dark matter or neutrino mass? No firm or compelling connections have been made to either, although the possibilities are intriguing. Unlike vacuum energy, which must be spatially uniform, scalar-field dark energy can clump, providing a possible new observational feature, but in most cases is only expected to do so on the largest observable scales today, which are challenging to probe.

Scalar-field dark energy opens up new phenomenology, but does not address the cosmological-constant problem because these theories simply assume that the minimum value of $V(\phi)$ is very small or zero. Scalar-field models also pose new challenges: in order to roll slowly enough to produce accelerated expansion, the effective mass of the scalar field must be very light compared to other mass scales in particle physics:

$$m_\phi \equiv \sqrt{V''(\phi)} \lesssim 3H_0 \simeq 10^{-33}\,\text{eV}, \tag{12.21}$$

even though the field amplitude is typically of order the Planck scale, $\phi \sim 10^{19}$ GeV. The phenomenally small mass seen in Eq. (12.21) is another reflection of the fine-tunings that a successful dark-energy theory needs to incorporate and, eventually, explain.

12.3.4 Modified Gravity

A very different approach holds that cosmic acceleration is a manifestation of new gravitational physics rather than dark energy (i.e., that it involves a modification of the geometric as opposed to the stress-tensor side of the Einstein equations). Assuming that 4D spacetime can still be described by a metric, the operational changes are twofold: (1) a new version of the Friedmann equation governing the evolution of $a(t)$; (2) modifications to the equations that govern the growth of the density perturbations that evolve into large-scale structure. A number of ideas have been explored along these lines, from models motivated by higher-dimensional theories and string theory to phenomenological modifications of the general relativity theory.

Changes to the Friedmann equation are in principle easy to derive, discuss, and analyze. In order not to spoil the success of the standard cosmology at early times (from Big Bang nucleosynthesis to the CMB anisotropy to the formation of structure), the Friedmann equation must reduce to the GR form for $z \gg 1$. As a specific example, consider the model of Dvali, Gabadadze, and Porrati (DGP), which arises from a five-dimensional (5D) gravity theory and has a 4D Friedmann equation,

$$H^2 = \frac{8\pi G\rho}{3} + \frac{H}{r_c}, \tag{12.22}$$

where r_c is a length scale related to the 5D gravitational constant. As the energy density in matter and radiation, ρ, becomes small, there is an accelerating solution, with $H = 1/r_c$. From the viewpoint of expansion, the additional term in the Friedmann equation has the same effect as dark energy that has an equation-of-state parameter which evolves from $w = -1/2$ (for $z \gg 1$) to $w = -1$ in the distant future. While attractive, it is not clear that a consistent model with this dynamical behavior exists.

Modified-gravity explanations for the accelerating universe lead to much new phenomenology as well as new signatures and predictions that are observable with current and future cosmological and astrophysical observations. We do not discuss them here any more, but direct the reader to the rich literature on the subject, as well as Box 12.1 where we discuss ways to distinguish between modified gravity and "normal" (fluid) dark energy.

12.3.5 Anthropic Principle

The challenge presented by the cosmological-constant problem has led some in the cosmology community to desperate measures. This refers to their subscription to the **anthropic principle** which says, roughly: things are the way they are, because if they were different, life as we know it (or, less dramatically, galaxies with stars in them) wouldn't exist.

For example, why is Earth about one astronomical unit (AU) distant from the Sun? The anthropic answer is: if it were much closer, it would be too hot for life to form; if it were much farther, it would be too cold. You get the idea.

The anthropic explanation for the cosmological-constant problem starts from the postulate that the value of the vacuum energy is a random variable which can take on different values in different causally disconnected pieces of the universe. Because a value much larger than needed to explain the observed cosmic acceleration would preclude the formation of galaxies, we could not find ourselves in such a region. Remarkably, Steven Weinberg used such an argument in 1989, well before cosmic acceleration was discovered, to predict a nonzero, positive cosmological constant of energy density $O(10 - 100)$ times the matter energy density.

This anthropic approach finds a possible home in the landscape version of string theory, in which the number of different vacuum states is very large and essentially all values of the cosmological constant are possible. Very roughly speaking, there may be $\sim 10^{500}$ so-called string vacua that correspond to different universes, but

Box 12.1	Dark Energy or Modified Gravity?

Is dark energy due to a new mass/energy component (e.g., a fluid) in the universe, or else due to a modification of gravity that leads to the apparent observed acceleration? Let us illustrate the fundamental difficulty of separating between these two possibilities. Consider the Friedmann I equation in two alternative forms:

$$H^2 - F(H) = \frac{8\pi G}{3}\rho_M \quad \text{vs.} \quad H^2 = \frac{8\pi G}{3}\left(\rho_M + \frac{3F(H)}{8\pi G}\right), \quad \text{(B1)}$$

where $F(H)$ is some function of the Hubble parameter and hence time. These two equations are obviously completely identical. Yet the first equation could be interpreted as modified gravity (The Friedmann equation is modified and there is no dark-energy fluid), while the second equation corresponds to standard gravity with the dark-energy term of the form $\rho_{\rm DE} \equiv 3F/(8\pi G)$. This illustrates the challenge in separating these two scenarios – based on measurements of the expansion rate $H(t)$ alone, modified gravity and standard gravity with a dark-energy component are completely indistinguishable.

The growth of cosmic structure can be used to break this degeneracy (using nevertheless some assumptions about the dark-energy sector that we won't go into here). Consider the linear growth equation, Eq. (9.37),

$$\ddot{\delta} + 2H\dot{\delta} - 4\pi G\rho_M(t)\delta = 0. \quad \text{(B2)}$$

A precise measurement of the expansion rate $H(t)$, along that of $\rho_M(t)$, could then be used to *predict* the growth of fluctuations in standard gravity by simply plugging these quantities into Eq. (B2) above. One could then go ahead and *measure* the growth of fluctuations $\delta(t)$. A comparison of the general-relativistic prediction and measurement would then be sensitive to departures from general relativity, and hence the presence of modified gravity.

The recipe outlined above is pedagogical but impractical in detail. Instead, one typically separates the mutual dependence of the expansion and growth history, given by Eqs. (B1) and (B2), by introducing new parameters. For example, one can define $\Omega_{\rm DE}^{\rm geom}$ and $\Omega_{\rm DE}^{\rm grow}$, where the former is fed into all theory equations for quantities that depend on geometry (e.g., SN Ia distances), while the latter is fed into theory equations for quantities that encode growth (e.g., the growth function that enters the abundance of galaxy clusters). This is the basis of the so-called geometry–growth split technique, which is one of the methods to test the *internal consistency* of the standard ΛCDM model and, specifically, not only search for departures from the cosmological-constant scenario, but also look for signatures of modified gravity.

only one out of every $\sim 10^{100}$ of them – so the still respectable $\sim 10^{400}$ vacua – is suitable for us to be here and talk about it.

As is probably clear from the discussion above, the anthropic principle is very controversial. To most cosmologists, it is nothing short of non-scientific, as such anthropic explanations are simply not how we usually explain natural phenomena.

While the present author agrees with such sentiments, it is also true that we do not know for a fact that the anthropic reasoning is wrong. It will be interesting to see whether a more physics-oriented explanation of dark energy and/or the cosmological-constant problem emerges over the next few decades.

12.3.6 Fate of the Universe with Dark Energy

The universe with dark energy exhibits some surprising behavior.

Recall that, without dark energy, a spatially closed universe recollapses, while the flat and spatially open universes expand forever, albeit slower and slower. With dark energy, this nice separation is not valid any more, and the results depend wholly on the past, present, and future properties of dark energy.

Assuming, however, that dark energy behaves something like vacuum energy ($w \simeq -1$), as current data indicate, and that the universe is spatially flat, the prediction is unequivocal: the universe will expand ever more rapidly in the future. This is illustrated in Fig. 12.5, which shows the past and future behavior of the scale factor in the standard cosmological model with three different values of the equation of state, as well as the matter-dominated models (without dark energy)

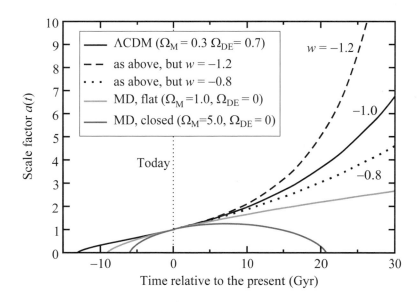

Fig. 12.5 Evolution of the FLRW scale factor in models with and without dark energy. The upper three curves show the model with dark energy ($\Omega_M = 1 - \Omega_{\mathrm{DE}} = 0.3$) for three values of the equation of state parameter: $w = -0.8$, $w = -1$ (ΛCDM), and $w = -1.2$. The fourth curve from the top shows the EdS model ($\Omega_M = 1$, flat). The curve on the bottom shows a closed matter-dominated model (which results in a Big Crunch), with $\Omega_M = 5$. Adopted from Frieman *et al.* (2008).

with $\Omega_M = 1$ and $\Omega_M = 5$. And if the equation of state is "phantom" ($w < -1$), even more extreme behavior is predicted, as you can work out in Problem 12.5.

Regardless of the precise value of w, in an accelerating universe galaxies are slowly leaving our Hubble volume. This can be understood by the fact that the comoving Hubble radius shrinks when $\ddot{a} > 0$; recall a similar discussion for inflation around Eq. (8.27) in Chapter 8. Therefore, in something like a hundred billion light-years, our sky will be empty.[5]

Some of the foremost upcoming cosmological telescopes and observatories – notably Rubin Observatory (and its Legacy Survey of Space and Time, or LSST), Dark Energy Spectroscopic Instrument (DESI), Euclid Space Telescope, and Roman Space Telescope – were designed principally with the idea to provide measurements that will help understand the physical nature of dark energy.

Bibliographical Notes

There are many excellent reviews of dark energy. For the big picture, history, phenomenology, and connection to observations, we have relied on the author's own reviews (Frieman *et al.*, 2008; Huterer and Shafer, 2018). The cosmological constant is reviewed in Carroll (2001), and the cosmological-constant *problem* is best understood from the legendary review of the subject by Steven Weinberg (1989). Dynamical (scalar field, etc.) models for dark-energy are reviewed in Linder (2008), and dark-energy theory is covered in great detail in Copeland *et al.* (2006). Basic phenomenology and connection to observations is given in Huterer and Turner (2001), while an up-to-date and very detailed analysis of how observations can be used to probe dark energy can be found in Weinberg *et al.* (2013). Finally, the original discovery papers of dark energy (Riess *et al.*, 1998; Perlmutter *et al.*, 1999) are very clearly laid out and well worth reading.

Problems

12.1 Dark energy for everyday life? The general public is fascinated by dark energy, and often asks the question of whether one can use it as a source of energy on Earth. Let us investigate this in some detail (without being worried about the practicalities of the dark-energy extraction process itself). Say you would like to power a 60 W bulb for one day. What volume of space contains the amount of dark energy required to do this? Illustrate your answer

---

[5] Although other calamities will have arrived by then – the Sun will have scorched Earth, and Andromeda will have collided with the Milky Way.

by linking it to a typical corresponding distance/volume on Earth or in the solar system, whichever is more applicable.

12.2 Understanding the offset in the Hubble diagram. Recall from the text that the normalization of theory curves in the SN Ia Hubble diagram is free to vary, corresponding to the fact that there is a variable nuisance parameter $\mathcal{M}$. Briefly demonstrate that this is equivalent to the statement that SN Ia observations effectively constrain *relative* distances.

12.3 Analytically marginalizing over script-M. The offset in the Hubble diagram, $\mathcal{M}$, is a nuisance parameter that needs to be marginalized over to obtain the likelihood (or, in Bayesian analysis, the posterior) on the cosmological parameters. Assume N apparent magnitude measurements m_i ($i = 1, 2, \ldots, N$) that are uncorrelated, with individual errors σ_i. Assume also a set of cosmological parameters $\mathbf{p}$ that does not include $\mathcal{M}$. Then the marginalization is

$$\mathcal{L}(\mathbf{p}) = \int \mathcal{L}(\mathbf{p}, \mathcal{M}) \, d\mathcal{M},$$

where

$$m^{\text{th}}(z_j, \mathbf{p}) = 5 \log_{10}[H_0 d_L(z_j, \mathbf{p})] + \mathcal{M}$$

$$\mathcal{L}(\mathbf{p}, \mathcal{M}) \propto \exp\left[-\sum_{j=1}^{N} \frac{[m_j - m^{\text{th}}(z_j, \mathbf{p})]^2}{2\sigma_j^2}\right]$$

are the theoretical magnitude and the likelihood, respectively.
(a) Show that the marginalization above can be done analytically, and that the result is simple:

$$\chi(\mathbf{p})^2 \equiv -2 \ln \mathcal{L}(\mathbf{p}) = A - \frac{B^2}{C}, \qquad \text{where}$$

$$A \equiv \sum_{j=1}^{N} \frac{[m_j - \widetilde{m}^{\text{th}}(z_j, \mathbf{p})]^2}{\sigma_j^2}$$

$$B \equiv \sum_{j=1}^{N} \frac{m_j - \widetilde{m}^{\text{th}}(z_j, \mathbf{p})}{\sigma_j^2}$$

$$C \equiv \sum_{j=1}^{N} \frac{1}{\sigma_j^2},$$

where $\widetilde{m}^{\text{th}}$ is just m^{th} with $\mathcal{M}$ ignored:

$$\widetilde{m}^{\text{th}} \equiv 5 \log_{10}[H_0 d_L(z_j, \mathbf{p})].$$

Hint: Complete the square in the exponent, so that you have a Gaussian integral in $\mathcal{M}$.

(b) How does the expression above change when we assume correlated SN Ia errors described by an $N \times N$ covariance matrix $\mathbf{C}$? *Hint:* This can be done by repeating the calculation with $\mathbf{C}$, but also probably by inspection of how A, B, and C in the final result above, which retains the same form, must generalize.

12.4 Standard sirens. A promising new probe of dark energy is the standard *siren* method, where the distance to galaxies is obtained from the waveform of gravitational waves (GWs) of a black-hole or neutron-star inspiral event in that galaxy. This scenario, proposed in 1986, was first applied in 2017 to one of the earliest detected GW events.

The challenge with the method is to get the redshift of the source, as that in turn requires an identification of the GW host galaxy which is not always available. [Note that this is the converse to the situation with standard candles, where the redshift is "easy" and the distance is "hard."] Hence, cosmologists have considered the scenario of using the standard sirens statistically, by averaging over all *potential* host galaxies in volume of the universe to which a given GW event has been localized by triangulation using GW detectors on Earth.

(a) Do the standard sirens provide an absolute or relative measure of distance?

(b) Imagine you had a *single* standard siren with very accurately measured distance, and an accurate corresponding redshift from an electromagnetic follow-up. If the event is at a moderate to high redshift ($z \sim 1$), could you obtain an interesting measurement of the Hubble constant from this single event? What about if the event is at a very low redshift ($z \ll 1$)? Explain.

(c) Consider now a more realistic scenario where the GW host has not been identified. Imagine instead that the volume to which the host has been localized corresponds to radial range $z = 0.5 \pm 0.05$ over 1000 square degrees on the sky. How many galaxies – potential GW hosts – are in this volume? Assume optimistically a low value of the number density of potential hosts, $n \sim 10^{-4}(h^{-1}\mathrm{Mpc})^{-3}$. You can simplify your formulae by assuming $\Delta z \ll z$ and $\Delta\Omega \ll 4\pi$, and feel free to use the $z \ll 1$ formulae (even though $z \simeq 0.5$), specifically $r(z) \simeq z/H_0$.

12.5 The Big Rip. When the equation of state is less than the vacuum-energy value – the so-called *phantom* dark-energy scenario with $w < -1$ – the universe's fate is strange and possibly upsetting (Caldwell *et al.*, 2003). It turns out that the scale factor becomes infinite in *finite* time, resulting in the so-called Big Rip ending, where space itself rips itself apart.

Assume the cosmological model with Ω_{DE} and a constant phantom equation of state $w < -1$. For simplicity, entirely ignore the dark-matter term in the Hubble parameter, that is, assume that $H^2(a) \simeq H_0^2\Omega_{\mathrm{DE}}a^{-3(1+w)}$ (you can keep the flat-universe formulae; we are being approximate here).

Demonstrate that indeed $a \to \infty$ as $\Delta t \to \Delta t_{\rm rip}$, and find $\Delta t_{\rm rip}$, the time to the Big Rip starting from the present time. Evaluate $\Delta t_{\rm rip}$ for two scenarios: $w = -1.5$ and $w = -1.05$.

12.6 **Reconstructed** $w(z)$**.** Derive Eq. (12.11) which gives a general, time-dependent dark-energy equation of state $w(z)$ as a function of comoving distance $r(z)$ and its derivatives. Assume a flat, matter-dominated universe with matter density Ω_M and dark-energy density $1 - \Omega_M$ and equation of state $w(z)$.

Hint: Start by taking the second derivative of $r(z)$ with respect to redshift in order to isolate $w(z)$, and make use of Eq. (12.10) in the process.

12.7 **[Computational] Evidence for** $w \neq -1$**?** In this problem, you will analyze the Pantheon SN Ia dataset and use two statistical methods discussed in Chapter 10 to investigate if the data prefer the flat ΛCDM model with Ω_M as the only free parameter, or flat wCDM with (Ω_M, w). Along the way, you will get practice producing credible intervals in 2D parameter space.

Download the Pantheon SN Ia data from `https://github.com/dscol nic/Pantheon`; the file you want is `lcparam_full_long_zhel.txt`. The relevant columns are redshift (`zcmb`), magnitude (`mb`), and its error (`dmb`). For simplicity, assume the "diagonal" statistical-only magnitude error, with no correlations between measurements or systematic-error contributions.

Set up a grid in Ω_M (for ΛCDM) and Ω_M–w (for wCDM) space. Start with a rough grid to establish your code, then refine the grid until you get reasonably well-converged results. For example, a grid sufficient for this purpose would be $\Omega_M \in [0.2, 0.4]$ with steps of $d\Omega_M = 0.01$ and $w \in [-1.5, -0.5]$ with steps $dw = 0.02$. Note that a single run over the grid (for each model) can be used to answer all parts of this problem. Note also that the computation in this problem is reasonably light given that the grids are in one and two parameters, respectively, so that you should not need to tabulate computation of the luminosity distance (i.e., find a lookup table for $z \to m(z, \Omega_M, w)$ for each model, where m is the theoretical apparent magnitude) – but you certainly *can* go with establishing a lookup table at each step in your parameter loops, as it is a useful tool often used in cosmology to speed up numerical computation.

Then compute the answers to the following questions:

(a) Marginalizing analytically over the Hubble-diagram offset $\mathcal{M}$ (see Problem 12.3), find the best-fit value of Ω_M in the ΛCDM model and of (Ω_M, w) in wCDM. Report these values. *Hint:* You will be presented with very small numbers in the likelihood calculation, because $\mathcal{L} \propto e^{-\chi^2/2}$, where $\chi^2 \sim N_{\rm SN} \sim 1000$. To avoid them and the numerical instabilities they may cause, you can simply subtract a constant value from each χ^2, say $\chi^2 \to \chi^2 - 1000$. This will not affect anything in the statistical analysis (being a renormalization of all likelihoods by a constant), but will solve the aforementioned problem.

(b) Find

$$\Delta\chi^2 \equiv \chi^2_{\text{best,wCDM}} - \chi^2_{\text{best,}\Lambda\text{CDM}},$$

that is, the difference in the best-fit chi-squared values in the two models. Report $\Delta\chi^2$, and comment on its sign. Then answer the question we are after: according to the usual $\Delta\chi^2$ criteria (e.g., the Wilks' theorem discussed in Sec. 10.4.2), do the data favor the wCDM model over ΛCDM?

(c) Now perform a Bayesian calculation to answer the same question: calculate the Bayes factor

$$B_{21} = \frac{P(D|\text{wCDM})}{P(D|\Lambda\text{CDM})} = \frac{\int \mathcal{L}(D|\Omega_M, w) P(\Omega_M, w) \, d\Omega_M dw}{\int \mathcal{L}(D|\Omega_M) P(\Omega_M) \, d\Omega_M},$$

where the numerator and the denominator are the Bayesian evidences for the two models, respectively, and are equal to integrals of likelihood times the prior evaluated over the parameter space. Do not forget to normalize the (flat) priors as per Eq. (10.33). With your result in hand, use the Jeffreys' scale (Table 10.1) to assess if the data prefer wCDM over ΛCDM.

(d) Calculate and plot the 68 percent credible interval contour in the (Ω_M, w) space (in the wCDM model), marginalized over $\mathcal{M}$ as usual. To get a better idea of what SN data constrain, you can adopt a wider parameter grid here (with a correspondingly coarser spacing), say $\Omega_M \in [0, 0.5]; w \in [-2, 0]$. Do not worry about producing a nice-looking contour if that is hard; you can simply show the set of points on your 2D parameter grid that lie within the contour.

12.8 [Computational] **Early dark energy.** Is it possible that dark energy not only dominates today, but has also appeared (once, or even periodically) in the past, at $z \gg 1$? Prolonged eras with significant dark energy are ruled out as they would spoil predictions of the ΛCDM model which are in agreement with modern observations. But periods with a small (percent-level contribution to ρ_{TOT}) dark-energy component are in principle allowed. Such scenarios go under the name of **early dark energy**.

To model a combination of early and late dark energy, we model the total dark-energy density as (Doran and Robbers, 2006)

$$\Omega_{\text{DE}}(a) = \frac{\Omega_{\text{DE}}^0 - \Omega_{\text{early}} \left(1 - a^{-3w_0}\right)}{\Omega_{\text{DE}}^0 + \Omega_M^0 a^{3w_0}} + \Omega_{\text{early}} \left(1 - a^{-3w_0}\right),$$

where Ω_{early} is the early dark-energy component density and $\Omega_{\text{DE}}^0 \equiv \Omega_{\text{DE}}$ and $w_0 \equiv w(a = 1)$ are the dark-energy density and equation of state today. Note that $\Omega_{\text{DE}}(a)$, along with w_0, is sufficient to calculate the distances; in other words, the dark-energy sector is specified by

$$\mathbf{p}^{\text{dark}} \in \{\Omega_{\text{DE}}^0, \Omega_{\text{early}}, w_0\}.$$

While not required for the computations that follow, it can be shown that the corresponding time-dependent dark-energy equation of state is

$$w(a) = -\frac{1}{3[1 - \Omega_{\mathrm{DE}}(a)]}\frac{d\ln\Omega_{\mathrm{DE}}(a)}{d\ln a} + \frac{a_{eq}}{3(a + a_{eq})},$$

where a_{eq} is the scale factor at matter–radiation equality. For nonzero Ω_{early}, the equation of state goes to values slightly larger than *zero* at high redshift in order to ensure the period of non-negligible early dark energy.

In the following, assume $\Omega_{\mathrm{early}} = 0.03$, $w_0 = -1$, and fiducial model (Table 3.1) for the other parameters, including $\Omega_{\mathrm{DE}}^0 = 0.7$. You may ignore the radiation contribution.

(a) Plot $\Omega_{\mathrm{DE}}(a)$ for $10^{-3} \leq a \leq 1$. Make the x-axis logarithmic.

(b) Show that the Hubble-parameter distance in the model with present-day matter density Ω_M and the time-dependent dark-energy density $\Omega_{\mathrm{DE}}(a)$ is given by

$$H(a) = H_0\sqrt{\frac{\Omega_M a^{-3}}{1 - \Omega_{\mathrm{DE}}(a)}}.$$

(c) What is the relative difference in $d_A(a = 0.5)$ (so at $z = 1$) between this model with $\Omega_{\mathrm{early}} = 0.03$ and the fiducial model with $\Omega_{\mathrm{early}} = 0$?

(d) What is the relative difference in $d_A(a = 10^{-3})$ (so at $z \simeq 1000$) between this model with $\Omega_{\mathrm{early}} = 0.03$ and the fiducial model with $\Omega_{\mathrm{early}} = 0$?

13   Cosmic Microwave Background

The cosmic microwave background (CMB) is relic radiation left over from the very early universe and observed today. It is probably the most familiar cosmological probe, as it lends itself to beautiful images of the CMB sky *and* produces impressively precise constraints on some of the most important physical processes in the early universe. We try to do justice to this enormously successful subfield of cosmology by outlining its observations, theory, and applications in this chapter.

13.1   CMB Observed

The discovery of the CMB is a story of missed opportunities followed by a great groundbreaking finding. Subsequent ever-higher-resolution measurements of the CMB paralleled the increased mathematical sophistication used in describing it. We now review these basics.

13.1.1   The Discovery of the CMB

In 1965, two Bell Labs engineers, Arno Penzias and Robert Wilson, were setting up a horn antenna in Crawford Hill, New Jersey. They were finding persistent noise that wouldn't go away even after they cleaned the pigeon droppings from the antenna. The noise seemed to be isotropic on the sky, that is, not coming from one source (they were working at radio frequencies, and there are many sources of radio noise on Earth). The radiation corresponded to a temperature of about 3 K.

News about the Penzias–Wilson detection trickled to astrophysicist Bernard Burke, who happened to know about a parallel effort at Princeton to search for leftover radiation from the hot early universe. He put the two teams in touch. It soon became apparent to everyone involved that Penzias and Wilson had made one of the biggest discoveries in physics in a generation: the cosmic microwave background (CMB) radiation.

The CMB is a relic from the early universe. At this early stage, photons were numerous enough and energetic enough to keep hydrogen ionized, so that electrons and protons were not bound together. During this period, photons were constantly scattering off the electrons. The CMB carries information about the end of this era, around 380,000 years after the Big Bang, when the photons were finally released and became free to propagate in space.

Incredibly, the CMB had actually been predicted and even *detected* before the famous Penzias–Wilson "official discovery" in 1965 – the predictions were unfortunately promptly forgotten, while the early discoverers did not realize the cosmic importance of their results. In particular:

- In 1948, Ralph Alpher and Robert Herman, closely following the work of George Gamow, theoretically predicted a relic radiation of temperature 5 K, but the significance of this prediction was not subsequently appreciated.

- Also in the 1940s, astronomer Andrew McKellar measured the ambient temperature by interpreting the observations of the absorption spectra of cyanogen (CN) observed toward the star Zeta Opiuchi. The ratio of the abundances of the excited and ground levels is known to be given by the Boltzmann factor, $n_1/n_0 = \exp[-(E_1 - E_0)/T]$. This ratio had been observed to be roughly 1/5 and, along with the knowledge of the energy levels (E_1, E_0), revealed the ambient temperature $T = 2$–3K. The discovery was not recognized, presumably because it was thought that molecular scattering, rather than background photons, were causing the excitations.

- In the 1950s, Russian radio astronomer Tigran Shmaonov, then a graduate student, operated an antenna sensitive to radiation of 3.2 cm wavelength, and measured a signal corresponding to an ambient temperature $T \simeq 4\,\mathrm{K}$. Because he and his advisors were radio observers and not in contact with cosmology theorists, their discovery did not attract attention.

The CMB was actively sought at the same time that Penzias and Wilson stumbled on their discovery: Jim Peebles and Bob Dicke, both at Princeton, had been building an instrument to detect the CMB. Having seen the discovery that the former two made, the Princeton group contacted them and they published their respective papers concurrently. The Nobel Prize (in 1978) went to Penzias and Wilson alone, though Jim Peebles received one very recently (in 2019) as well.

The CMB is observed to be remarkably uniform across the sky. The very precisely measured temperature of the CMB has been established by the COBE satellite:

$$T_0 = (2.7255 \pm 0.0006)\,\mathrm{K}. \tag{13.1}$$

This average temperature is called the CMB temperature **monopole**. [This monopole is not to be confused with *magnetic* monopoles, which are particles that may or may not exist, and which we discussed in Chapter 8.] We will discuss the **anisotropies** in the CMB – fluctuations around the monopole – further below.

13.1.2 CMB Acts as a Blackbody

A blackbody is an idealized object that absorbs all electromagnetic radiation that falls on it. No electromagnetic radiation passes through it and none is reflected.

However, a blackbody emits a temperature-dependent spectrum of light. This thermal radiation from a blackbody is termed the **blackbody radiation**.

Here we review a few key results about the blackbody radiation that we derived in Chapter 4. The energy of a blackbody at frequency ν is[1]

$$\rho_{\mathrm{CMB}}(\nu)d\nu = 16\pi \frac{\nu^3 d\nu}{\exp\left(2\pi\nu/T\right) - 1}.\tag{13.2}$$

As you may recall from your first quantum-mechanics course, the energy density at frequency ν depends on the temperature of the radiation T. At low frequencies we have the classical (Rayleigh–Jeans) regime; $\rho(\nu) \propto \nu^2$. The Rayleigh–Jeans formula alone would lead to **ultraviolet catastrophe** – infinite energy density of radiation at high energies (high ν). This classical disaster is saved by the quantization of light which produces the denominator in Eq. (13.2), and an exponential suppression at $\nu \gg T$. The peak of the blackbody energy-density curve is at the frequency

$$\nu_{\mathrm{CMB,peak}} \approx 2.8T.\tag{13.3}$$

The CMB has a temperature of about 3 K; therefore, its spectrum peaks at $\lambda \approx 2$ mm, or $\nu \simeq 150\,\mathrm{GHz}$.

The energy density of the blackbody radiation can be computed by integrating Eq. (13.2). The result has been derived in Eq. (4.20) in Chapter 4; for $g = 2$ degrees of freedom for photons, it evaluates to

$$\rho_{\mathrm{CMB}} = \frac{\pi^2}{15}T^4.\tag{13.4}$$

[In MKS units, $\rho_{\mathrm{CMB}} = \alpha T^4$ with $\alpha = \pi^2 k^4/(15\hbar^3 c^3) = 7.56 \times 10^{-16}$ J/m^3/K^4.] The blackbody number density is

$$n_{\mathrm{CMB}} = \frac{2\zeta(3)}{\pi^2}T^3 = 0.243\,T^3.\tag{13.5}$$

Evaluated today, at $T = T_0 = 2.725\,\mathrm{K}$, the number density of CMB photons is $n_{\mathrm{CMB,0}} = 4.11 \times 10^8\mathrm{m}^{-3}$, or

$$n_{\mathrm{CMB,0}} = 411\,\mathrm{cm}^{-3},\tag{13.6}$$

so 411 CMB photons per cubic centimeter. These CMB photons are "everywhere" but, unlike neutrinos, not in the room where you are reading this, as a significant fraction of them are stopped by Earth's atmosphere (except at some of the best CMB-viewing sites, such as the South Pole or the Atacama Desert in Chile). And all microwave photons are stopped by building walls, of course.

---

[1] To restore MKS units, the prefactor to Eq. (13.2) would be $16\pi\hbar/c^3$ and the argument of the exponent would be $h\nu/k_B T$. Also, note that the energy density of radiation ρ (sometimes labeled u in the literature) is related to the specific intensity $I(\lambda) \equiv E/(\Delta t \delta A_\perp \delta\lambda)$ via $\rho = (4\pi/c)I(\lambda)$.

The CMB is a very nearly perfect blackbody. The blackbodiness of the CMB was verified to high precision by NASA's COBE (COsmic Background Explorer) experiment in 1992. Figure 13.1 shows historical measurements of the CMB spectrum at various frequencies. It also shows the measured spectrum as measured with COBE's FIRAS (Far InfraRed Absolute Spectrophotometer) that verified the blackbodiness of the CMB in the wavelength range $0.1\,\mathrm{mm} < \lambda < 10\,\mathrm{mm}$. The COBE FIRAS errors are actually much smaller than the size of the corresponding points in Fig. 13.1, showing how remarkably close to blackbody the measurements lie. No departures from the blackbody spectrum of the primordial CMB have been detected to date, although very small deviations are expected; we will discuss this further in Sec. 13.6.

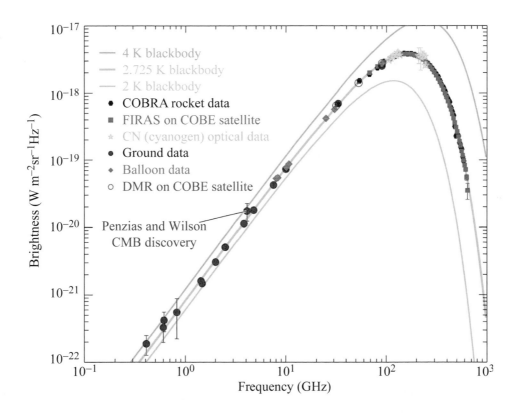

Fig. 13.1   CMB spectrum as measured by various early experiments. The best constraints to date have been provided by COBE FIRAS, whose actual error bars are much smaller than the size of the corresponding points in this figure. Reprinted with permission of *Annual Reviews*, from Samtleben *et al.* (2007); permission conveyed through Copyright Clearance Center Inc. A black and white version of this figure will appear in some formats. For the color version, please refer to the plate section.

13.1.3 CMB Dipole

In the early 1970s, another milestone was achieved: researchers detected a **dipole** in the CMB temperature. A dipole is a pattern where the temperature in one direction of the sky is colder, while the temperature in the opposite direction is hotter (by the same amount) than the mean. The CMB dipole is measured to be about one thousandth of the monopole (the mean), that is

$$\delta T_{\rm dip} \simeq 10^{-3}\, T_0. \tag{13.7}$$

However the observed CMB dipole is not primordial, but rather:

> The observed CMB dipole is due to our (i.e., the Sun's) motion through the CMB rest frame.

Intuitively, this is very much like driving a motorcycle on a cold day. Your face feels colder and the back of your head feels hotter, with a $\cos\theta$ pattern in between. A full analysis is a little more involved, but this roughly illustrates why you expect a dipole temperature pattern due to your motion.

The dipole is therefore not cosmological, but it is very useful in one regard: it can be used to determine the direction and velocity of our motion through the universe, where the "rest frame" is specified by the CMB rest frame. Note that this motion is the (vector) sum of individual motions: Earth around the Sun; Sun around the Milky Way galaxy; Milky Way relative to our local group of galaxies; and local group of galaxies relative to even larger structures. Some of the individual contributions are poorly measured, but their *sum* is extremely well known, thanks to the CMB dipole.

The velocity of the Sun with respect to the CMB rest frame, and direction of motion, can thus be determined precisely as

$$v_{\rm Sun-CMB} = 369.8 {\rm km/s} \quad \text{in direction of} \quad (l, b) = (264.02°, 48.25°), \tag{13.8}$$

where l and b are the galactic latitude and longitude, respectively. This is the direction in which we (i.e., the solar system) are hurtling through the CMB rest frame.

The dipole velocity in Eq. (13.8) can be decomposed into two physically interesting components: the Sun's velocity towards the local group of galaxies and the local group velocity through the CMB rest frame. These two are less accurately measured, and take the values $v_{\rm Sun-LG} \simeq 300 {\rm km/s}$ and $v_{\rm LG-CMB} \simeq 620 {\rm km/s}$; they point in nearly opposite directions and add (vectorially) so as to produce the Sun–CMB velocity shown in Eq. (13.8).

13.1.4 CMB Anisotropy

A momentous discovery that Stephen Hawking called "the greatest discovery of the millennium, if not all time," took place in 1992.

The COBE experiment was the first satellite launched to map the CMB above the atmosphere. One of COBE's instruments was FIRAS, which we already mentioned in the context of the CMB blackbody measurements. Another was DIRBE (Diffuse InfraRed Background Experiment), which opereted in the shorter wavelength range than the peak of the CMB blackbody spectrum, $0.001\,\text{mm} < \lambda < 0.24\,\text{mm}$, where it was mainly detecting light and heated dust from stars. However, the most important instrument on COBE was the third one, DMR (Differential Microwave Radiometer) whose job was to measure temperature *differences* across the sky.

The DMR discovered temperature fluctuations all across the sky, all the way down to COBE's resolution of 7 degrees; see Fig. 13.2(a). In particular, it found the **quadrupole**, which is the largest primordial fluctuation mode (recall that the dipole is not primordial). This great discovery brought the 2006 Nobel Prize to COBE's John Mather and George Smoot. The size of CMB fluctuations that COBE discovered is

$$\left\langle \left(\frac{\delta T}{T}\right)^2 \right\rangle^{1/2} \simeq 10^{-5}. \tag{13.9}$$

(a)   (b)

Fig. 13.2 (a) COBE DMR 90 GHz map of CMB fluctuations (credit: NASA / COBE Science Team; see also Bennett *et al.*, 1996). (b) WMAP's W-band (roughly also around 90 GHz) map (credit: NASA / LAMBDA Archive Team; see also Bennett *et al.*, 2013). Both maps show significant galactic contamination near the equator; the primordial CMB anisotropies are the features away from it. The resolution of the COBE map is about $7°$, while that of the WMAP map is about $15'$. A black and white version of this figure will appear in some formats. For the color version, please refer to the plate section.

Let us immediately point out that

$\delta T/T \simeq 10^{-5}$ is arguably the single most important number in cosmology. It is the initial condition, laid down by inflation, for the CMB temperature and matter density fluctuations.

Conversely, note that the mean CMB temperature measured today, $T_0 \simeq 2.725\,\text{K}$, is *not* as fundamentally an important a number, simply because it is changing in time; we just happen to live when it is around 3 K.

It is worth pausing for a moment and contemplating the significance of the CMB anisotropies first seen by COBE and later mapped out in impressive detail by WMAP and Planck; see Figs. 13.2(b) and 13.3. These temperature anisotropies are the original perturbation seeds of structures that later, on smaller scales, grew to become objects in the universe (galaxies and clusters of galaxies). As we learned in Chapter 9, dark matter is free to fall into these early potential wells, while baryons initially did not as they were tightly coupled to photons. At recombination, at $z \sim 1000$, photons and baryons decouple, and baryons can finally follow dark matter and start clustering. The clustering continues with cosmic time, up to the present day, though it is suppressed by the onset of dark energy at $z \lesssim 1$.

Fig. 13.3 Planck experiment's full-sky map in Galactic coordinates. Galactic contamination has been cleaned using multi-frequency data using the so-called SMICA technique. The resolution of the map is approximately 5 arcmin. Adopted from Akrami *et al.* (2020a) and ESA and the Planck Collaboration. A black and white version of this figure will appear in some formats. For the color version, please refer to the plate section.

13.1.5 Angular Power Spectrum of the CMB

We will thoroughly explore how to statistically describe the CMB later on in this chapter, but for now we *briefly* introduce the harmonic-space description of the CMB anisotropies. Let us start with the two-point correlation function of CMB temperature, $C(\theta)$, which is the CMB equivalent of the two-point angular correlation function of galaxies, $w(\theta)$, introduced in Chapter 9 (see Sec. 9.2.4). We can expand the angular two-point correlation function in Legendre series,

$$C(\theta) \equiv \left\langle \frac{\delta T}{T}(\hat{\mathbf{n}}) \frac{\delta T}{T}(\hat{\mathbf{n}}') \right\rangle_{\hat{\mathbf{n}} \cdot \hat{\mathbf{n}}' = \cos\theta} = \sum_{\ell=2}^{\infty} \frac{2\ell+1}{4\pi} C_\ell P_\ell(\cos\theta). \tag{13.10}$$

Here the multipoles C_ℓ are simply the harmonic coefficients of the CMB temperature's two-point correlation function $C(\theta)$. Note that the C_ℓ are dimensionless when defined as above. [Cosmologists sometimes choose to expand δT, rather than $\delta T/T$, and in those cases the C_ℓ have units of temperature squared.] Finally, the correspondence between multipoles and angles is $\ell \simeq \pi/\theta$, with θ in radians. We remind the reader that, for a Gaussian random field, either the correlation function $C(\theta)$ or the angular power spectrum C_ℓ contains all statistical information. The C_ℓ and $C(\theta)$ in the fiducial cosmological model are shown in Fig. 13.6 later.

13.2 Recombination and Photon Decoupling

We next describe the physics of how photons become decoupled from baryons and free-stream to us, hitting our detectors today.

13.2.1 Tightly Coupled Limit

At early times in the history of the universe (redshift $\gg 1000$), photons and baryons are "**tightly coupled**." Photons scatter off electrons,

$$\gamma + e^- \rightleftharpoons \gamma + e^-, \tag{13.11}$$

via Thomson scattering, which has a cross-section (restoring MKS units briefly)

$$\sigma_{\mathrm{T}} = \frac{8\pi}{3} \left(\frac{\alpha h}{m_e c} \right)^2 = 6.65 \times 10^{-29}\,\mathrm{m}^2. \tag{13.12}$$

Meanwhile, electrons and protons are coupled via the Coulomb interaction. Therefore, electrons, photons, and protons are all coupled and behave as a fluid, which we call the baryon–photon fluid (or the baryon–photon plasma).

Let us estimate some parameters in the tightly coupled regime. The mean distance that a photon travels before scattering from an electron – the **mean free path** – is

$$\lambda = \frac{1}{n_e \sigma_{\mathrm{T}}}, \tag{13.13}$$

where n_e is the ambient electron density. Using this, and the fact that the photons are moving with speed c, we can calculate the scattering rate (defined really as the inverse of the mean time between scatters)

$$\Gamma = c n_e \sigma_{\mathrm{T}}. \tag{13.14}$$

In the fully ionized tightly coupled regime, $n_e = n_p = n_B$, where n_p and n_B are the photon and baryon densities. Moreover, at the present time, $n_{B,0} = \rho_{\mathrm{crit},0} \Omega_B / m_p \simeq 0.25 \, \mathrm{m}^{-3}$ and

$$n_e = n_B = \frac{n_{B,0}}{a^3}, \tag{13.15}$$

so that

$$\Gamma = \frac{c n_{B,0} \sigma_{\mathrm{T}}}{a^3} \simeq \frac{5.0 \times 10^{-21} \, \mathrm{s}^{-1}}{a^3}. \tag{13.16}$$

The photons and baryons will be "tightly coupled" (scattering off each other) if $\Gamma > H$, whereas we know

$$H = H_0 \left(\Omega_M a^{-3} + \Omega_R a^{-4} \right)^{1/2}, \tag{13.17}$$

where we ignored the dark-energy contribution to H since it is insignificant at redshifts greater than about one.

However, setting H and Γ equal to each other to find when the tightly coupled era ends would produce a grossly incorrect result because we assumed a fully ionized universe. A more accurate estimate follows in the subsections below. First, however, we summarize the three important processes happening at $z \gtrsim 1000$:

1. **Recombination.** The baryonic component in the universe, consisting basically of protons, goes from being ionized to being neutral.

2, 3. **Decoupling and last scattering.** The rate at which photons scatter off free electrons goes down (because the number density of the latter decreases sharply) and at some point becomes smaller than the Hubble parameter at the time. We say that photons and baryons become decoupled. Photons are now free to propagate, and they free-stream toward us with almost no further scattering.[2] Hence one also talks about the "last scattering" (of photons). The photon temperature pattern that we observe today from the epoch of last scattering is called the last-scattering surface.

We now discuss the recombination and decoupling/last scattering in that order.

13.2.2 Recombination

Recombination – which probably should be called *combination* – is the process where electrons and protons combine into hydrogen atoms; see Fig. 13.4. It signifies the process in the history of the universe when atoms formed.

---

[2] About 5 percent of the photons scatter much later, at redshift $z \sim 10$, during the process of **reionization**, which we discuss in Sec. 13.2.4.

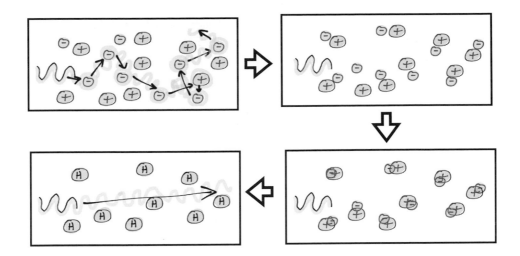

Fig. 13.4 Sketch of the recombination process, where protons and electrons combine into hydrogen atoms. The radiation (shown with wiggly lines), previously coupled to the electrons, is subsequently released. Drawing made by Jessica Muir.

Let us do some energetics to see how the hydrogen recombination proceeds. The reaction we are talking about is

$$H + \gamma \rightleftharpoons p + e^-. \tag{13.18}$$

As long as its rate is much larger than the Hubble rate, this reaction will be in statistical equilibrium. In that case, we can determine the number density of particles i via the Maxwell–Boltzmann distribution (see Eq. (4.23))

$$n_i = g_i \left(\frac{m_i T}{2\pi} \right)^{3/2} e^{-(m_i - \mu_i)/T}, \tag{13.19}$$

where g_i, m_i, and μ_i are the statistical weight, mass, and chemical potential for particles i. For electrons, protons, and neutrons this number is $g_e = g_p = g_n = 2$, corresponding to two spin states, while for the hydrogen atom, $g_H = 4$. Moreover, the chemical potentials cannot be ignored, but they are related by the chemical-equilibrium relation

$$\mu_H + \mu_\gamma = \mu_p + \mu_e, \tag{13.20}$$

with $\mu_\gamma = 0$ as usual. Taking the following ratio, the chemical potentials cancel out and we have

$$\frac{n_H}{n_p n_e} = \frac{g_H}{g_p g_e} \left(\frac{m_H}{m_p m_e} \right)^{3/2} \left(\frac{T}{2\pi} \right)^{-3/2} \exp\left(\frac{m_p + m_e - m_H}{T} \right). \tag{13.21}$$

Defining the binding energy of hydrogen,

$$Q \equiv m_p + m_e - m_H \simeq 13.6\,\mathrm{eV}, \tag{13.22}$$

using $m_p \approx m_H \gg m_e$, and realizing that the weight prefactor in Eq. (13.21) evaluates to one, we end up with the **Saha equation**

$$\frac{n_H}{n_p n_e} = \left(\frac{m_e T}{2\pi}\right)^{-3/2} \exp\left(\frac{Q}{T}\right) \qquad \text{(Saha equation)}. \qquad (13.23)$$

Now we would like to determine how the ionization fraction, defined as

$$\frac{n_p}{n_p + n_H} = \frac{n_p}{n_B} = \frac{n_e}{n_B} \equiv x_e, \qquad (13.24)$$

varies over time. First, rewrite it as

$$\frac{1 - x_e}{x_e} = \frac{n_H}{n_p} = n_e \left(\frac{m_e T}{2\pi}\right)^{-3/2} \exp\left(\frac{Q}{T}\right)$$
$$= n_p \left(\frac{m_e T}{2\pi}\right)^{-3/2} \exp\left(\frac{Q}{T}\right), \qquad (13.25)$$

where in the last line we used $n_e = n_p$. Now we want to relate this to the baryon-to-photon ratio η. Since

$$\eta = \frac{n_B}{n_\gamma} = \frac{n_p}{x_e n_\gamma}, \qquad (13.26)$$

we have

$$\frac{1 - x_e}{x_e^2} = n_\gamma \eta \left(\frac{m_e T}{2\pi}\right)^{-3/2} \exp\left(\frac{Q}{T}\right). \qquad (13.27)$$

For a blackbody the number density of particles is directly related to temperature, $n_\gamma = (2\zeta(3)/\pi^2)T^3$ (see Eq. (4.19)), so that

$$\frac{1 - x_e}{x_e^2} \simeq 3.84\eta \left(\frac{T}{m_e}\right)^{3/2} \exp\left(\frac{Q}{T}\right). \qquad (13.28)$$

To see how ionization varies as a function of scale factor, it is useful to remember that

$$Ta = \text{const.}, \qquad (13.29)$$

so that one can simply replace $T = T_0/a$ in Eq. (13.28). This equation can be solved numerically to obtain either $x_e(T)$ or $x_e(a)$.

The results of the Saha calculation for $x_e(a)$ are shown as the dashed line in Fig. 13.5(a). Clearly, the ionization fraction decreases from unity to very small values, indicating that hydrogen in the universe goes from being completely ionized to being neutral.

In the same figure, we show the results from the three-level (Peebles) recombination calculation (see Box 13.1), as well as exact results that assume many-level

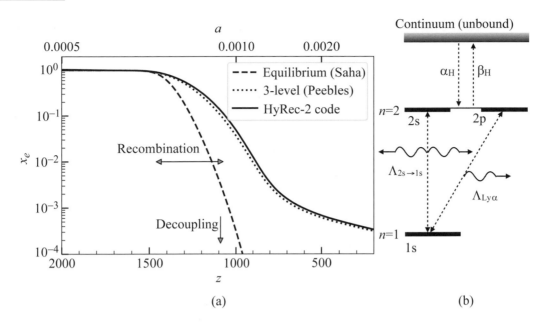

Fig. 13.5 (a) Hydrogen recombination in the standard cosmological model. We show the ionization fraction x_e vs. redshift (lower axis) or scale factor (upper axis). The dashed line shows the Saha (equilibrium) approximation, the dotted line the "Peebles" three-level recombination, while the solid line is a precise, many-level result obtained using HyRec-2 computer code (Lee and Ali-Haïmoud, 2020). Note that the Saha approximation already informs us about the time when recombination takes place, while the three-level result is close to the exact one. (b) Illustration of the three-level recombination model discussed in Box 13.1.

hydrogen atoms obtained using HyRec-2 code (Ali-Haimoud and Hirata, 2011; Lee and Ali-Haïmoud, 2020). Saha approximation gives a good estimate of the redshift of recombination, but its x_e decreases exponentially at later times. In contrast, the three-level (Peebles) approximation does an excellent job of tracking the late-time ionized fraction, which decreases much more slowly than an exponential would indicate.

The moment when half of the hydrogen in the universe is ionized ($x_e = 1/2$) is reached at $z = 1275$ or temperature of $\sim$3500 K (referring to the exact HyRec-2 result). This could perhaps be called the "moment of recombination," but this would be slightly misleading as recombination is a gradual process. Consider that a 90 percent ionization is reached at $z = 1450$, and it falls to 10 percent ($x_e = 0.1$) only at $z = 1060$. The time period between 90 percent and 10 percent reionization corresponds to about 300,000 years, which is about 40 percent of the age of the universe at the end of this period (when $z = 1060$). This gives you an idea of the leisurely pace of recombination.

Box 13.1   **Three-Level (Peebles) Approximation for Decoupling**

In the main text of this chapter, we covered the Saha approximation for decoupling that assumes equilibrium throughout. Here we outline the three-level atom approximation, pioneered by Peebles (1968), which gives the ionization fraction $x_e(z)$ that agrees much better with the exact result; see Fig. 13.5(a). This derivation follows Ali-Haimoud and Hirata (2011).

Direct recombinations to the ground level of hydrogen are inefficient, as the resulting photon immediately ionizes another hydrogen atom. The path forward is provided by recombinations where an electron and a proton combine to an excited state of hydrogen, then cascade down to the ground state. The ground-state level of hydrogen is denoted $1s$, while the first excited levels are $2s$ and $2p$; see Fig. 13.5(b). Let $x_1 = n_{H(n=1)}/n_H$ be the fraction of hydrogen in the $n = 1$ state, and similarly for x_2. Because atomic rates are much faster than the overall timescale of recombination, it is a good approximation to assume a steady-state situation for the $n = 2$ level where the rate of recombinations is equal to the rate of transitions to the $n = 1$ ground state, that is, $\dot{x}_2 = 0$. The ionization rate, $\dot{x}_e$, is just the negative of the rate of change in the $1s$ level and is given by

$$\dot{x}_e = -\dot{x}_{1s} = -(\dot{x}_{1s}|_{2s} + \dot{x}_{1s}|_{2p}), \tag{B1}$$

where $\dot{x}_{1s}|_{2s}$ and $\dot{x}_{1s}|_{2p}$ are contributions to $1s$ from the $2s$ and $2p$ levels, respectively. The former decays occur via a two-photon process (to conserve angular momentum since both the $2s$ and $1s$ state have angular momentum of zero); the rate is

$$\dot{x}_{1s}|_{2s} = \Lambda_{2s \to 1s}\left(x_{2p} - x_{1s}e^{-B_{21}/T}\right), \tag{B2}$$

where $\Lambda_{2s \to 1s} = 8.22\,\mathrm{s}^{-1}$ is the rate of the (slow) two-photon decays from the $2s$ state and $B_{21} = 10.2\,\mathrm{eV}$ is the energy difference between the $n = 1$ and $n = 2$ states of hydrogen. The first term on the right-hand side of Eq. (B2) accounts for the decays from $2s$, while the second term accounts for two-photon absorptions.

Similarly, the rate of electrons cascading from level $2p$ can be calculated considering that those transitions release a Lyman-α photon. In a static universe, every such photon would go ahead to reionize another H atom leading to no net recombination. In the expanding universe, a fraction of Lyman-α photons will be redshifted out of the natural width of the Lyman-α line, leading to net recombinations. This rate, $\Lambda_{\mathrm{Ly}\alpha}$, will clearly be proportional to the characteristic expansion time, that is, proportional to the Hubble parameter H. A similar reasoning as in Eq. (B2) gives

$$\dot{x}_{1s}|_{2p} = \Lambda_{\mathrm{Ly}\alpha}\left(x_{2p} - 3x_{1s}e^{-B_{21}/T}\right), \tag{B3}$$

where the factor of 3 accounts for three orbital angular momentum states in the $2p$ atomic level.

Box 13.1   **Three-Level (Peebles) Approximation for Decoupling (continued)**

Moreover, the occupancy of the $2p$ level is three times that of the $1s$ level, so that $x_{2s} = (1/3)x_{2p} = x_2$. Replacing x_{2p} and x_{2s} in favor of x_2 and combining Eqs. (B2) and (B3), we get

$$\dot{x}_e = -\Lambda_{2s \to 1s}\left(\frac{1}{4}x_2 - x_{1s}e^{-B_{21}/T}\right) - 3\Lambda_{\text{Ly}\alpha}\left(\frac{1}{4}x_2 - 3x_{1s}e^{-B_{21}/T}\right). \quad (B4)$$

Finally, the net rate of *recombinations* to/from level 2 (from electrons at higher levels) is

$$x_2|_{\text{rec}} = -\dot{x}_e = \alpha_H n_H x_e^2 - x_2\beta_H, \quad (B5)$$

where the first term gives the rate from the $n > 2$ states to $n = 2$ and vice versa for the second term. Here $\alpha_H(T)$ and $\beta_H(T)$ are two known temperature-dependent rates that are related by a detailed-balance relation (see Problem 13.12). Combining Eqs. (B4) and (B5) and eliminating x_2 from them, we finally get our expression for the rate of change of x_e in the two-level approximation:

$$\dot{x}_e = -C\left(n_H x_e^2 \alpha_H - 4x_1\beta_H e^{-B_{21}/T}\right),$$

where

$$C \equiv \frac{\frac{3}{4}\Lambda_{\text{Ly}\alpha} + \frac{1}{4}\Lambda_{2s \to 1s}}{\beta_H + \frac{3}{4}\Lambda_{\text{Ly}\alpha} + \frac{1}{4}\Lambda_{2s \to 1s}}. \quad (B6)$$

The equation for $\dot{x}_e$ is stiff and somewhat difficult to integrate. The result of that calculation is shown as the dotted line in Fig. 13.5(a), and shows very good agreement with the exact result obtained using many-level numerical calculation.

13.2.3 Decoupling of Photons and Baryons

Not coincidentally, around the time of recombination photons and baryons decouple – that is, they stop behaving as a single fluid where photons scatter off electrons which are in turn coupled to protons via Coulomb interactions. Earlier we saw that the photon–electron scattering rate is

$$\Gamma = \frac{cn_{B,0}\sigma_{\text{T}}}{a^3} = \frac{5.0 \times 10^{-21}\,\text{s}^{-1}}{a^3}, \quad (13.30)$$

where $n_{B,0}/a^3$ is the baryon number density at scale factor a and σ_{T} is the Thomson cross-section. This needs to be modified once recombination starts taking place, since $n_e(z) = x_e(z)n_B$, so that we need to multiply the rate (which is proportional to the number of free electrons) by $x_e(z)$. This results in

$$\Gamma = 5.0 \times 10^{-21}\,\text{s}^{-1}x_e(z)(1 + z)^3, \quad (13.31)$$

which should be compared to the Hubble rate in the (then, at $z = 1000$) matter-dominated universe:

$$H(z) \simeq H_0 \sqrt{\Omega_M}(1+z)^{3/2} \simeq 1.19 \times 10^{-18}\,\mathrm{s}^{-1}(1+z)^{3/2}. \tag{13.32}$$

Requiring

$$\Gamma(z_{\mathrm{dec}}) = H(z_{\mathrm{dec}}) \quad \text{(condition for decoupling)}, \tag{13.33}$$

we get

$$1 + z_{\mathrm{dec}} = \frac{38.3}{x_e(z_{\mathrm{dec}})^{2/3}}. \tag{13.34}$$

Using the results for $x_e(z)$ obtained from the Saha equation, we get $z_{\mathrm{dec}} \simeq 1120$. A more complete analysis gives

$$z_{\mathrm{dec}} \simeq 1090. \tag{13.35}$$

This is the redshift at which photons and baryons formally decouple. This epoch therefore signifies the **last scattering** of photons[3] (hence frequent references to the "last-scattering surface" in cosmology). From then on, photons free-stream, that is, propagate freely without interactions. The next thing some of them experience, about 13.5 Gyr later, is – splat! – to hit our CMB detectors.

Note that decoupling does not happen at the same time for all photons – some last scatter earlier than others. So while the mean decoupling (and last scattering) redshift is 1090, the rms variation in this redshift is about 100; that is, $z_{\mathrm{dec}} \equiv z_{\mathrm{ls}} \simeq 1090 \pm 100$.

Concluding, we highlight the following important epochs at $z \sim 1000$:

- Matter–radiation equality ($z \approx 3250$, $t \approx 50{,}000$ yrs, $T \approx 9000\,\mathrm{K}$).

- Recombination ($z \approx 1275$, $t \approx 290{,}000$ yrs, $T \approx 3500\,\mathrm{K}$).

- Decoupling and last scattering ($z \approx 1090$, $t \approx 380{,}000$ yrs, $T \approx 3000\,\mathrm{K}$).

At a much later time, one more important process affects CMB photons, which we now discuss.

---

[3] More precisely, this is the redshift when photons, but not baryons, last scatter. Because there are so many fewer baryons than photons, the baryons are actually released slightly later, at the so-called "drag epoch," characterized by the drag horizon r_d. In the standard cosmological model, $z_d \simeq 1060$ (compared to $z_{\mathrm{dec}} \simeq 1090$), with r_d correspondingly (~ 2 percent) larger than the (photon) sound horizon r_s.

13.2.4 Reionization

Following recombination and decoupling, the universe is largely neutral – recall from Fig. 13.5 that the ionization fraction x_e falls sharply to become $\ll 1$, so that most of the baryonic material in the universe is in neutral hydrogen and helium. Nevertheless, the universe today is highly ionized. How did this come about?

At redshift $z \sim 10$, roughly a billion years after decoupling, the smooth and uneventful travel of *some* CMB photons is interrupted. These photons scatter off the electrons that had just been freed. The immediate question is how exactly were these electrons disassociated from the hydrogen atoms in the first place – what freed them? The answer lies in the radiation emitted by some of the **first objects** in the universe – early stars and quasars. These first objects form at redshift $8 \lesssim z \lesssim 30$, and some fraction of the outgoing radiation – the so-called **ionizing photons**, those of energy $E > 13.6\,\text{eV}$ – ionizes neutral gas (mainly hydrogen) around them. The newly ionized regions are isolated, but they grow with time and eventually merge. These regions contain electrons that serve as targets for Thomson scattering of CMB photons in the process called **reionization**.

The key quantity describing reionization is its **optical depth** τ; it gives the probability of the CMB photon scattering off a free electron, $P = 1 - e^{-\tau} \simeq \tau$ (assuming $\tau \ll 1$ in the latter expression). The CMB is very sensitive to τ, as only the surviving photons, which are a fraction $e^{-\tau}$ of all photons, live to tell about the CMB anisotropy. Hence, the CMB *power* spectrum is proportional to the quantity $A_s e^{-2\tau}$, where A_s is the power spectrum normalization and $e^{-2\tau}$ is the new term that encodes the suppression due to reionization. The Planck constraint is

$$\tau = 0.054 \pm 0.008 \qquad \text{(68 percent C.L., Planck).} \qquad (13.36)$$

This means that about one CMB photon in 20 was scattered in the process of reionization.

We can work out the basics of the physics of reionization as follows. Assume for simplicity that reionization is *instantaneous*. Then we need to sum up the collective effect of Thomson scattering of the CMB between the time of reionization t_{reion} until the present time t_0. Our arguments for *recombination* will be useful here; Eq. (13.30) states that the Thomson scattering rate of the CMB photons is

$$\Gamma = \frac{c n_{e,0} \sigma_{\text{T}}}{a^3} \simeq \frac{c n_{B,0} \sigma_{\text{T}}}{a^3}, \qquad (13.37)$$

where σ_T is the Thomson scattering cross-section. Again, we have assumed that the reionization is complete and instantaneous, so that the universe is 100 percent ionized at $t_{\text{reion}} < t < t_0$, and the free electron number density in that time interval is $n_{e,0}/a^3$ (and $n_{e,0} \simeq n_{B,0} \simeq 0.25\,\text{m}^{-3}$). The optical depth is just the integral of the scattering rate:

$$\tau = \int_{t_{\text{reion}}}^{t_0} \Gamma(t)dt = c n_{e,0} \sigma_{\text{T}} \int_{t_{\text{reion}}}^{t_0} \frac{dt}{a(t)^3}. \qquad (13.38)$$

This expression can be straightforwardly evaluated, and from it the assumed-instantaneous reionization redshift z_{reion} can be determined; see Problem 13.3.

Reionization is studied in more detail using numerical simulations that incorporate star formation and include atomic-physics treatment of the interaction of ionizing photons and the surrounding gas. While the Planck measurement of the optical depth in Eq. (13.36) is very precise, the details of the rate at which reionization progressed in redshift, as well as its spatial dependence (its "patchiness," in the language used in the field), are still largely unexplored. There is also great interest in directly observing the earliest stars and quasars using, for example, the recently launched James Webb Space Telescope in order to learn more about the physical processes that started reionization.

13.3 Physics Behind the Power Spectrum

As we already hinted, the physics behind all structures in the angular power spectrum (see Fig. 13.6) is well understood, but fairly complicated. We will not discuss it in great detail here; see the Dodelson–Schmidt (2020) or Weinberg (2008) books for technical treatments. Instead, we highlight the most important physical processes, and provide a few simple, pedagogic derivations.

The key physical phenomena that determine the power spectrum are:

- The Sachs–Wolfe effect, which is due to the combination of intrinsic fluctuations at the last scattering surface, and the subsequent photons' climbing out of potential wells. Sachs–Wolfe is the dominant effect at large angular scales ($\ell \lesssim 20$) in the CMB angular power spectrum.

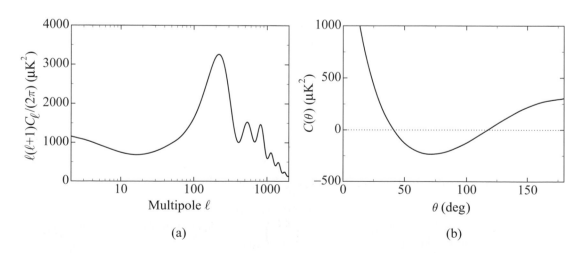

(a)   (b)

Fig. 13.6 The angular power spectrum $\ell(\ell+1)C_\ell/(2\pi)$ (a) and the angular two-point correlation function $C(\theta)$ (b) of CMB temperature in the fiducial LCDM model.

- The acoustic oscillations observed at intermediate scales ($100 \lesssim \ell \lesssim 1000$), with multiple peaks and valleys in the angular power spectrum spaced out regularly in multipole.

- Damping of the angular power spectrum at small angular scales ($\ell \gtrsim 1000$), where the angular power exponentially goes to zero.

We devote the rest of this section to describing these physical effects in some detail.

13.3.1 Sachs–Wolfe Effect

Let us take a closer look at the Sachs–Wolfe effect in the Standard Model with adiabatic density fluctuations. We are considering the temperature of photons that are coming out of the hot and cold spots on the last scattering surface, "climbing out" of potential wells, and making their way to us. We assume that the perturbed metric takes the form

$$ds^2 = -(1 + 2\Psi)dt^2 + a^2(1 + 2\Phi)d\mathbf{x}^2, \tag{13.39}$$

where Ψ and Φ are two metric potentials. For the applications of our interest, $\Psi \simeq -\Phi$, which is what we assume in what follows. Moreover, for brevity we drop the sky direction in our notation, so for example $\delta T/T(\hat{\mathbf{n}}) \equiv \delta T/T$.

The Sachs–Wolfe temperature shift is

$$\left(\frac{\delta T}{T}\right)_{\mathrm{SW}} = \left.\frac{\delta T}{T}\right|_i - \Phi. \tag{13.40}$$

The first term on the right-hand side indicates the intrinsic temperature fluctuation at the last scattering surface. The second term just denotes the familiar gravitational redshift of photons – they are "climbing out of potential wells" – whereby they lose energy.

The first term, $\delta T/T|_i$, can be evaluated as follows. The gravitational potential affects the scale factor of the universe $a(t)$ because of time dilation; note that $ds = (1 + 2\Psi)^{1/2}\, dt \simeq (1 - 2\Phi)^{1/2}\, dt \simeq (1 - \Phi)\, dt$, so that $\delta t/t \simeq -\Phi$. Moreover, $aT = \mathrm{const.}$, so that a fluctuation in the scale factor induces a fluctuation in the temperature. Therefore

$$\left.\frac{\delta T}{T}\right|_i = -\frac{\delta a}{a} = -\frac{2}{3}\frac{\delta t}{t} = +\frac{2}{3}\Phi, \tag{13.41}$$

where we also assumed that $a(t) \propto t^{2/3}$ since the last scattering is safely in the matter-dominated era. Equations (13.40) and (13.41) imply

$$\left(\frac{\delta T}{T}\right)_{\mathrm{SW}} = -\frac{1}{3}\Phi. \tag{13.42}$$

The minus sign in the equation above signifies that, on large angular scales, potential *overdensities* register as *cold* spots on the sky. This is because hotter-than-average photons coming from overdense areas more than lose their excess energy climbing out of the potential wells.

We can calculate the Sachs–Wolfe signature in the angular power spectrum as follows (we omit constants of proportionality for simplicity). The projection of any 3D field to the angular power spectrum is given by the following expression that features the comoving distance r_* to recombination (see Problem 13.8):

$$C_\ell \propto \int \Delta_\Phi^2(k) j_\ell^2(kr_*) d\ln k, \tag{13.43}$$

where we already specialized in the case of gravitational potential (hence the subscript $_\Phi$). From the Poisson equation $-k^2\Phi = 4\pi G\rho_M\delta$, we see that $\Phi \propto k^{-2}\delta$. Since $\Delta^2 \equiv k^3 P(k)/(2\pi^2) \propto k^{n_s+3}$, the contribution of *potential* power per unit log interval in k goes as $\Delta_\Phi^2 \propto k^{-4}\Delta^2 \propto k^{n_s-1}$. Therefore

$$C_\ell \propto \int k^{n_s-1} j_\ell^2(kr_*) d\ln k = \int_0^\infty k^{n_s-2} j_\ell^2(kr_*) dk. \tag{13.44}$$

The integral can be done analytically, and is equal to

$$C_\ell \propto \frac{\Gamma\left(\ell + \dfrac{n_s-1}{2}\right)\Gamma(3-n_s)}{\Gamma\left(\ell + \dfrac{5-n_s}{2}\right)\Gamma^2\left(2-\dfrac{n_s}{2}\right)} \xrightarrow{n_s=1} \frac{\Gamma(\ell)\,\Gamma(2)}{\Gamma(\ell+2)\,\Gamma^2\left(\dfrac{3}{2}\right)}, \tag{13.45}$$

where the latter expression assumes $n_s = 1$. Noting that $\Gamma(3/2) = \sqrt{\pi}/2$ and that $\Gamma(x+1) = x\Gamma(x)$, the right-hand side above is proportional to $1/(\ell(\ell+1))$, so that

$$\ell(\ell+1)C_\ell = \text{const.} \qquad \text{(Sachs–Wolfe, for } n_s = 1). \tag{13.46}$$

Therefore, the Sachs–Wolfe effect leads, for a perfectly scale-invariant ($n_s = 1$) primordial spectrum, to a flat relation between $\ell(\ell+1)C_\ell$ vs. ℓ. This is why the CMB angular power spectrum is generally presented as a plot of $\ell(\ell+1)/(2\pi)C_\ell$ vs. ℓ, which in turn explains why people refer to the Sachs–Wolfe "plateau" at these large scales.

In the presence of dark energy, the **integrated Sachs–Wolfe effect** (ISW) also contributes. The ISW contribution occurs whenever the universe is *not* matter-dominated (more precisely, not EdS with $\Omega_{\rm TOT} = \Omega_M = 1$). Then it turns out that the Newtonian gravitational potential Φ decays in time, and the photons' gain in energy when entering a potential well is not matched by their loss of energy when they come out of the well, as the potential itself becomes shallower during the time of a photon's traversal. Because dark energy dominates in the late universe, the ISW plays an important role at large scales (multipoles $\ell \lesssim 10$), corresponding to modes that enter the horizon at late times (redshift $\lesssim 1$) when dark energy dominates.

The ISW can be seen as a sloping-down large-angle anisotropy that is added to a mostly flat Sachs–Wolfe-only plateau in Fig. 13.6(a).[4]

We now turn to CMB anisotropy at smaller angular scales (degree and below), at which a series of peaks in the angular power spectrum is observed. We start by discussing the so-called sound horizon.

13.3.2 Role of the Sound Horizon

It turns out that a typical comoving separation between the hot and cold spots is given by the **sound horizon** – the distance that sound can travel between the Big Bang and recombination. The speed of sound is given by $c_s = (\delta P/\delta \rho)^{1/2}$ in the early epoch when the photons and baryons are strongly coupled in a plasma. The energy density and pressure of the photons are

$$\rho_\gamma = \alpha T^4; \qquad P_\gamma = \frac{1}{3}\alpha T^4, \tag{13.47}$$

so that the familiar $P_\gamma/\rho_\gamma = 1/3$ obtains. Then

$$c_s = \sqrt{\frac{\delta P_\gamma}{\delta \rho_B + \delta \rho_\gamma}} = \sqrt{\frac{(1/3)\,\delta \rho_\gamma}{\delta \rho_B + \delta \rho_\gamma}} = \frac{1}{\sqrt{3(1+R)}}, \tag{13.48}$$

where $R \equiv \delta \rho_B/\delta \rho_\gamma$. To evaluate R, we assume that the fluctuations are adiabatic, that is, that each species' density fluctuation separately obeys the continuity equation $\dot\rho_i + 3H(\rho_i + P_i) = 0$. Rearranging we get $\delta \rho_i/[\rho_i(1+w_i)] = -3H\delta t$, where the right-hand side is clearly species independent, so that, for any two species i and j:

$$\frac{\delta \rho_i}{(1+w_i)\rho_i} = \frac{\delta \rho_j}{(1+w_j)\rho_j} \quad \text{(adiabatic condition for any } i, j). \tag{13.49}$$

For baryons and matter we have

$$\frac{\delta \rho_B}{\rho_B} = \frac{\delta \rho_\gamma}{(4/3)\rho_\gamma} \quad \text{(adiabatic condition for photons, baryons)} \tag{13.50}$$

so that $\delta \rho_B/\delta \rho_\gamma = (3/4)\rho_B/\rho_\gamma$. Therefore

$$R \equiv \frac{\delta \rho_B}{\delta \rho_\gamma} = \frac{3\rho_B}{4\rho_\gamma} = \frac{3\Omega_B}{4\Omega_\gamma}\,a \tag{13.51}$$

which, note, scales as a, or $(1+z)^{-1}$. Therefore, the speed of sound is $1/\sqrt{3}$ of the speed of light, corrected for the non-negligible presence of baryons. The sound horizon is just defined as the coordinate distance from the Big Bang to recombination, $r_s = c_s \int dt/a(t) = c_s \int da/(a^2 H(a))$ or, adopting Eqs. (13.49) and (13.51), we find the sound horizon to be

---

[4] In addition to this so-called *late*-time ISW effect due to the presence of dark energy, there is also the so-called *early*-time ISW coming from around the time of recombination, at $z \sim 1000$, when the universe was also not totally matter dominated but had a $\sim$10 percent contribution from radiation. The early-time ISW leaves signatures in the CMB power spectrum at scales comparable to the horizon at recombination, at $\ell \sim 200$.

$$r_s = \frac{1}{\sqrt{3}} \int_0^{a_*} \frac{da}{a^2 H(a) \sqrt{1 + \frac{3\Omega_B}{4\Omega_\gamma} a}} \simeq 100 \, h^{-1} \text{Mpc}, \qquad (13.52)$$

where $a_* \simeq 0.001$ is the scale factor at recombination. [Note that this is the expression for the *comoving* sound horizon; the physical sound horizon at $a_* \simeq 0.001$ is $r_s^{\text{phys}} = a_* r_s \simeq 100 \, h^{-1} \text{kpc}$.]

The sound horizon plays an important role in cosmology, as it gives a clear scale in the universe. This scale is imprinted in the distribution of hot and cold spots of the CMB as we discuss just below, and also in the distribution of galaxies, where it registers as the BAO feature discussed in Sec. 9.5.5.

13.3.3 Physics of the Bumps and Wiggles

Understanding and deriving the full structure of the CMB anisotropy theory is a formidable task that was carried out in the pioneering works of physical cosmology, notably Peebles and Yu (1970), Sunyaev and Zeldovich (1970), and others in the 1980s. One starts from the coupled Einstein equation of general relativity and the Boltzmann equation, perturbs both, and goes from there. Such an analysis is well beyond the scope of this book; it is covered in more advanced texts and it involves considerable acrobatics with the perturbed Boltzmann equation. Here we proceed to merely sketch the essence of the origin of the acoustic oscillations seen in the CMB power spectrum.

Let us first consider the continuity and Euler equations for the photon fluid, ignoring for the moment the effects of both baryons and gravity. Let $\delta T/T(\hat{\mathbf{n}}) \equiv \Theta(\hat{\mathbf{n}})$ be the temperature fluctuation, and Θ_k be its Fourier transform. The fully relativistic Euler equation encodes momentum conservation, and for photons in an expanding universe reads

$$\frac{d}{dt}\left[(P_\gamma + \rho_\gamma)\mathbf{v}_\gamma\right] = -4\frac{\dot{a}}{a}\left(P_\gamma + \rho_\gamma\right)\mathbf{v}_\gamma - \nabla P_\gamma, \qquad (13.53)$$

where the factor of four on the right-hand side accounts for the redshifting power law for photons. Because for radiation $P = \rho/3$, $\rho + P = 4/3\rho$. Moreover, as $\rho_\gamma \propto a^{-4}$ we have $d\rho_\gamma/dt = -4(\dot{a}/a)\rho_\gamma$. The Euler equation thus simplifies to

$$\frac{4}{3}\rho_\gamma \dot{\mathbf{v}}_\gamma = -\nabla P_\gamma. \qquad (13.54)$$

Further,

$$\nabla P_\gamma = \frac{1}{3}\nabla \rho_\gamma = \frac{4}{3}\bar{\rho}_\gamma \nabla \Theta, \qquad (13.55)$$

where we used the fact that $\rho_\gamma \propto T^4$ and also that $\nabla(T + \delta T) = \nabla(\delta T)$. Fourier transforming both sides, so that $\mathbf{v} \rightarrow i\mathbf{v}_k$ (where the imaginary factor i is added here for convenience) and $\nabla \to -i\mathbf{k}$, Eqs. (13.54) and (13.55) lead to

$$\dot{\mathbf{v}}_k = \mathbf{k}\Theta_k. \tag{13.56}$$

Next, the continuity equation for the photon number in an expanding universe,

$$\frac{dn_\gamma}{dt} + 3\frac{\dot{a}}{a}n_\gamma + \nabla(n_\gamma \cdot \mathbf{v}) = 0, \tag{13.57}$$

can be perturbed by taking $n_\gamma \to \bar{n}_\gamma + \delta n_\gamma$ and recognizing that $\mathbf{v}_\gamma$ is the same order as δn. At zeroth order, we get $d/dt(\bar{n}) + 3(\dot{a}/a)\bar{n} = 0$. At first order, we have $d/dt(\delta n/n) + \nabla \cdot \mathbf{v} = 0$. Recalling that $n_\gamma \propto T^3$, Fourier transforming, and making use again of $\mathbf{v} \to i\mathbf{v}_k$ and $\nabla \to -i\mathbf{k}$, we get

$$\dot{\Theta} = -\frac{1}{3}\mathbf{k}\mathbf{v}_k. \tag{13.58}$$

Taking the derivative of Eq. (13.58) and plugging into Eq. (13.56), we finally get

$$\ddot{\Theta} + c_s^2 k^2 \Theta = 0, \tag{13.59}$$

with $c_s^2 \simeq 1/3$ at this level of approximation.

Phew, done. Equation (13.59) already illustrates a key feature of the Fourier modes of the CMB – Fourier-mode oscillations. The solution to Eq. (13.59) is[5]

$$\Theta(t_*) = \Theta_0 \cos(kr_s), \tag{13.60}$$

where t_* is the time between the Big Bang and recombination, and $r_s = \int_0^{t_*} c_s(t)dt$ is the corresponding comoving distance, usually referred to as the sound horizon and defined in Eq. (13.52). It is clear that there are oscillations, with peaks located at $k_n = n\pi/r_s$, with n an integer. Because we will be interested in the power spectrum – the "square" – of temperature fluctuations, both minima and maxima of the oscillations of Θ will register as peaks. And because the distance that sound can travel plays a fundamental role in this, the oscillations are called **acoustic oscillations**, while the peaks in the CMB power spectrum are called the acoustic peaks. Here, "acoustic" refers to the cosmic sound, which propagates with $c_s \simeq c/\sqrt{3}$.

Since the CMB photons are coming to us from the last scattering surface, we observe these acoustic oscillations in projection. The sound-horizon scale is the transverse side of a triangle, opposite to an observer and at a distance equal to the distance to the surface of last scattering (or the epoch of photon decoupling), $r(z_{\mathrm{dec}})$. The *angle* on which we observe the first peak, $\theta_{\mathrm{1st\,peak}}$, is then equal to the ratio $r_s/r(z_{\mathrm{dec}})$; see Fig. 13.7. It is often more convenient to work in quantities conjugate to angles – the multipoles $\ell \simeq \pi/\theta$. Then the peaks in the CMB angular power spectrum are

---

[5] There is also a $\sin(kr_s)$ term, corresponding to initial velocity perturbation; we ignore it in our simplified discussion.

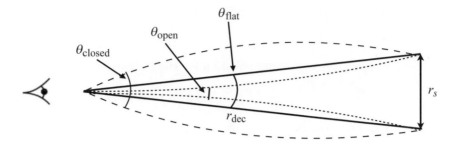

Illustration of how the angle at which we observe the acoustic peak – that is, the sound horizon – helps measure the curvature of the universe, showing three (exaggerated) cases of positive-, negative-, and zero-curvature space. Further, by *assuming* a flat universe the sound-horizon measurement can give a precise measurement of the distance to decoupling, $r_{\rm dec} = r(z_{\rm dec})$.

$$\ell_{\rm peak} = n\ell_{\rm 1st\,peak}, \qquad \text{where} \qquad \ell_{\rm 1st\,peak} \simeq \frac{\pi}{\theta_{\rm 1st\,peak}} = \pi \frac{r(z_{\rm dec})}{r_s}, \qquad (13.61)$$

where $r(z_{\rm dec})$ is the (comoving) distance from us to the epoch of decoupling. For the cosmological parameters from Table 3.1 and $z_{\rm dec} = 1090$, $r_s \simeq 110\,h^{-1}{\rm Mpc}$, $r(z_{\rm dec}) \simeq 9500\,h^{-1}{\rm Mpc}$, indicating that the first peak should be at $\ell_{\rm 1st\,peak} \simeq 270$. A more complete theoretical analysis shows, and data confirm, that

$$\ell_{\rm 1st\,peak} \simeq 220. \qquad (13.62)$$

Because the CMB angular power peaks at this scale (corresponding to $\theta \simeq \pi/\ell \simeq 1°$), we say that the "typical" separation between hot and cold spots in the CMB is about one degree.

The peak locations can be determined to a fabulously good accuracy from the CMB data; the Planck experiment reports a 0.03 percent accuracy for what we call $\ell_{\rm 1st\,peak}$. The sound horizon r_s is also determined very well by the morphology of the acoustic peaks (their absolute and relative heights); Planck reports a 0.4 percent accuracy in it and this measurement is largely independent of the information encoded in peak locations. With measurements of $\ell_{\rm 1st\,peak}$ and r_s in hand, we have a sub-percent measurement of either one of these quantities (see Fig. 13.7):

1. We have cosmology's most sensitive measure of the geometry of the universe. Recall that the comoving distance formula contains a sin or a sinh for closed and open universes, respectively: an open universe leads to a smaller angle θ (or larger $\ell_{\rm 1st\,peak}$) for a fixed r_s measurement, and vice versa for a closed universe. The measurement of the first-peak location in multipole, $\ell_{\rm 1st\,peak} \simeq 220$, is in excellent agreement with a flat-universe scenario even after other cosmological parameters are varied; see Table 3.1.

2. *Assuming* that the universe is flat, we can solve for the distance to recombination r_{dec}, which in turn depends on the dark-energy parameters Ω_{DE} and w. We effectively get a single, very precise, measurement of a combination of these dark-energy parameters.

We conclude:

> The locations of the acoustic peaks in the CMB power spectrum provide an extremely sensitive measure of the geometry of the universe. In a flat universe, they provide a valuable measurement of the distance to the last-scattering surface, $r(z \simeq 1000)$.

13.3.4 CMB Physics: Adding Gravity and Baryons

In our introduction to the origin of the acoustic peaks above, we have ignored the effects of baryons and gravity. The overall result of including these two is that the Fourier-space oscillations are actually not in Θ, but in $\Theta + (1 + R)\Psi$, where $\Psi \simeq -\Phi$ is one of the gravitational potentials, and R is the baryon-to-photon density ratio defined in Eq. (13.51). Baryons also change the sound speed from $1/\sqrt{3}$ to $1/\sqrt{3(1 + R)}$, as described in Eq. (13.48), and thus modify the value of r_s to what we found in Eq. (13.52).

Moreover, baryons act as an additional source of inertia in the Euler equation: the pressure and density terms get multiplied by $1 + R$, and pressure and potential gradients have more intertia to overcome. As a consequence, the zero point of the oscillations is shifted from $\Theta \simeq -\Psi$ to $\Theta \simeq -(1+R)\Psi$. This is very much equivalent to an oscillating mass m on a spring under the influence of gravity. If we now add an additional mass so that $m \rightarrow m(1 + R)$, the zero point of the oscillations will be shifted. The amplitude *squared* of the oscillations will therefore exhibit a pattern where every odd peak (first, third, etc.) is enhanced relative to every even peak (second, fourth, etc.). Therefore

> The relative amplitudes of the consecutive peaks in the CMB angular power spectrum are sensitive to the baryon density in the universe.

13.3.5 Silk Damping

On small angular scales, the CMB angular power spectrum is damped. The effect can be phenomenologically described as

$$\ell(\ell + 1)C_\ell \propto e^{-(\ell/\ell_D)^{1.25}}, \tag{13.63}$$

with $\ell_D \simeq 1000$ the damping scale; this is called **Silk damping**. The reason for the damping is photon *diffusion*, and it can be thought of as a random walk of photons as they cross from hot to cold regions if these regions are small enough (roughly

$\lesssim 10'$) – that is, when observed at small angular scales. Because the medium goes from being optically thick to optically thin during recombination when the photons' rate of scattering dramatically decreases, a full treatment of Silk damping involves numerical evaluation of perturbation equations. We will not say more about it here, but you will have a chance to make some estimates of Silk damping in Problem 13.1.

13.3.6 CMB: The Cosmological Rosetta Stone

The CMB deserves this name (Bennett *et al.*, 1997), since it encodes precise information about a wealth of cosmological parameters. Very briefly:

- The amplitude of the primordial power spectrum, A_s, can be determined from the overall amplitude of the CMB angular power spectrum C_ℓ.

- The spectral index n_s can be determined from the "tilt" of C_ℓ as a function of ℓ.

- The physical matter and baryon densities $\Omega_M h^2$ and $\Omega_B h^2$ (why do I refer to these as "physical" densities?) can be determined from the morphology of the peaks.

- One linear combination of the dark-energy parameters Ω_{DE} and w, together with Ω_M, can be constrained very accurately, as these parameters enter the angular diameter distance to last scattering, and thus $\ell_{\mathrm{1st\,peak}}$.

We now move to the statistical description of CMB anisotropies, as well as connection to experiments.

13.4 Describing CMB Anisotropies

We now turn to the mathematical description of the CMB temperature anisotropies, and the statistical and computational methodology of how to measure the anisotropies on the sky and compare them to theory. This has been one of the most successful areas of cosmology: over the period of about three decades (from the early 1990s until the present day), it enabled us to reach previously unfathomable levels of precision in both measurements and theory, allowing percent-level constraints on a number of key cosmological parameters. This data–statistics–theory triple-headed monster made the CMB arguably the most important cosmological probe, and played a large role in the advance of precision cosmology.

13.4.1 Harmonic Decomposition of the CMB Temperature

The temperature of the CMB can be expanded in terms of spherical harmonics $Y_{\ell m}$:

$$\frac{\delta T}{T}(\hat{\mathbf{n}}) = \sum_{\ell=2}^{\infty} \sum_{m=-\ell}^{\ell} a_{\ell m} Y_{\ell m}(\hat{\mathbf{n}}), \tag{13.64}$$

where $a_{\ell m}$ are complex coefficients (because the $Y_{\ell m}$ are complex and $\delta T/T$ is real). The description of the temperature anisotropy in Eq. (13.64) is completely general and, as $\ell \to \infty$, can describe any pattern on the sky, no matter how complicated. Note that in the expansion above we skipped the monopole term $\ell = 0$ since we are expanding *around* the mean, and the dipole term $\ell = 1$ since the dipole is, as discussed around Eq. (13.8), due to our local motion and is not cosmological.

Working with spherical harmonics has many nice properties, but the main thing to know is that they enable separating scales – each ℓ corresponds to a characteristic variation on a scale of

$$\theta \simeq \frac{\pi}{\ell} \, \text{rad} = \frac{180°}{\ell}. \tag{13.65}$$

Typically, the sum in Eq. (13.64) only needs to be carried out to the maximum multipole $\ell_{\max} \simeq \pi/\theta_{\text{res}}$, where θ_{res} is the resolution limit of the map that we are analyzing. In Problems 13.4 and 13.6, you will further investigate how this correspondence works.

Note that it is very easy to determine the coefficients $a_{\ell m}$ from the CMB map, by inverting Eq. (13.64) using orthonormality of spherical harmonics:

$$a_{\ell m} = \int \frac{\delta T}{T}(\hat{\mathbf{n}}) Y_{\ell m}^{*}(\hat{\mathbf{n}}) d\Omega, \tag{13.66}$$

where the asterisk denotes the complex conjugate as usual. So far we are assuming a full-sky map of CMB anisotropy so that the orthogonality of spherical harmonics holds. The cut-sky case is discussed further below.

There are a total of $2\ell + 1$ coefficients $a_{\ell m}$ for each ℓ, since m ranges from $-\ell$ to ℓ. Since the $a_{\ell m}$ are complex numbers, they can be written as $a_{\ell m} = a_{\ell m}^{\text{RE}} + i a_{\ell m}^{\text{IM}}$. Because the CMB temperature is described by purely real numbers, the $a_{\ell m}$ coefficients with positive and negative values of the azimuthal number m are mutually related:

$$a_{\ell(-m)} = (-1)^m a_{\ell m}^{*}. \tag{13.67}$$

Therefore, the $(2\ell + 1)$ degrees of freedom in an $a_{\ell m}$ at a fixed ℓ consist of a total of $2 \times (2\ell + 1)$ real and imaginary parts of each $a_{\ell m}$, subject to $2\ell + 1$ relations in Eq. (13.67).

One simple choice for specifying these degrees of freedom is to work with these $2\ell + 1$ numbers: $a_{\ell 0}$, $a_{\ell 1}^{\text{RE}}$, $a_{\ell 1}^{\text{IM}}$, ..., $a_{\ell \ell}^{\text{RE}}$, $a_{\ell \ell}^{\text{IM}}$. A visual representation of such a decomposition of a CMB map is shown in Fig. 13.8. Assuming real and imaginary

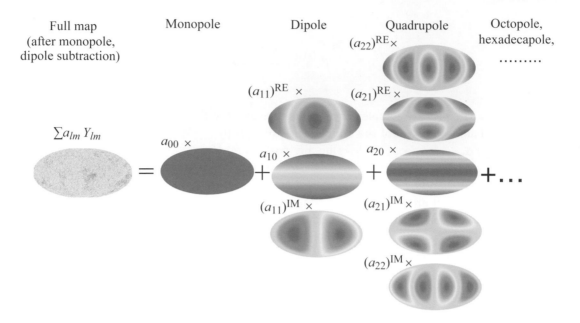

Fig. 13.8 A sketch of how individual multipoles are combined to make the CMB map. In the sum, each $Y_{\ell m}$ pattern (individual ovals) is multiplied by a coefficient ($a_{\ell m}$). Note that the monopole and dipole are not used in the map of CMB anisotropy because the monopole is the mean temperature (around which we are expanding), while the dipole is due to our local motion through the CMB rest frame, and thus not cosmological. Note that the sum over the real and imaginary parts of $a_{\ell m}$ is a bit simplified in this sketch; see Eq. (13.68) for the exact expression. A black and white version of this figure will appear in some formats. For the color version, please refer to the plate section.

parts of $a_{\ell m}$ as base parameters, and accounting for the relation in Eq. (13.67), Eq. (13.64) can be written purely in terms of real numbers $Y_{\ell m}^{\mathrm{RE,IM}}$ and $a_{\ell m}^{\mathrm{RE,IM}}$:

$$\frac{\delta T}{T}(\hat{\mathbf{n}}) = \sum_{\ell=2}^{\infty} \left[a_{\ell 0} Y_{\ell 0}(\hat{\mathbf{n}}) + 2 \sum_{m=1}^{\ell} \left(a_{\ell m}^{\mathrm{RE}} Y_{\ell m}^{\mathrm{RE}}(\hat{\mathbf{n}}) - a_{\ell m}^{\mathrm{IM}} Y_{\ell m}^{\mathrm{IM}}(\hat{\mathbf{n}}) \right) \right]. \qquad (13.68)$$

This form is useful for numerical work, as it explicitly avoids complex numbers.

Inflation predicts that the resulting CMB anisotropies are a Gaussian random field. This means that the coefficients $a_{\ell m}$ come from a Gaussian normal distribution with mean zero, and some variance, call it C_ℓ. More precisely, if $\mathcal{N}(\mu, \sigma^2)$ is a Gaussian random distribution with mean μ and variance σ^2, then the real and imaginary parts of the $a_{\ell m}$ coefficients are given via

$$a_{\ell 0} \in \mathcal{N}(0, C_\ell); \qquad a_{\ell m}^{\mathrm{RE}}, a_{\ell m}^{\mathrm{IM}} \in \mathcal{N}(0, C_\ell/2) \quad (m \neq 0). \qquad (13.69)$$

This is visually shown in Fig. 13.9. We will prove a bit further below that the variance C_ℓ doesn't depend on m if we assume statistical isotropy.

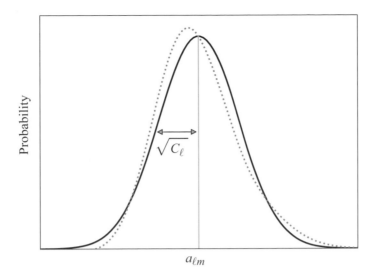

Fig. 13.9   A sketch of the probability distributions for the coefficients $a_{\ell m}$. Note that the distributions in $a_{\ell m}^{\mathrm{RE}}$ and $a_{\ell m}^{\mathrm{IM}}$ (for $m \neq 0$) separately have half the variance, that is, $C_\ell/2$, while $a_{\ell 0}$ has variance C_ℓ. The dotted line shows non-Gaussianity of the so-called local type about 2000 times larger (!) than the current upper limit imposed by the Planck experiment's data; this will be discussed in Sec. 13.7.

13.4.2 Two-Point Correlation Function and Angular Power Spectrum of the CMB

We continue our discussion of CMB angular clustering by considering the angular correlation function, $C(\theta)$. The angular two-point function, introduced briefly earlier in Eq. (13.10), is the exact equivalent of the galaxy-clustering angular two-point function $w(\theta)$ that we introduced in Chapter 9. Recall that the two-point function, at any separation θ, can be positive (excess probability of finding a hot spot) or negative (decreased probability), or else zero (random probability of finding a spot of the same temperature sign – this is the uncorrelated case). The angular two-point correlation function is defined as

$$C(\theta) \equiv \left\langle \frac{\delta T}{T}(\hat{\mathbf{n}}) \frac{\delta T}{T}(\hat{\mathbf{n}}') \right\rangle_{\hat{\mathbf{n}}\cdot\hat{\mathbf{n}}'=\cos\theta}, \qquad (13.70)$$

so that it is the average over the observed sky of products of temperatures at all points (or in all pixels in a map) separated by an angle θ.

It is useful to expand the correlation function $C(\theta)$ as follows:

$$C(\theta) \equiv \sum_{\ell=2}^{\infty} \frac{2\ell+1}{4\pi} C_\ell \, P_\ell(\cos\theta), \tag{13.71}$$

where $P_\ell(\cos\theta)$ are Legendre polynomials. For now the C_ℓ can just be considered some arbitrary expansion coefficients. Note that we can always do such an expansion (provided statistical isotropy holds), since the Legendre polynomials $P_\ell(x)$ form a complete basis on the interval $x \in [-1, 1]$.

Starting from Eq. (13.71), we will now prove the relation between the coefficients $a_{\ell m}$ and the coefficients C_ℓ that was gently introduced in Eq. (13.69). More exactly, the relation that we will prove is a key result in the mathematical description of CMB anisotropy:

$$\langle a_{\ell m} a_{\ell' m'}^* \rangle = C_\ell \delta_{\ell\ell'} \delta_{mm'} \quad \text{(assuming statistical isotropy)}. \tag{13.72}$$

In other words, the coefficients $a_{\ell m}$ are uncorrelated except for the same ℓ and m, and in *that* case their variance is given by C_ℓ as defined in Eq. (13.71). Two assumptions go into the validity of the remarkably simple relation in Eq. (13.72):

- Statistical isotropy of the CMB (i.e., the fact that anisotropies are distributed statistically the same in every direction on the sky).

- Full-sky coverage.

The first assumption above is very important, and holds apparently to an excellent approximation in our universe – we discussed this in Chapter 1. The second assumption trivially breaks on an incomplete sky (i.e., observations that cover less than 4π steradians on the sky), as the spherical harmonics are not orthogonal any more. In this case, neighboring multipoles are coupled over a coherence length of roughly $\Delta\ell_{\rm coupled} \simeq \pi/\theta_{\rm survey}$, where $\theta_{\rm survey}$ is (again roughly) the linear size of the survey in radians. This coupling can be mathematically calculated for a given sky footprint of the survey and taken into account in data analysis; we will not discuss it here any more.

Because Eq. (13.72) is central to the description and analysis of the CMB, let us derive it. We start from Eq. (13.66):

$$a_{\ell m} = \int \frac{\delta T}{T}(\hat{\mathbf{n}}) Y_{\ell m}^*(\hat{\mathbf{n}}) d\Omega. \tag{13.73}$$

The two-point correlation function of $a_{\ell m}$ is given by

$$\langle a_{\ell m} a_{\ell' m'}^* \rangle = \int\int \left\langle \frac{\delta T}{T}(\hat{\mathbf{n}}) \frac{\delta T}{T}(\hat{\mathbf{n}}') \right\rangle Y_{\ell m}^*(\hat{\mathbf{n}}) Y_{\ell' m'}(\hat{\mathbf{n}}') d\Omega \, d\Omega'$$

$$\equiv \int\int C(\hat{\mathbf{n}} \cdot \hat{\mathbf{n}}') Y_{\ell m}^*(\hat{\mathbf{n}}) Y_{\ell' m'}(\hat{\mathbf{n}}') d\Omega \, d\Omega'. \tag{13.74}$$

Note that we have assumed statisical isotropy in these expressions, since we assumed that the correlation function only depends on the angular separation between the two points, so that $\langle \delta T/T(\hat{\mathbf{n}}) \delta T/T(\hat{\mathbf{n}}') \rangle \equiv C(\hat{\mathbf{n}} \cdot \hat{\mathbf{n}}')$.

Adopting now Eq. (13.71) to express $C(\hat{\mathbf{n}} \cdot \hat{\mathbf{n}}')$ as a function of C_ℓ, and then using the addition theorem for spherical harmonics,

$$P_\ell(\hat{\mathbf{n}} \cdot \hat{\mathbf{n}}') = \frac{4\pi}{2\ell+1} \sum_{m=-\ell}^{\ell} Y_{\ell m}(\hat{\mathbf{n}}) Y_{\ell m}^*(\hat{\mathbf{n}}') \qquad \text{(addition theorem)}, \qquad (13.75)$$

Eq. (13.74) becomes, after cancellation of $(2\ell+1)/4\pi$ and $4\pi/(2\ell+1)$,

$$\langle a_{\ell m} a_{\ell' m'}^* \rangle = \int \int \left(\sum_{\ell''} C_{\ell''} \sum_{m''=-\ell''}^{\ell''} Y_{\ell'' m''}(\hat{\mathbf{n}}) Y_{\ell'' m''}^*(\hat{\mathbf{n}}') \right) Y_{\ell m}^*(\hat{\mathbf{n}}) Y_{\ell' m'}(\hat{\mathbf{n}}') \, d\Omega \, d\Omega'$$

$$= \sum_{\ell''} C_{\ell''} \sum_{m''=-\ell''}^{\ell''} \int (Y_{\ell'' m''}(\hat{\mathbf{n}}) Y_{\ell m}^*(\hat{\mathbf{n}}) \, d\Omega) \int (Y_{\ell'' m''}^*(\hat{\mathbf{n}}') Y_{\ell' m'}(\hat{\mathbf{n}}') \, d\Omega')$$

$$= \sum_{\ell''} C_{\ell''} \sum_{m''=-\ell''}^{\ell''} \delta_{\ell \ell''} \delta_{m m''} \delta_{\ell' \ell''} \delta_{m' m''}$$

$$= C_\ell \, \delta_{\ell \ell'} \, \delta_{m m'}, \qquad (13.76)$$

which proves the desired expression in Eq. (13.72). Its simplicity can be intuitively understood: rotational invariance dictates that the azimuthal indices m and m' are not featured in the two-point function amplitude C_ℓ.

Statistical isotropy clearly makes the life of a CMB cosmologist much easier. The angular power spectrum, instead of being a function of as many as four indices ($C_{\ell \ell' m m'}$), has just one index (C_ℓ). This makes the angular power spectrum much easier to measure (since one can do, roughly speaking, a lot of averaging across the sky), as well as to compare to theory.

Figure 13.10 shows the measured angular power spectrum of the CMB by the Planck satellite, as well as the theoretical prediction from the best-fit ΛCDM model. Note the excellent agreement between the two. The impressive fact that the "bumps and wiggles" in the power spectrum were predicted, based on known physics in the early universe, about three decades before they were confirmed with measurements is one of the triumphs of modern cosmology.

13.4.3 Estimators for the C_ℓ

It should be clear by this point that extracting the angular power spectrum C_ℓ from the observed CMB anisotropy is of fundamental importance, as it links observations and theory.

Let us write the relation between the harmonic coefficients $a_{\ell m}^{\text{map}}$ measured in a CMB map and the actual C_ℓ realization *in that map*; this is

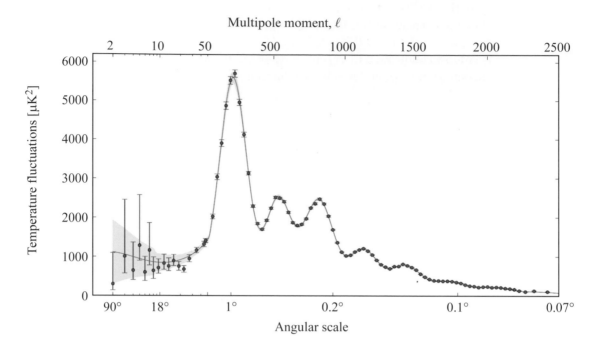

Multipole moment, ℓ

Fig. 13.10 Angular power spectrum (more precisely, $\ell(\ell+1)C_\ell/(2\pi)$) as a function of multipole ℓ, measured by the Planck experiment. Points are the Planck measurements, while the line is the best-fit ΛCDM cosmological model. The band around theory is the cosmic variance error. The errors on the measurements include the measurement error (which is typically small and barely visible in this graph) and the cosmic variance as well (which dominates on large scales and has, somewhat confusingly, been added to both the theory and measurements in this plot). Adopted from Ade *et al.* (2014b), reproduced with permission Ⓒ ESO.

$$C_\ell^{\text{map}} = \frac{\sum_{m=-\ell}^{\ell} |a_{\ell m}^{\text{map}}|^2}{2\ell + 1}. \qquad (13.77)$$

In other words, the measured C_ℓ is just the average of the squares of the measured harmonic coefficients.

The C_ℓ^{map} coefficients extracted from a map are typically noisy, but much less so than the individual $a_{\ell m}$ over which we effectively average in Eq. (13.77). We are able to do this averaging thanks to statistical isotropy, which stipulates that $a_{\ell m}$ for each azimuthal number m comes from the same underlying C_ℓ.

Equation (13.77) suggests a (rather naive, as it turns out) approach to extract the angular power spectrum from a CMB map: simply measure the $a_{\ell m}^{\text{map}}$ from the map using Eq. (13.73), then get the C_ℓ^{map} using Eq. (13.77). This works well on the full sky but, if implemented without any further corrections, fails miserably on the cut sky. If we only cover 25 percent of the sky, for example, each $a_{\ell m}$ obtained from Eq. (13.73) will be, on average, 50 percent of the true, full-sky value,[6] and the resulting C_ℓ^{map} (which are basically the squares of the $a_{\ell m}$) will therefore be 25 percent of the true value. One could apply a corresponding normalization correction for f_{sky}, but that still leaves a coupling of the extracted harmonics, which is very similar to the correlation of Fourier components of a function $f(x)$ that has been measured on a fraction of a full interval of x. One can make a correction for that as well (see the pseudo-C_ℓ estimator discussed just below), but the takeaway message is – when estimating the angular power spectrum, careful analysis is needed!

Box 13.2 describes two principal estimators for the angular power spectrum of the CMB. Here we assume a realistic case with partial sky coverage, that is, sky which is **masked**. The pseudo-C_ℓ estimator is the easier one of the two to implement, but is not optimal (unbiased and minimum variance). The maximum-likelihood method, on the other hand, *is* optimal, but is also numerically challenging to implement at smaller scales because of the required inversion of the $q \times q$ covariance matrix $\mathbf{C}$ in Eq. (B5). Consider, for example, that even at the relatively large scales with $\ell \leq \ell_{\max} \simeq 30$, the required pixel size in order to resolve the temperature anisotropy features is of order $\theta_{\text{pix}} \simeq \pi/\ell_{\max} \simeq 0.1\,\text{rad}$, and the total number of pixels in a full-sky map is $q \simeq 4\pi/\theta_{\text{pix}}^2 \simeq 1000$. Inverting a 1000×1000 matrix is not hard these days, but this has to be done repeatedly in the analysis. And if we wish to go to smaller scales, the matrix size increases sharply, scaling as $\ell_{\max}^2$ in each dimension. Therefore, the maximum-likelihood method is only practical at relatively large scales where a crude pixelization of the map is sufficient. At smaller scales, the pseudo-C_ℓ estimator is numerically much less demanding. Fortunately, it is precisely at smaller angular scales where the pseudo-C_ℓ estimator performs about as well as the maximum-likelihood one.

13.4.4 Cosmic Variance

The variance of the C_ℓ is another important quantity, as it tells us about the statistical error on this key quantity that links CMB observations to theory. The angular power spectrum C_ℓ is itself a two-point correlation function (i.e., a quadratic quantity in the temperature overdensity field), so its variance will be a *four*-point correlation function. To evaluate it, we will use Wick's theorem (see Box 13.3).

---

[6] The reader may wonder why this is, given that the ensemble average of any given $a_{\ell m}$ is zero. A good way to think about Eq. (13.73) is to see it as a random walk in the temperature field, with positive and negative contributions to $a_{\ell m}$ from $\delta T/T$ in pixels. The expected value of a random-walk variable after N steps goes as $\sqrt{N}$. Similarly, the expected value of $a_{\ell m}$ with f_{sky} goes as $f_{\text{sky}}^{1/2}$, so $f_{\text{sky}} = 0.25$ will lead to $a_{\ell m}$ that are, on average, 0.5 of their true, full-sky value.

| Box 13.2 | Angular Power Spectrum Estimators |

Here we describe two of the most commonly used estimators for the angular power spectrum of the CMB.

- **Pseudo-C_ℓ estimator**. Here we start from Eq. (13.73) and first calculate the $a_{\ell m}$ over the (observed sky, or masked) map

$$\hat{a}_{\ell m}^{\mathrm{masked}} = \int \left(\frac{\delta T}{T}\right)^{\mathrm{masked}}(\hat{\mathbf{n}}) Y_{\ell m}^*(\hat{\mathbf{n}}) d\Omega, \tag{B1}$$

where the hat indicates that $\hat{a}_{\ell m}^{\mathrm{masked}}$ is a statistical estimate. Then one gets the estimated masked C_ℓ:

$$\hat{C}_\ell^{\mathrm{masked}} = \frac{\sum_{m=-\ell}^{\ell} |\hat{a}_{\ell m}^{\mathrm{masked}}|^2}{2\ell + 1}. \tag{B2}$$

The full-sky (true) pseudo-C_ℓ angular power spectrum is related to the masked one by applying Eq. (13.77):

$$\hat{C}_\ell^{\mathrm{masked}} = \sum_{\ell'} G_{\ell\ell'} \hat{C}_{\ell'}^{\mathrm{pseudo}}, \tag{B3}$$

where $G_{\ell\ell'}$ is a fairly complicated matrix that can be calculated analytically given the geometry of the mask – see the appendix of the Hinshaw *et al.* (2003) WMAP paper for details and further references. The main effect of the matrix G is to correct the overall power for the masked sky so, roughly speaking, $G \propto f_{\mathrm{sky}}$. We can then invert this to get the true underlying C_ℓ as

$$\hat{C}_\ell^{\mathrm{pseudo}} = \sum_{\ell'} G_{\ell\ell'}^{-1} \hat{C}_{\ell'}^{\mathrm{masked}}. \tag{B4}$$

Note that the G^{-1} inversion is often ill behaved (the matrix G is nearly singular, especially if $f_{\mathrm{sky}} \ll 1$), and the inversion may require some further tricks.

- **Maximum-likelihood estimator**. This is the best estimator of the C_ℓ but, as we explain below, it is unfeasible to apply it on small angular scales. Assume that the probability distribution of the temperature data vector $\mathbf{T} = (\delta T/T(\hat{\mathbf{n}}_1), \ldots, \delta T/T(\hat{\mathbf{n}}_q))$ is Gaussian:

$$\mathcal{L} = \frac{1}{(2\pi)^{n/2}[\det(\mathbf{C})^{1/2}]} \exp\left(-\frac{1}{2}\mathbf{T}^T \mathbf{C}^{-1} \mathbf{T}\right), \tag{B5}$$

where q is the number of pixels in the sky, and $\mathbf{C}$ is the $q \times q$ pixel covariance matrix with elements

$$C_{ij} = S_{ij} + N_{ij} = \sum_\ell \frac{2\ell + 1}{4\pi} C_\ell P_\ell(\cos(\hat{\mathbf{n}}_i \cdot \hat{\mathbf{n}}_j)) + N_{ij}, \tag{B6}$$

where S_{ij} is the signal (compare to Eq. (13.71)) and N_{ij} is noise (usually unimportant at low ℓ, but important at high ℓ). Here C_ℓ is the *a priori* unknown angular power spectrum that we would like to determine. Because the likelihood $\mathcal{L}$ is the probability of data ($\mathbf{T}$) given the model ($\mathbf{C}$) and we would like the probability of the model given the data, one can use a Bayesian analysis to determine the covariance $\mathbf{C}$, and then the angular power spectrum C_ℓ.

We carry out the evaluation explicitly:

$$
\langle \hat{C}_\ell \hat{C}_{\ell'} \rangle - \langle \hat{C}_\ell \rangle \langle \hat{C}_{\ell'} \rangle = \frac{1}{2\ell+1} \frac{1}{2\ell'+1} \left\langle \sum_m a_{\ell m} a_{\ell m}^* \sum_{m'} a_{\ell' m'} a_{\ell' m'}^* \right\rangle - C_\ell C_{\ell'}
$$

$$
= \frac{1}{2\ell+1} \frac{1}{2\ell'+1} \sum_{m,m'} \left(\langle a_{\ell m} a_{\ell m}^* \rangle \langle a_{\ell' m'} a_{\ell' m'}^* \rangle + \langle a_{\ell m} a_{\ell' m'} \rangle \langle a_{\ell m}^* a_{\ell' m'}^* \rangle \right.
$$

$$
\left. + \langle a_{\ell m} a_{\ell' m'}^* \rangle \langle a_{\ell m}^* a_{\ell' m'} \rangle \right) - C_\ell C_{\ell'} \tag{13.78}
$$

$$
= \frac{1}{2\ell+1} \frac{1}{2\ell'+1} \sum_{m,m'} \langle a_{\ell m} a_{\ell' m'} \rangle \langle a_{\ell m}^* a_{\ell' m'}^* \rangle + \langle a_{\ell m} a_{\ell' m'}^* \rangle \langle a_{\ell m}^* a_{\ell' m'} \rangle
$$

$$
= \frac{1}{2\ell+1} \frac{1}{2\ell'+1} 2 \sum_{m,m'} C_\ell C_{\ell'} \delta_{\ell\ell'} \delta_{mm'}
$$

$$
= \frac{2}{2\ell+1} C_\ell^2 \, \delta_{\ell\ell'}
$$

where, going to the second line, we applied Wick's theorem to evaluate the four-point function of the $a_{\ell m}$, and in the third line we have shuffled the conjugate signs by using $a_{\ell m} = (-1)^m a_{\ell -m}^*$ and the fact that a sum over m gives the same answer as the same sum over $-m$. So the **cosmic variance** is given by

$$
\sigma(\hat{C}_\ell) = \sqrt{\frac{2}{2\ell+1}} C_\ell \qquad \text{(cosmic variance when } f_{\text{sky}} = 1\text{)}. \tag{13.79}
$$

In other words, at a fixed ℓ, we have $(2\ell + 1)$ samples $a_{\ell m}$ from the distribution whose variance is C_ℓ. This matches the standard result in statistics that the variance decreases as the inverse number of samples, namely goes as $1/(2\ell + 1)$.

A characteristic feature of a sample variance-type error is the proportionality of the standard deviation in some quantity to the quantity itself – as in Eq. (13.79), where $\sigma(C_\ell) \propto C_\ell$. Specifically, in the case where the true angular power spectrum is zero, its variance is also zero.

13.4.5 Total Error in the C_ℓ

How does the variance change if sky coverage is partial, so that $f_{\text{sky}} < 1$? In that case, we can simply think about the number of *modes* – $a_{\ell m}$ coefficients or fractions thereof – sampled on the sky. Clearly, $N_{\text{modes}} \propto f_{\text{sky}}$. Moreover, in Problem 10.1 we learn that the variance of the mean of a random variable goes as $1/N_{\text{modes}}$ (e.g., if our sample size decreases by a factor of two, the variance of its mean halves as well). Hence $\sigma(\hat{C}_\ell) \propto N_{\text{modes}}^{-1/2} \propto f_{\text{sky}}^{-1/2}$.

Moreover, instrumental noise coupled with a finite beam (resolution) of the survey introduces an additional error that goes as $(1/w)e^{\ell^2 \sigma_b^2}$, where w is the noise per pixel

Box 13.3   Wick's Theorem

It is generally very difficult to evaluate high-order correlation functions of random variables. For example, modeling even the three-point correlation function (of, say, the density field δ) is a full-time job for a number of cosmologists worldwide. However, things simplify dramatically if the random variable is Gaussian (with mean zero in typical LSS and CMB applications to cosmological density/temperature fields). In that case, a useful result called Wick's theorem applies. Named after the Italian physicist Gian Carlo Wick, it actually comes from quantum field theory, where it is an especially useful result when evaluating the expectation values of quantum-field correlators.

For a Gaussian distribution, all *odd*-order correlation functions are zero, so that we don't need to worry about them. [As a simple example, consider the fact that the first moment of a Gaussian variable, its mean $\langle X \rangle$, is zero, and so is its third moment $\langle X^3 \rangle$.] And the *even*-order correlation functions are given in terms of the two-point function.

Let us therefore consider the even-order correlation function of n Gaussian random variables $X_1, X_2, \ldots, X_n$. Wick's theorem states that their expectation value is given by

$$\langle X_1 X_2 \ldots X_n \rangle = \sum_{\text{all pair associations}} \ \prod_{\text{all pairs } (i,j)} \langle X_i X_j \rangle. \tag{B1}$$

The key point is that the right-hand side contains only the products of the two-point correlation functions for various pairings of the variables. It encodes the fact that all the information about the (high-order) correlation function on the left is given by the two-point correlation functions.

It is a little difficult to parse the right-hand side of the expression above, so let us take an example. Consider the general four-point correlation function of a Gaussian field. Wick's theorem states that this correlation function can be expressed as

$$\langle X_1 X_2 X_3 X_4 \rangle = \langle X_1\ X_2\ X_3\ X_4 \rangle + \langle X_1\ X_2\ X_3\ X_4 \rangle + \langle X_1\ X_2\ X_3\ X_4 \rangle$$

$$\equiv \langle X_1 X_2 \rangle \langle X_3 X_4 \rangle + \langle X_1 X_3 \rangle \langle X_2 X_4 \rangle + \langle X_1 X_4 \rangle \langle X_2 X_3 \rangle, \tag{B2}$$

where the overlines show pairings of the variables. For a special case $X_1 = X_2 = X_3 = X_4 \equiv X$, the fourth moment simply evaluates to

$$\langle X^4 \rangle = 3 \langle X^2 \rangle^2. \tag{B3}$$

One can similarly evaluate a six-point correlation function of Gaussian fields, eight-point correlation function, etc.

The expression in Eq. (B1) is the so-called *unconnected* term (again using common quantum-field-theory language). If the variables X_i are non-Gaussian, then there is an additional, *connected* term. The connected term is sometimes simply denoted as $T(X_1, \ldots, X_n)$, and it encodes all additional contributions to the Wick formula due to non-Gaussianity. There exist a number of approaches to evaluate the connected term analytically given mildly non-Gaussian variables or fields. The CMB anisotropy is Gaussian to a good approximation, and so are the $a_{\ell m}$ coefficients that describe it, so that the unconnected term suffices in our applications.

and σ_b is the characteristic width of the beam (in radians). The exponential term is fairly easy to understand: as $\ell \gg \sigma_b^{-1}$, that is, for scales smaller than the beam, the C_ℓ error blows up as we simply cannot measure CMB anisotropy on scales below the resolution of the experiment.

Including the finite sky coverage and finite resolution of an experiment, the total error in the C_ℓ is thus given by the sum of cosmic variance and the noise term

$$\sigma(\hat{C}_\ell) = \sqrt{\frac{2}{(2\ell + 1)f_{\text{sky}}}} \left(C_\ell + \frac{1}{w} e^{\ell^2 \sigma_b^2} \right) \qquad \text{(CV plus noise).} \qquad (13.80)$$

In conclusion, two kinds of statistical errors are important in the CMB analysis:

1. Errors due to the finite number of samples on the sky – dominant at large scales (low ℓ); these are referred to as the cosmic variance.

2. Errors due to the noise and finite resolution of the experiment – dominant at small scales (high ℓ); these are referred to as the shot noise.

We now turn to the study of polarization of the CMB.

13.5 CMB Polarization

CMB radiation is linearly polarized. This conclusion comes from two basic facts: (1) Thomson scattering that characterizes the interaction between photons and electrons leads to linearly polarized outgoing radiation, and (2) the existence of the quadrupolar anisotropy in the CMB radiation. We now explain this in a little more detail.

Why is the local temperature quadrupole required for the polarization to be generated? This is illustrated in Fig. 13.11, where we show two incoming photons, one moving in the $+\hat{\mathbf{x}}$ direction and one in the $+\hat{\mathbf{y}}$ direction. Their electromagnetic waves oscillate in the directions perpendicular to the direction of propagation: $\pm\hat{\mathbf{y}}$ and $\pm\hat{\mathbf{z}}$ for the first photon, and $\pm\hat{\mathbf{x}}$, $\pm\hat{\mathbf{z}}$ for the second. Only the polarizations perpendicular to the incoming direction of motion survive in the outgoing wave; they are shown as solid lines ($\pm\hat{\mathbf{y}}$ and $\pm\hat{\mathbf{x}}$, respectively) in the figure. If their amplitudes were the same, the outgoing wave would have no net polarization. The CMB quadrupole ensures that the amplitudes of the two incoming waves (which are 90° apart) differ, leading to net outgoing linear polarization.

The CMB polarization is conventionally represented by Stokes parameters: I denotes the total intensity of light, while Q and U describe linear and V circular polarization. The relevant Stokes parameters for the (linearly polarized) CMB polarization are therefore Q and U. Because polarization at any given point on the sky is described by a direction – a "stick" – it formally forms a spin-2 field whose main property is that it is unchanged after a rotation by 180° (rather than 360°

Thomson scattering

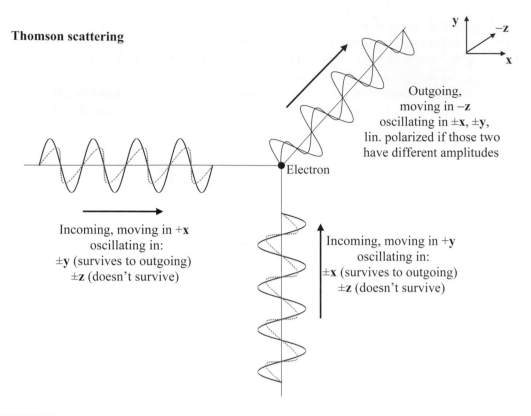

Fig. 13.11 Illustration of how a quadrupole is responsible for the polarization of the CMB. We consider two mutually perpendicular photons hitting an electron. Each photon's electromagnetic field oscillates in directions perpendicular to the direction of propagation of the respective photon. The polarizations that survive in the outgoing direction are denoted by solid sine waves; the ones that do not are dashed. A quadrupole – difference in amplitude between the two incoming waves, and thus surviving outgoing polarizations – is required in order to generate net outgoing polarization.

for a spin-1, vector field). Intuitively, two numbers are required to describe polarization at any given point in the sky, since we are talking about a length and (say) azimuthal angle of each polarization "stick." The fields $Q(\hat{\mathbf{n}})$ and $U(\hat{\mathbf{n}})$ are one such description.

Because $Q(\hat{\mathbf{n}})$ and $U(\hat{\mathbf{n}})$ each require two numbers to be described, the (Q, U) pattern at some location $\hat{\mathbf{n}}$ on the sky is conventionally described by a 2×2 matrix, with elements

$$\begin{pmatrix} Q(\hat{\mathbf{n}}) & U(\hat{\mathbf{n}}) \\ U(\hat{\mathbf{n}}) & -Q(\hat{\mathbf{n}}) \end{pmatrix}, \tag{13.81}$$

which can roughly be thought of as a stretch along the x-direction ($Q > 0$), or the y-direction ($Q < 0$), or the $x = y$ direction ($U > 0$), or else along the $x = -y$ direction ($U < 0$); see Fig. 13.12.

Figure 13.12 also shows that the (Q, U) description of polarization unfortunately depends on the choice of the coordinate system. If we go from one coordinate system to another (unprimed to primed in the figure), Q and U change. This is not good – it means that CMB observers would have to agree every time they measure or report the Q, U maps of CMB polarization. Worse, they would also have to agree on the coordinate system convention when they report the *power spectra* of Q and U.

The coordinate dependence of Q and U can be resolved in an elegant way, by defining a coordinate-*in*dependent representation of the CMB polarization. This results in the so-called E and B modes of CMB polarization, which we now describe. First transform Q and U into harmonic space:[7]

$$Q(\mathbf{l}) = \int Q(\hat{\mathbf{n}})\, e^{i\mathbf{l}\cdot\hat{\mathbf{n}}}\, d^2\hat{\mathbf{n}}; \qquad U(\mathbf{l}) = \int U(\hat{\mathbf{n}})\, e^{i\mathbf{l}\cdot\hat{\mathbf{n}}} d^2\hat{\mathbf{n}}. \qquad (13.82)$$

Then E and B fields in harmonic space are given by

$$\begin{aligned} E(\mathbf{l}) &= \cos(2\phi_l)U(\mathbf{l}) + \sin(2\phi_l)U(\mathbf{l}) \\ B(\mathbf{l}) &= -\sin(2\phi_l)U(\mathbf{l}) + \cos(2\phi_l)U(\mathbf{l}), \end{aligned} \qquad (13.83)$$

where $\mathbf{l} = (l_x, l_y) = l(\cos\phi_l, \sin\phi_l)$; here $\mathbf{l}$ is the 2D wavenumber described by the azimuthal angle ϕ_l. You can check that E and B are invariant with a change of the coordinate system.

Figure 13.12 illustrates the physical meaning of E and B fields. Around an overdensity (underdensity), the E-field is tangential (radial). The B-field, conversely, points at 45° clockwise (anti-clockwise), relative to the direction of overdensity (underdensity). Under a parity flip $\mathbf{x} \to -\mathbf{x}$, $E \to E$, while $B \to -B$ (see again Fig. 13.12). This mathematically makes E a scalar and B a *pseudo-scalar*.

Density perturbations can create only the E modes. This can be intuitively understood yet again by the inspection of Fig. 13.12, where it is clear that any physical effect must be parallel to the direction of the density perturbation wavenumber (e.g., velocities of test particles), or else perpendicular to it (e.g., potential iso-contours). In our example, the polarization from density perturbations must be either radial or tangential relative to the center of the over/underdensity, corresponding to pure E modes, and none is expected to be at 45° relative to it (B modes).

---

[7] Here we are assuming the so-called flat-sky approximation where, instead of the usual spherical-harmonic transform, we have a Fourier transform on the 2D sky. The flat-sky approximation would be accurate when the analysis is done on small areas of the sky, $\Omega \ll 4\pi$. The general, harmonic analysis is more complicated as it introduces, for a spin-2 field such as polarization, scary-looking spin-2 spherical harmonics $_{\pm}Y_{\ell m}$. However all of the physical arguments in the full-sky analysis are the same as in the simpler flat-sky limit that we assume.

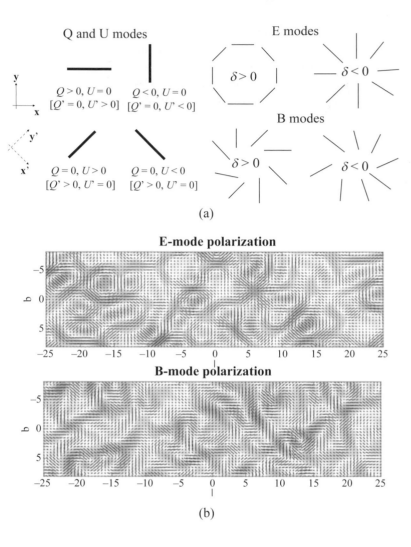

(a)

(b)

Fig. 13.12 (a) Q and U modes, and the fact that they depend on the coordinate system. Each polarization direction is reported in either the unprimed or the primed coordinate system shown on the left. E and B modes around a spherical region which is overdense ($\delta > 0$) or underdense ($\delta < 0$) are shown on the right. (b) (Reprinted with permission of *Annual Reviews*, from Kamionkowski and Kovetz, 2016; permission conveyed through Copyright Clearance Center Inc.) Synthetic polarization map with pure E (top) and pure B (bottom) modes. The central and right illustrations both illustrate that the E modes have polarization direction which is tangential to a temperature overdensity and radial around a temperature underdensity, while the B-mode pattern is at $45°$ relative to an under/overdensity.

There are in fact three reasons why one *could* nevertheless measure a nonzero B-mode signal:

- astrophysical foregrounds or instrumental systematic errors, which can in principle produce *any* kind of pattern including the B modes;
- gravitational lensing of the CMB by large-scale structure – a guaranteed signal which turns E modes into B modes as seen in Fig. 13.13 and, most fascinatingly,
- primordial *tensor* fluctuations – meaning gravitational waves – which also generate B modes. In single-field models of inflation, the amplitude of the gravitational-wave B-mode signal is directly proportional to the scale of inflation, as discussed in Chapter 8.

Because the detection and amplitude of the primordial tensor fluctuations would directly inform us about the physical conditions during inflation, they constitute a

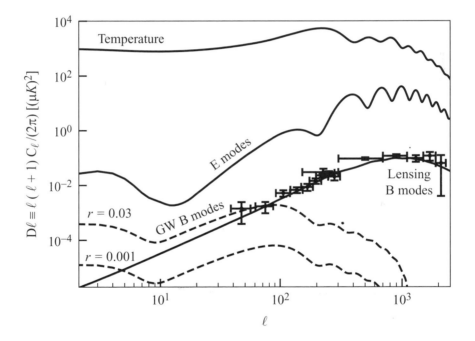

Fig. 13.13 Theoretically predicted angular power spectrum of CMB temperature, polarization E modes, and polarization B modes. The B-mode spectrum has contributions from gravitational lensing of the CMB and primordial gravitational waves (GW). The amplitude of the GW B-mode spectrum is characterized by parameter r, which measures the energy scale of inflation; see Eqs. (8.67) and (8.68) in Chapter 8. We show the theoretical prediction for the GW B modes for two models: $r = 0.03$, roughly consistent with the current upper bound from data, and $r = 0.001$, comparable to the lowest signal potentially detectable with future data. Current data from the BICEP/Keck (Ade *et al.*, 2021) and South Pole Telescope (Sayre *et al.*, 2020) experiments are also shown; note that they are consistent with the lensing-only contribution to B modes.

highly sought-after signal which is being eagerly pursued by the current and next generation of CMB experiments.

The theoretically expected CMB polarization power spectrum in the standard model is shown in Fig. 13.13. The CMB polarization was first detected in 2002 by the Degree Angular Scale Interferometer (DASI) experiment (Kovac *et al.*, 2002). Measurements of the E power spectrum (EE), as well as the cross-correlation between temperature and E-mode polarization (TE) have steadily improved over time, and polarization information now significantly helps constrain the cosmological parameters. Measurements of B modes, on the other hand, have thus far detected the guaranteed lensing B-mode signature, but have not yet detected the holy-grail gravitational-wave B-mode signature. In 2013, the BICEP2 collaboration did in fact announce a discovery of the gravitational-wave signature in B modes (implying $r \simeq 0.2$), but that claim was soon shown to be spurious and due to B modes produced by magnetized dust particles. Nevertheless, it remains the case that:

> Detection of gravitational wave-induced B modes in the CMB polarization is one of the most hotly pursued goals in cosmology. The amplitude of this signal would inform us about the energy scale of inflation.

Primordial gravitational waves is just one example of fascinating early-universe physics that could potentially be observed in the CMB. Another such example, that of cosmic strings, is discussed in Box 13.4.

13.6 The Sunyaev–Zeldovich Effect

The Sunyaev–Zeldovich (SZ) effect uses the CMB as a *backlight* that effectively illuminates the largest objects in the universe – clusters of galaxies – and tells us about a variety of interesting physical processes.

13.6.1 Thermal SZ Effect

In this chapter we have seen that the CMB is a nearly perfect blackbody. This is generally true for the CMB observed over most of the sky. However, when viewed in the direction of galaxy clusters, the CMB spectrum is slightly distorted. This is because the CMB photons scatter off hot ($T \sim 10\,\mathrm{keV}$) electrons present in the cluster; we say that the photons are "upscattered" since their energy typically increases. Some of the photons therefore acquire more energy, and they shift the blackbody spectrum, borrowing from the low-frequency (low-energy) end and giving to the high frequencies. The probability of the scattering – that is, the electron optical depth in the cluster – is of order $\tau_e \simeq 0.01$.

Box 13.4	Cosmic Strings and the CMB

One of the more spectacular predicted signatures of primordial physics in the CMB is that of cosmic strings. Cosmic strings are long 1D bundles of energy that can be created during phase transitions in the early universe. They are a fairly generic prediction of quantum field theory, and finding them would open a dramatic new window to the physics of the early universe. These strings are not the tiny ones from string theory, but rather very long (length $\simeq H_0^{-1}$) though very thin and with a huge linear density. For phase transitions at the Grand Unified Theory (GUT) symmetry-breaking energy scale $m_{\rm GUT} \sim 10^{16}\,{\rm GeV}$, the thickness of the string is $\sim m_{\rm GUT}^{-1} \sim 10^{-18}\,{\rm m}$, and the mass per length is $\mu \equiv M/L \simeq m^2 \simeq 10^{20}\,{\rm kg/m}$. Equivalently, a GUT string would be characterized by the dimensionless parameter $G\mu \simeq (m/m_{\rm Pl})^2 \simeq 10^{-6}$.

The metric of the cosmic string is given in polar coordinates (z, ρ, θ) as

$$ds^2 = dt^2 - dz^2 - d\rho^2 - \rho^2 d\theta^2, \qquad \text{where} \qquad 0 \le \theta < 2\pi(1 - 4G\mu). \qquad \text{(B1)}$$

The metric is completely flat *except in the angular coordinate* θ which covers an interval less than 2π. Thus, the metric exhibits a "deficit angle" $(\Delta\theta)_{\rm deficit} = 8\pi G\mu$.

Cosmic strings leave a clear signature in the CMB temperature anisotropy via the Kaiser–Stebbins effect. This occurs when the string has a component of motion perpendicular to our line of sight to the CMB. The photons on the way to us passing in front of the string are undeflected. The ones passing behind the string bend toward us at an angle $(\Delta\theta)_{\rm deficit}$, and therefore have a small component of motion toward us in the plane of the sky – and are therefore Doppler shifted. The Doppler shift is proportional to the aforementioned velocity component which is equal to $v\,(\Delta\theta)_{\rm deficit}$. As a consequence, the string makes a sharp, thin line feature in the CMB with temperature gradient

$$\left(\frac{\delta T}{T}\right)_{\rm string} = 8\pi G\mu v\, \gamma(v), \qquad \text{(B2)}$$

where $v \simeq 1$ is the velocity of the string perpendicular to the line of sight, and the relativistic factor $\gamma(v)$ is also $\simeq 1$ for the typical not-very-relativistic case. Since $G\mu \lesssim 10^{-6}$ or so, the cosmic string signal is roughly comparable to the primordial CMB anisotropy, and may be detectable. Higher-resolution maps have a better chance of detecting cosmic strings, since they can reach down to a smaller $G\mu$. The figure on the right, adopted from Ringeval (2010), shows a numerical simulation of how a network of cosmic strings (right) would show up in the CMB (left); note the barely perceptible sharp edges in the CMB map.

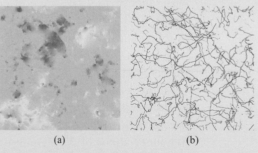

(a)   (b)

© 2010 Christophe Ringeval. CC BY 3.0.

The current constraint on the parameter $G\mu$ from Planck is

$$G\mu \lesssim 10^{-7}.$$

Future, higher-resolution CMB experiments may further improve upon this bound.

This effect is called the Sunyaev–Zeldovich effect after the renowned cosmologists who predicted it about half a century ago (Sunyaev and Zeldovich, 1972), based on analytically solving the so-called Kompaneets equation, which is a special case of the Boltzmann equation for photons interacting with electrons. More precisely, we will refer to this effect as the **thermal Sunyaev–Zeldovich effect** (or tSZ), to distinguish it from the kinetic SZ effect discussed just below.

It is useful to consider the effect of tSZ on the *intensity* of the CMB radiation, which is typically what the radio observations measure at the frequencies relevant here. The fiducial blackbody intensity, which takes the familiar form (restoring MKS units temporarily in the next few equations)

$$I_\nu = \frac{2h}{c^2} \frac{\nu^3}{\exp(h\nu/k_B T) - 1} \tag{13.84}$$

is shifted by the frequency-dependent amount given by

$$\Delta I_\nu^{\text{tSZ}} = g(x) I_0 y, \quad \text{with}$$

$$g(x) = \frac{x^4 e^x}{(e^x - 1)^2} \left[x \frac{e^x + 1}{e^x - 1} - 4 \right]; \quad I_0 = 2(k_B T)^3 / (hc)^2, \tag{13.85}$$

where $x \equiv h\nu/(k_B T)$, and y is defined just below. The expression for the temperature shift is similar, $(\delta T/T)_{\text{tSZ}} = f(x) y$, where $f(x)$ is the part of $g(x)$ in square brackets in Eq. (13.85). A key quantity for the tSZ effect is the so-called y-parameter, defined as

$$y \equiv \sigma_{\text{T}} \int n_e \frac{k_B T_e}{m_e c^2} d\ell, \tag{13.86}$$

where σ_{T} is the Thomson cross-section, n_e is the electron number density, T_e is the electron temperature, k_B is the Boltzmann constant, $m_e c^2$ is the electron rest mass energy, and the integration is along the line of sight. Because the pressure of an ideal gas can be written as $P = n k_B T$, we recognize the y-parameter as proportional to the pressure of the electron gas integrated along the line of sight. The integral in Eq. (13.86) goes from the observer to the CMB last-scattering surface, but essentially all of its contribution comes from within the galaxy cluster.

Figure 13.14(a) shows change in the blackbody spectrum I_ν due to the tSZ effect for two typical values of the y-parameter. The spectrum changes sign going from low to high frequency, with the y-independent null at $\nu = 217\,\text{GHz}$. The specific frequency shape of $\Delta I_\nu^{\text{tSZ}}$ makes the tSZ detection relatively easy. By the 1990s, the tSZ effect was being reliably detected in numerous clusters. Typical observations (shown in Fig. 13.14(b)) are made at frequencies lower than $217\,\text{GHz}$, meaning that one looks for, roughly speaking, "holes" in the CMB sky.

A tSZ signal can also be used to *detect* clusters of galaxies. Provided the clusters can be resolved and their surface brightness measured, the tSZ signal is redshift

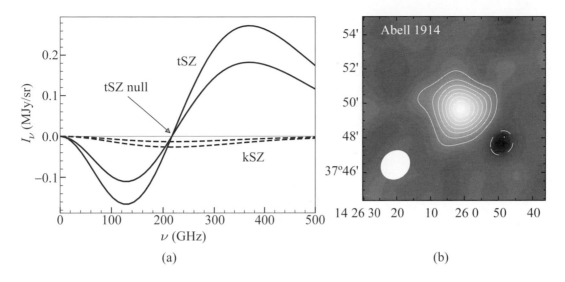

Fig. 13.14 (a) Distortion in the spectrum of CMB radiation due to the SZ effect, showing the change in intensity ΔI_ν as a function of frequency. The units of intensity are shown in megaJanskys per steradian, where $1\,\mathrm{Jy} = 10^{-26}\,\mathrm{W/m^2}$. We show the tSZ effect for two values of the y-parameter ($y = 0.00010$ and 0.00015) and the kSZ effect for two values of peculiar velocity along the line of sight ($v_\mathrm{pec} = 300$ and $600\mathrm{km/s}$) and an optical depth $\tau_e = 0.01$. (b) A galaxy cluster at $z = 0.17$ seen via its thermal SZ signature. The contours show decrement, while the round white blob shows the beam (i.e., the resolution of the instrument). The cluster has been detected with the 30 GHz BIMA/OVRO array in California. Reprinted by permission from Springer Nature Customer Service Centre GmbH: Mohr *et al.* (2000), copyright (2000).

independent, as can be seen by the inspection of Eqs. (13.85) and (13.86). Contrast this with cluster detection in X-rays or in the optical, where the flux falls dramatically with increasing redshift. Several major surveys that currently operate – notably, South Pole Telescope (SPT) and Atacama Cosmology Telescope (ACT) – aim to find thousands of clusters via their tSZ signature.

In addition to detecting clusters, the y-parameter integrated through the cluster turns out to tightly correlate with the cluster mass, and is therefore an excellent mass proxy. One can thus perform a classic test of counting clusters as a function of the mass proxy and redshift to constrain the cosmological model. Finally, there is interest in making full-sky *maps* of tSZ, without reference to individual objects. Because tSZ is sensitive to the thermal energy of the medium from which the signal comes, it is an excellent probe of astrophysical processes where energy injection plays a role, such as mergers or feedback from supermassive black holes and supernovae.

13.6.2 Kinetic SZ Effect

Figure 13.14(a) also shows another, smaller effect called the **kinetic SZ effect** (kSZ). The kSZ is caused by galaxy clusters' peculiar velocities in the CMB rest frame. A cluster with peculiar velocity $v_{\rm pec}$ along the line of sight will incur a density shift of the CMB temperature in the direction of the cluster of

$$\left(\frac{\delta T}{T}\right)_{\rm kSZ} \simeq -\tau_e \frac{v_{\rm pec}}{c}, \tag{13.87}$$

where τ_e is the optical depth in the cluster. The spectrum therefore looks like the blackbody spectrum (so, following Eq. (13.84)), but with the temperature shifted as in Eq. (13.87). For typical values $\tau_e \simeq 0.01$ and $v_{\rm pec} \simeq 500{\rm km/s}$, the kSZ is quite small, as Fig. 13.14(a) shows. The kSZ effect was first detected about a decade ago (Hand *et al.*, 2012), and its better mapping will allow more information about the velocity field of clusters of galaxies. Given independent velocity information, the kSZ effect can also be used to map out the baryonic density of individual galaxy clusters to fairly large radii around them.

13.6.3 More General Spectral Distortions

The SZ effect is a special case of spectral distortions of the CMB spectrum. A spectral distortion is a departure from the perfect blackbody form captured in Eq. (13.84). Such an effect takes place whenever a departure from thermodynamic equilibrium between photons and baryons occurs in such a way that the new thermodynamic equilibrium (i.e., *thermalization* – arrival at a blackbody form with some new temperature) cannot be reached. Various physical processes – notably, particle–antiparticle annihilations, particle decays, magnetic fields, topological defects, or primordial black holes – are expected to leave their signature by depositing energy that would register as a spectral distortion.

If the energy deposition occurs at a redshift larger than about two million, perfect thermalization takes place and no spectral distortions are observed. Energy released at $10^4 \lesssim z \lesssim 10^6$ leads to the so-called μ-distortions, while energy injected at $10^3 \lesssim z \lesssim 10^4$ leads to the so-called y-distortions. The SZ effect, which occurs at very low redshift as described above, also registers as a y-distortion to the CMB blackbody spectrum.

So far no spectral distortion has been observed. The best limits are obtained with the COBE experiment's FIRAS instrument, roughly $|\mu| \lesssim 10^{-5}$ and $|y| \lesssim 10^{-4}$; here μ is a dimensionless parameter describing μ-distortions, and y has been defined in Eq. (13.86). There is, however, considerable interest in building a new instrument to operate in space, more sensitive than COBE FIRAS, in order to search for spectral distortion signatures of annihilating particles and other early-universe artifacts. Moreover, there is a guaranteed spectral-distortion signature, at a level of $\sim 10^{-8}$ (roughly equal to the ratio of the number density of atoms to photons), due to recombination of hydrogen and helium. Detecting the spectral-distortion signatures of these recombination lines could be used to further probe the thermal history of the universe.

13.7 Primordial Non-Gaussianity

The simplest models of inflation, where a single field slow-rolls with no interactions and assuming standard Einstein gravity, produce primordial density fluctuations that are *almost* perfectly Gaussian. More specifically, one can study the so-called **local model of non-Gaussianity**

$$\Phi = \Phi_G + f_{NL}\Phi_G^2. \tag{13.88}$$

Here Φ is the Newtonian gravitational potential, and Φ_G is its Gaussian part. For $f_{NL} = 0$ we have a perfectly Gaussian potential (and hence CMB anisotropy field that follows from it), and vice versa. Note that, because $\Phi \simeq 10^{-5}$, quite a large value of f_{NL} ($\sim$30,000) is required to make visible, order-unity non-Gaussianity in the CMB temperature field. For example, the non-Gaussian probability density distribution of $a_{\ell m}$, shown in Fig. 13.9, was shown for $f_{NL} = 10,000$. The aforementioned simplest models of inflation predict tiny levels of primordial non-Gaussianity, $|f_{NL}| \simeq |n_s - 1| \simeq 0.03$, but models with multiple fields or interactions can produce a much higher signal.

There are multiple ways to search for non-Gaussianity in CMB temperature (and polarization) maps. One could, very naively, start by plotting a histogram of temperature (or of $a_{\ell m}$, as in Fig. 13.9), but this is certainly not the most sensitive way to search for non-Gaussianity. Instead, we rely on the fact that the three-point correlation function of the CMB, in either real space with $\delta T/T$ or else harmonic space with $a_{\ell m}$, is linearly proportional to f_{NL}. To illustrate this, consider that taking the third moment of Eq. (13.88) gives an expression proportional to f_{NL}:

$$\langle \Phi^3 \rangle \simeq 3f_{NL}\langle \Phi_G^2 \rangle^2, \tag{13.89}$$

where we make use of the fact that Φ_G is Gaussian so its three-point function is zero, and apply Wick's theorem. The equation above is simplified because one usually calculates the three-point correlation of temperature (not Φ), in harmonic space (not real space), and evaluated at three vertices of a triangle (and not a single point), but it still communicates an important point: the three-point correlation function (or its harmonic cousin, the bispectrum) of the CMB temperature carries key information about primordial non-Gaussianity.

The constraints from Planck on the local model from Eq. (13.88) are extremely tight and fully consistent with Gaussianity:

$$f_{NL} = -0.9 \pm 5.1. \tag{13.90}$$

This constraint is at least a factor of 100 better than you could do "by eye," scrutinizing a difference between a Gaussian and a non-Gaussian map even given

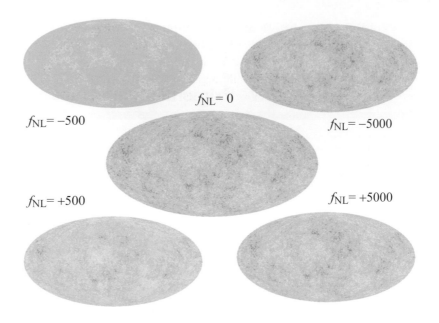

$f_{\rm NL} = 0$

$f_{\rm NL} = -500$

$f_{\rm NL} = -5000$

$f_{\rm NL} = +500$

$f_{\rm NL} = +5000$

Fig. 13.15 Synthetic maps of the Gaussian sky (center), and with non-Gaussianity of the local type for $f_{\rm NL} = \pm 500$, and $f_{\rm NL} = \pm 5000$. Current limit from Planck is $f_{\rm NL} = -0.9 \pm 5.1$. A black and white version of this figure will appear in some formats. For the color version, please refer to the plate section.

the same phases; see Fig. 13.15. It showcases the awesome power of statistics applied to cosmology.

Primordial non-Gaussianity is a large and active area of research for theorists, data-analysts, and experimentalists alike, as it pursues a direct signature of the physics of the early universe. Besides the local model in Eq. (13.88), there are many other models and parameterizations of primordial non-Gaussianity that have been both motivated by theory and constrained by data. One can also use large-scale structure – the distribution of galaxies or other tracers – to test primordial non-Gaussianity (Dalal *et al.*, 2008). Such probes are more challenging than the CMB as they contain significant *non*-primordial non-Gaussianity (due to nonlinearity of the gravitational collapse), but they offer the benefit of probing non-Gaussianity on different scales than the CMB.

Bibliographical Notes

CMB is covered in most modern textbooks on cosmology, but none of them covers both theory and data analysis at an intermediate level as we do here. A recommended clear introduction is the review by Hu and Dodelson (2002). At a more advanced level, the topic is covered by Dodelson and Schmidt (2020). At a very technical level, it is covered in Weinberg (2008) and Durrer (2020). The CMB theory presented in this chapter follows Hu (2008), while the description of the Sachs–Wolfe effect follows White and Hu (1997). A fascinating collection of articles about the history of cosmology leading up to, and including, the detection of the CMB is given in Peebles *et al.* (2009).

Problems

13.1 **Silk damping scale.** Estimate the Silk damping scale – the distance that photons effectively traverse during the period of decoupling. The significance of the Silk scales comes from the fact that, on scales smaller that r_{Silk}, the CMB photons diffuse and the observed CMB fluctuations are suppressed. I will now guide you through a very rough estimate of r_{Silk}.

A photon at the last-scattering surface random-walks (diffuses) a distance $r_{\text{Silk}} = \sqrt{N} r_{\text{diff}}$, where $r_{\text{diff}} = 1/(n_e \sigma_e)$ is the diffusion length (with n_e the electron density and σ_e the Thomson scattering cross-section), and $N = \Delta t_{\text{dec}}/\Delta t_{\text{diff}}$ is the number of times the photon scatters during the period of decoupling. Therefore, $r_{\text{Silk}} \simeq \sqrt{\Delta t_{\text{dec}} r_{\text{diff}}}$ (working as usual in units with $c = 1$) is the geometric mean between the time available for diffusion and the diffusion length. Given that $\Delta t_{\text{dec}} \simeq (\Delta a/a) t_{\text{Hubble}} \simeq (|\Delta z|/z) H^{-1}$, and adopting $\Delta z \sim 200$ (at $z \simeq 1000$), assume that the time available for diffusion is about a fifth of the age of the universe. Adopt $n_e = (0.25 \text{m}^{-3}) a_{\text{dec}}^{-3}$ (this is really the expression for the baryon density, which is related to n_e via $n_e(z) = x_e(z) n_{\text{bary}}(z)$, but ignore departures of the ionization fraction x_e from unity in this approximate analysis). Then estimate the Silk scale, and report it in comoving megaparsecs.

13.2 **Precision in N_{eff}.** In this problem you will understand how the CMB constrains the number of relativistic species, N_{eff}, and estimate the precision in it. First, recall from Chapter 5 that N_{eff} parameterizes the number of relativistic species (other than photons), that $N_{\text{eff}} \simeq 3$ is the Standard-Model value for three neutrino species, and that a substantially higher value of N_{eff} would indicate the presence of additional relativistic species – the "dark radiation." Then, carry out the following estimates (which loosely follow Hou *et al.*, 2013).

(a) Writing $N_{\text{eff}} = 3 + \Delta N_{\text{eff}}$ (where 3 corresponds to neutrinos and ΔN_{eff} to dark radiation), show that the effective number of relativistic degrees of freedom (for neutrinos/antineutrinos and any dark radiation) can be written as

$$g_\nu + g_{\text{DR}} = g_\nu \left(1 + \frac{\Delta N_{\text{eff}}}{3}\right),$$

where g_ν and g_{DR} are the effective numbers of relativistic neutrinos/antineutrinos and dark radiation, respectively.

(b) Given the result in part (a), show that the change in the number of relativistic species of ΔN_{eff} leads to the change in the Hubble parameter at recombination ($z \sim 1000$) of

$$2\frac{\Delta H}{H} = \frac{\Delta \rho}{\rho_{TOT}} \simeq 0.04\,\Delta N_{eff},$$

where $\Delta\rho$ corresponds to dark radiation. In deriving the latter expression, you will want to recall that the universe is mostly matter dominated at recombination, so that $\rho_{TOT}(z = 1100) \simeq \rho_M(z = 1100)$. Do not forget that matter and radiation scale differently with redshift. And make use of the present-day density of matter and massless neutrinos that you can find in Table 3.1.

(c) Consider now two characteristic angular scales: the damping/diffusion scale θ_{diff}, which is the angle at which the Silk damping scale (discussed in Problem 13.1) is observed, and the sound horizon scale θ_s, which is the angle at which the sound horizon r_s is observed. Each of these angles gives information about N_{eff}, but also includes the angular diameter distance to recombination (e.g., $\theta_{diff} = r_{diff}/r(z_{dec})$ and $\theta_s = r_s/r(z_{dec})$), and the distance depends on the amounts of dark matter and dark energy. In order to isolate the dependence on N_{eff}, we can take the ratio θ_{diff}/θ_s in which $r(z_{dec})$ cancels out. Because damping is fundamentally diffusion (see again Problem 13.1), it scales as a square root of the age of the universe, $\theta_{diff} \propto \sqrt{t}$. The sound horizon, on the other hand, increases linearly with time (since $r_s = \int_0^t c_s(t')dt'$), $\theta_s \propto t$.

With all that, find the (simple!) relation between $\Delta(\theta_{diff}/\theta_s)/(\theta_{diff}/\theta_s)$ and $\Delta H/H$.

(d) Combine now the results of parts (b) and (c) to answer the following question: given a half-percent precision in θ_{diff}/θ_s from Planck, what precision in N_{eff} can be obtained? That is, find $\sigma_{N_{eff}}$ in the relation

$$\frac{\Delta(\theta_{diff}/\theta_s)}{\theta_{diff}/\theta_s} = 0.005\frac{\Delta N_{eff}}{\sigma_{N_{eff}}}.$$

13.3 **Redshift of reionization.** Starting from Eq. (13.38) and comparing it to the CMB measurement of optical depth in Eq. (13.36), calculate the redshift of reionization, z_{reion}. Assume that reionization is instantaneous. Also assume our fiducial flat ΛCDM model with $\Omega_M = 0.3 = 1 - \Omega_\Lambda$. *Hint:* Convert the time integral into a redshift integral. For the flat ΛCDM model, the integral is completely analytic and you will not need numerical integration.

13.4 **Relation between the multipoles and angles.** The angular two-point correlation function of the CMB is defined as $C(\theta) \equiv \langle (T_i - \bar{T})(T_j - \bar{T}) \rangle_\theta$, where i and j run over all pixels in the map separated by θ (in practice, the separation interval is $[\theta, \theta + d\theta]$, where $d\theta \ll \theta$). The real-space two-point function can be transformed to harmonic space where it is called the angular power spectrum, C_ℓ. The relation between the real space and harmonic space two-point function is given by Eq. (13.71).

(a) The approximate relation between the angles and multipoles is $\theta \simeq \pi/\ell$, where θ is in radians. Motivate this relation between θ and ℓ by studying the structure of Legendre polynomials $P_\ell(\cos\theta)$ for $\ell = 2, 3, 4$. *Hint:* Take the derivative of Legendre polynomials with respect to their argument to find out at which angle they peak, and compare to $\theta \simeq \pi/\ell$. You may need the full form of these Legendre polynomials – consult Wikipedia, for example.

(b) Recall that the dominant hot and cold spots in the CMB have a characteristic size of about one degree on the sky. What multipole (approximately!) does this correspond to?

13.5 **Angular power spectrum of a simple map.** Consider an (admittedly artificial in the context of the CMB) map defined as

$$\frac{\delta T}{T}(\theta, \phi) = \begin{cases} 1 & \text{for } 0 \leq \theta \leq \pi/2 \\ 0 & \text{for } \pi/2 < \theta \leq \pi. \end{cases}$$

That is, the map is unity above the equator and zero below. Note that the monopole has *not* been subtracted from the map.

(a) Which single multipole does the map above "most look like"?

(b) Calculate analytically the angular power spectrum, C_ℓ, of this map up to the octopole (so for multipoles $\ell = 0, 1, 2, 3$). You may want to recognize and use the symmetries of this map – this can significantly simplify the calculation.

(c) Imagine now that you have a realistic map (so not the one above) which has a resolution of $5'$, similar to the Planck satellite. Above which multipole ℓ_{res} is the angular power close to zero?

13.6 C_ℓ **in terms of** $C(\theta)$**.** Using orthonormality of the Legendre polynomials, analytically invert Eq. (13.71),

$$C(\theta) \equiv \sum_{\ell=2}^{\infty} \frac{2\ell + 1}{4\pi} C_\ell \, P_\ell(\cos\theta),$$

to get C_ℓ in terms of $C(\theta)$.

13.7 **Variance in $C(\theta)$.** Derive analytically the variance in $C^{\mathrm{th}}(\theta)$,

$$\mathrm{Var}[C(\theta)] \equiv \left\langle [C(\theta) - \langle C(\theta)\rangle]^2 \right\rangle.$$

[There is also a nonzero *co*variance in the angular correlation function between different angles θ and θ', but do not worry about it here.] Your answer should be a function of the C_ℓ and the Legendre polynomials. You will want to make use of Eq. (13.78). Ignore shot noise, but do keep the f_{sky} term which goes in the denominator (see, e.g., Eq. (13.79)).

13.8 **Projections of statistically homogeneous and isotropic random fields.** Say we have a 3D field (e.g., the correlation function of density perturbations $\xi(r)$) and you would like to calculate its 2D projection on the sky (e.g., the angular power spectrum C_ℓ). How do we calculate the latter, given the former?

You will successfully complete this task! Consider a time-independent random field $f(x)$ which is statistically homogeneous and isotropic and has zero mean. The projection of this field on the surface of the sphere $|\mathbf{x}| = r$ defines a field $f(\hat{\mathbf{n}}) \equiv f(r\hat{\mathbf{n}})$. Making use of the Rayleigh plane-wave expansion

$$e^{i\mathbf{k}\cdot\mathbf{r}} = 4\pi \sum_{\ell,m} i^\ell j_\ell(kr) Y_{\ell m}^*(\hat{\mathbf{k}}) Y_{\ell m}(\hat{\mathbf{n}}),$$

complete the following tasks:

(a) Express $f(\mathbf{n})$ in terms of the coefficients $f_{\ell m} = \int f(\hat{\mathbf{n}}) Y_{\ell m}^*(\hat{\mathbf{n}}) d\Omega$ (Ω is the solid angle pointing in direction $\hat{\mathbf{n}}$, and $d\Omega = d^2\hat{\mathbf{n}}$). Then Fourier-transform $f(\hat{\mathbf{n}})$,

$$f(\hat{\mathbf{n}}) = \frac{1}{(2\pi)^3} \int f(\mathbf{k}) e^{i\mathbf{k}\cdot\mathbf{r}} d^3\mathbf{k},$$

to obtain $f_{\ell m}$ in terms of $f(\mathbf{k})$. Use the Rayleigh expansion, and the orthogonality relation for spherical harmonics, both of which should help simplify the expression.

(b) Calculate the angular power spectrum of the field f, defined as

$$C_\ell \equiv \frac{1}{2\ell+1} \sum_{m=-\ell}^{\ell} \left\langle |f_{\ell m}|^2 \right\rangle.$$

In the process, use the fact that the 3D power spectrum of the Fourier-transformed field is given by

$$\langle f(\mathbf{k}) f^*(\mathbf{k}') \rangle = (2\pi)^3 P(k) \delta^{(3)}(\mathbf{k} - \mathbf{k}').$$

Also use the spherical harmonic addition theorem, which for $\mathbf{k} = \mathbf{k}'$ reads

$$\sum_m Y_{\ell m}^*(\hat{\mathbf{k}}) Y_{\ell m}(\hat{\mathbf{k}}) = \frac{2\ell+1}{4\pi}.$$

Your answer should take a remarkably simple form:

$$C_\ell = \text{const.} \int (\text{function\_of\_}\ell\_\text{and\_k\_and\_}r)\, d\ln k,$$

where r can be considered known and fixed – it's the distance to the CMB last-scattering surface.

13.9 **[Computational] Theoretical angular power spectrum.** A time-honored task for a cosmologist working on the interface between theory and data is to calculate the angular power spectrum C_ℓ for a given cosmological model. This task requires solving the coupled Einstein–Boltzmann set of equations for the perturbed quantities (dark matter, photons, etc.), and would be far too demanding as a "from-scratch" exercise. Therefore, cosmologists typically resort to using specialized and highly efficient computer codes developed for this purpose.

Download (and install) such a code. The most widely used choices are CAMB (`https://camb.info/`) or CLASS (`https://lesgourg.github.io/class_public/class.html`). Both are very efficient and fairly easy to use, so the choice is yours.

(a) Compute and plot the angular power spectrum, C_ℓ as a function of ℓ, for the fiducial cosmological model from Table 3.1. [You may use the fiducial set of parameters that comes as a default in the parameter file in either code, since they are typically very close to our fiducial model.] Plot it for the range of multipoles $2 \leq \ell \leq 2000$. Note that you want to plot the quantity $D_\ell \equiv \ell(\ell+1)C_\ell/(2\pi)$ as a function of multipole ℓ, and that the Einstein–Boltzmann codes are likely reporting the D_ℓ as a default.

(b) Now plot the angular power spectrum with the following parameters changed one at a time:

- Amplitude A_s increased by 50 percent.
- Curvature density Ω_k increased from 0 to 0.5.
- Physical baryon density $\Omega_B h^2$ increased by 50 percent.
- Spectral index n_s increased by 50 percent.

(c) For each case in part (b), explain qualitatively the change that you observe relative to the fiducial power spectrum in part (a).

13.10 **[Computational] Angular power spectrum from a CMB map.** This is the first of the two exercises that will give you practice in using "raw," pixel-level CMB maps. Here, our goal is to calculate the angular power spectrum of the temperature anisotropy in a map and compare it to the theoretical prediction.

In order to perform the tasks in this problem and 13.11, you will use the software package Healpix (Górski *et al.*, 2005). Healpix is an ingenious algorithm to pixelize the sphere in equal-area pixels. This pixelization is hierarchical, meaning that a pixel in a higher-resolution setting neatly splits into four pixels at the resolution that is one step lower. The resolution is specified with an input parameter `NSIDE` which must be a power of 2 (so, 1, 2, 4, 8, etc.); as a useful rule of thumb `NSIDE = 64` roughly corresponds to pixels that are 1° on a side (and then `NSIDE = 128` is pixelization at about half a degree, etc.). Crucially, `Healpix` comes with routines for Python (called `healpy`), but also for C and other languages as well, which make the CMB analysis possible, even easy! Having installed `healpy` (or equivalent for C), you may first want to familiarize yourself with `Healpix`'s basic commands (`https://healpix.sourceforge.io/`).

Go to Planck experiment's Legacy Archive (`https://pla.esac.esa.int/#maps`) and find the SMICA full-sky CMB map (`COM_CMB_IQU-smica_2048_R3.00_full.fits`). Also download a mask, say `COM_Mask_CMB-common-Mask-Int_2048_R3.00.fits`. [I am providing both the map and the mask at an `NSIDE = 512` resolution in supplementary materials for this book.] The mask helps cover areas likely contaminated by galactic emission (and point sources and other unwanted artifacts), and consists of ones and zeros, where the zeros indicate areas to be masked. Then carry out the following tasks.

(a) Calculate the angular power spectrum of the (full-sky) CMB map. You will want to make use of the `Healpix` function `anafast`. Note that `anafast` returns the coefficients C_ℓ, while you will want to plot

$$D_\ell \equiv \frac{\ell(\ell+1)C_\ell}{2\pi}$$

as a function of multipole ℓ. This simple analysis is expected to accurately recover the angular power spectrum only on large and intermediate (but not small) scales, so only show this and the following results in the range of, say, $\ell \in [2, 300]$, so capturing the first acoustic peak.

(b) Compare this measurement with the best-fit theoretical model. You could calculate the latter yourself using some standard software (say `camb` or `class`); as you were asked to do in Problem 13.9. Alternatively, I am providing output of such a file in supplementary materials for this book. [Note that this output (produced by `camb`) already gives the D_ℓ coefficients, and in the same units as the map – so $(\mu K)^2$ for the D_ℓ, while the map was in μK.] Do the measurements from part (a) match the theory in part (b)? Why?

(c) Repeat the analysis from part (a), but now mask out the area around the galactic plane. To do that, apply the Planck mask that you downloaded to the SMICA map. There are a few ways to do this; it may be simplest to multiply the map pixel values with the corresponding mask values. The resulting map will have zero temperature in pixels to which the map has been applied. Now repeat the `anafast` analysis. Without further adjustments, you should find that, in contrast to part (b), the measured power is now somewhat smaller than the theory expectation. Why do you think that is?

(d) To fix up the result in part (c), it is useful to re-read the "A very naive approach" paragraph a bit below Eq. (13.77). Clearly, we need to correct the measured power for the missing, masked area. [Masking also introduces covariance between the different C_ℓ which can also be taken into account, and which we ignore here.] To do that, simply rescale all D_ℓ by $1/f_{\rm sky}$, where $f_{\rm sky}$ is the unmasked fraction of the sky area. The measured angular power spectrum should now match the theory quite well over the range of multipoles suggested in part (a).

13.11 **[Computational] Pixel-level analysis of a CMB map.** This is the second of the two exercises that will give you practice in using "raw," pixel-level CMB maps. The goal is to compute the angular two-point correlation function from its definition in Eq. (13.70).

To do this, follow the guidance from Problem 13.10 on using `Healpix`. The commands you will find useful in this problem are `read_map` (reads the map), and `ud_grade` (degrades the map from a higher resolution at some fiducial `NSIDE` to a lower resolution with a lower value of this parameter).

Use the same input map and mass as in Problem 13.10: go to Planck experiment's Legacy Archive and find the SMICA full-sky CMB map and the "common" mask. [I am providing both the map and the mask at an `NSIDE = 512` resolution in supplementary materials for this book.] Degrade both the map and the mask to a much lower resolution – while testing your code you can go to as low as `NSIDE = 1`, and for a production plot use whatever your algorithm will allow for the code to run in less than a minute or so; this will be in the ballpark of `NSIDE = 16` or 32.

Mask your map (see again Problem 13.10 for more on this). Then subtract the monopole – that is, the mean – from your cut-sky map. Note that even a zero-monopole *full-sky* map will have some remaining nonzero monopole after masking; removing it is important, as otherwise this single $\ell = 0$ mode in harmonic space would "contaminate" all θ in $C(\theta)$.

(a) Compute the angular two-point correlation function using its most obvious estimator,

$$ C(\theta) \equiv \frac{1}{N_{\rm pairs}} \sum_{a=1}^{N_{\rm pairs}} \delta T_i \delta T_j, $$

where the sum runs over all pairs where pixels i and j are separated by an angle θ (or in practice, in the interval $[\theta, \theta + d\theta]$). Here, it may be easier to work in equal intervals in $\cos \theta$; the choice is yours. Plot $C(\theta)$ vs. θ (or $\cos \theta$).

(b) Now compute the *theoretically* expected two-point function

$$C^{\mathrm{th}}(\theta) \equiv \sum_{\ell=2}^{\infty} \frac{2\ell+1}{4\pi} \, C_{\ell}^{\mathrm{th}} \, P_{\ell}(\cos(\theta)),$$

where the power spectrum coefficients C_{ℓ}^{th} can be obtained using either `camb` or `class`. [I am also providing them in supplementary materials.] Plot $C^{\mathrm{th}}(\theta)$ on the same graph as the measured $C(\theta)$. [Optional: If you have completed Problem 13.7, add the error bar around $C^{\mathrm{th}}(\theta)$; this would be the square root of the variance calculated in Problem 13.7.]

(c) Does any particular feature strike you about the measured $C(\theta)$ and, if so, what is it? [If you are stumped consult, e.g., Copi *et al.*, 2015.] What do you conclude about the amplitude of the measured angular power on the largest observable scales in the universe?

13.12 **[Computational] Integrating the three-level recombination equation.** Integrate the three-level (Peebles) recombination equation derived near the end of Box 13.1:

$$\dot{x}_e = -C \left(n_H x_e^2 \alpha_H - 4 x_1 \beta_H e^{-B_{21}/T} \right),$$

where

$$C \equiv \frac{\frac{3}{4}\Lambda_{\mathrm{Ly\alpha}} + \frac{1}{4}\Lambda_{2s \to 1s}}{\beta_H + \frac{3}{4}\Lambda_{\mathrm{Ly\alpha}} + \frac{1}{4}\Lambda_{2s \to 1s}}$$

and where the rates are given by

$$\Lambda_{\mathrm{Ly\alpha}} = \frac{8\pi H(T)}{3 n_H(T)(1-x_e)\lambda_{\mathrm{Ly\alpha}}^3}$$

$$\Lambda_{2s \to 1s} = 8.22\,\mathrm{s}^{-1}$$

$$\alpha_H(T) = 2.8 \times 10^{-17}\, T^{-1/2}\,\mathrm{s}^{-1}$$

$$\beta_H(T) = \frac{1}{4}\alpha_H(T) \left(\frac{m_e T}{2\pi} \right)^{1/2} e^{-B_{21}/T}$$

where $\lambda_{\mathrm{Ly\alpha}} = 1216\,\text{Å}$ is the wavelength of the Lyα line, $B_{21} = 3.4\,\mathrm{eV}$ is the energy split between $n = 1$ and $n = 2$ levels of hydrogen, and m_e is the electron mass. Your result should look like the dashed line in Fig. 13.5. Assume $n_H(a) = 0.25 a^{-3}\,\mathrm{m}^{-3}$, and the usual expression for $H(a)$. Don't forget the radiation contribution to the Hubble parameter.

Note that the $\dot{x}_e$ equation above is a little "stiff," as it includes large and mostly canceling terms on the right-hand side (at low a). The most principled way to address this is to use a stiff-equation solver (or so-called implicit integration), but a simpler way to circumvent this problem is simply to start the integration late, say at $a = 0.0005$.

14 Gravitational Lensing

The last topic discussed in this book is gravitational lensing, which entails some of the most fascinating connections between observations and theory in cosmology. Light from distant objects is deflected by mass that it encounters on its way to our telescopes. This bending of light can be measured, and provides a crucial cosmological probe which is sensitive to the distribution and amount of *all* mass – dark and visible – between us and the source. We review the history of our understanding of gravitational lensing, then cover the mathematical details of simple scenarios with a point-mass deflector. We continue by discussing strong and weak gravitational lensing, and use that as a segue to discuss how lensing helps us better understand dark matter and dark energy in the universe.

14.1 Gravitational Lensing Basics

14.1.1 Brief History of Gravitational Lensing

According to physicist John Archibald Wheeler, the general theory of relativity (GR) can be summarized by the statement

Space-time tells matter how to move
matter tells space-time how to curve.

This poetic 12-word description stands in marked contrast with the highly mathematical and technical nature of GR. The theory, developed by Einstein (and almost simultaneously by Hilbert) in 1915, describes the motion of objects in the presence of the gravitational field – that is, in the presence of mass. Einstein realized that it was possible for astronomical objects to bend light. In fact, GR was confirmed precisely in this way: in 1919, astronomers Arthur Eddington and Frank Dyson organized two teams to verify the GR prediction that the amount of bending of light around a massive body is twice what the non-relativistic physics predicts. The two teams went to two places simultaneously to observe, during a solar eclipse, the light from stars passing close to the Sun: Principe Island in West Africa, and Sobral in the northeast of Brazil. They were able to confirm Einstein's theory. This result was publicized worldwide, and almost overnight confirmed the validity of GR.

Already in 1912 Einstein realized that one could potentially observe multiple images of a single source, where light is bent by an object along the line of sight between the observer and the source – the gravitational lens. However, as he only considered gravitational lensing by single stars, Einstein concluded that the phenomenon would most likely remain unobserved for the foreseeable future. Next up, in 1937 Fritz Zwicky first considered the case where a galaxy could act as a gravitational lens, where observing multiple images would, according to his calculations, be well within the reach of observations.

It was not until 1979 that the first multiply imaged gravitational lens system was discovered accidentally by Walsh, Carswell, and Weymann using the Kitt Peak National Observatory 2.1 m telescope in Arizona (Walsh *et al.*, 1979). This system became known as the "Twin QSO" since it initially looked like two identical quasistellar objects. The source of light is the quasar Q0957+561 at redshift $z = 1.4$, and it appears as two images separated by $6''$ due to the gravitational lensing of a group of galaxies in the foreground. Many more of these multiply imaged – so-called **strongly lensed** – systems have been discovered since, with the current total at about 100.

The situation was made more interesting by a realization, starting in the 1960s, that weak distortion of background galaxy **shapes**, as their light passes near a foreground mass, is also of interest. In this so-called **weak gravitational lensing** regime, the lens mass is not sufficient to generate multiple images of the background source. Rather, the image of the background galaxy, in addition to being magnified, also gets slightly distorted ("sheared"), implying that the ratio of its long to short axis appears changed by $O(1)$ percent relative to the truth. While this is a very small signal given that we do not know the intrinsic shape of a source galaxy, the statistical *correlations* of source-galaxy shapes can be both measured and unambiguously related to theory. The race for the first detection of weak lensing culminated in the year 2000, with four teams near-simultaneously announcing first detections. Since that time, weak lensing has become a major tool of precision cosmology.

Before we start with the mathematical treatment, we emphasize the single most important thing about gravitational lensing:

> Unlike most other observational probes in astrophysics, gravitational lensing has the unique power of being sensitive to *all* matter – dark as well as baryonic.

This makes gravitational lensing a powerful probe of the amount and distribution of dark matter in the universe.

14.1.2 The Lens Equation

Gravitational lensing geometry is described in Fig. 14.1. There are three entities involved: the **source** of light (which is in the background), the **lens** that bends the source's light to us (which is in the foreground), and the **observer**. The angle that the observer would measure between the source and the lens, in the absence of lensing, is β. The angle that the observer actually measures, between the image (of the source) and the lens, is θ. The lens itself can often not be readily observed (i.e., it is simply not luminous enough), but the angle θ, and its generalizations in cases with a more complicated makeup of the lens, is still useful as it is related to the relative positions of images on the sky, which of course *are* observed.

Angular separations on the sky are distorted in the presence of the lens, so that $\theta \neq \beta$. The angle between the source and its image, when measured from the lens position (see Fig. 14.1), is called the **reduced lensing angle**, α. Note that a lensing analysis necessarily involves angles which indicate separations on the 2D sky surface. A position on the sky relative to some reference point can therefore be described by two coordinates, such as right ascension and declination. For convenience, we shall describe such sky positions as 2D vectors and denote them in boldface. We define the 2D angle between the source and the lens angular positions,

$$\boldsymbol{\beta} = \boldsymbol{x}_{\text{source}} - \boldsymbol{x}_{\text{lens}}, \tag{14.1}$$

and the 2D angle between the image and the lens positions,

$$\boldsymbol{\theta} = \boldsymbol{x}_{\text{image}} - \boldsymbol{x}_{\text{lens}}. \tag{14.2}$$

In terms of these two, the reduced lensing angle is given by

$$\boldsymbol{\alpha} = \boldsymbol{\theta} - \boldsymbol{\beta}. \tag{14.3}$$

Because both the reduced lensing angle $\boldsymbol{\alpha}$ and the actual deflection angle $\hat{\boldsymbol{\alpha}}$ span the arc length between the source and the image, albeit at different distances (see Fig. 14.1), they are related by

$$\boldsymbol{\alpha} = \frac{d_{LS}}{d_S} \hat{\boldsymbol{\alpha}}, \tag{14.4}$$

where d_{LS} is the distance between the lens and the source, and d_S is the distance to the source (from the observer). Combining these equations, we get the **lens equation** which reads $\boldsymbol{\beta} = \boldsymbol{\theta} - \boldsymbol{\alpha}$, or

$$\boldsymbol{\beta} = \boldsymbol{\theta} - \frac{d_{LS}}{d_S} \hat{\boldsymbol{\alpha}} \qquad \text{(lens equation)}. \tag{14.5}$$

Now we need to clarify what the distances d really are. All gravitational lensing observations necessarily involve measuring angles on the sky. Therefore, all distances are angular diameter distances $d_A(z) = r(z)/(1+z)$. Note that they are not additive; for example, $d_S \neq d_L + d_{LS}$ even in a flat universe.

14.1.3 Deflection Angle

The deflection angle can be derived from GR by studying the null geodesics (relevant for the propagation of light) in the presence of gravitational potential. Here we will

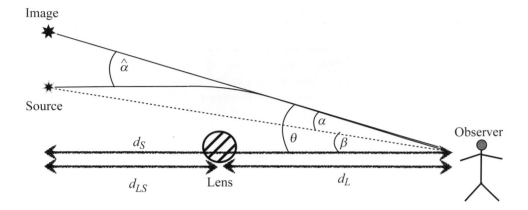

Fig. 14.1 Sketch of the gravitational lensing geometry. The deflection angle is $\hat{\alpha}$, while the reduced lensing angle is α. All distances are angular diameter distances.

just quote the result of that calculation. The deflection angle is

$$\hat{\boldsymbol{\alpha}} = 2 \int_0^{s_{\text{src}}} \boldsymbol{\nabla}_\perp \Phi \, ds, \tag{14.6}$$

where Φ is the (Newtonian) gravitational lensing potential ("the one from the metric"), $\boldsymbol{\nabla}_\perp$ is the gradient in the perpendicular direction to the photon's trajectory, ds is the radial path element, and s_{src} is the distance to the source. This result is higher than what we would get for a point mass in classical mechanics by the famous factor of two, which is precisely the difference from the classical prediction that Dyson and Eddington's teams looked for – and found – in 1919.

14.2 Point-Mass Lens

The simplest model of a lens assumes it is a point mass of mass M. To evaluate the deflection angle for a point-mass lens deflector, we need to evaluate the integral in Eq. (14.6). We work in cylindrical coordinates, as illustrated in Fig. 14.2. We do the analysis in the plane given by the direction of photon propagation and the point mass, so that all of our angles will be 1D (rather than the more general 2D). At an arbitrary distance r away from the point mass, the distance between the photon and the point mass is related to the impact parameter b via $b = r \cos \phi$. The gradient term in the perpendicular direction is $|\boldsymbol{\nabla}_\perp \Phi| = (GM/r^2) \cos \phi$. Because $ds \cos \phi = r \, d\phi$, the differential path length is given by $ds = r \, d\phi / \cos \phi = r^2 \, d\phi / b$. Then

$$\hat{\alpha} = 2 \int_{-\pi/2}^{\pi/2} \left(\frac{GM}{r^2} \cos \phi \right) \left(\frac{r^2 \, d\phi}{b} \right) = \frac{2GM}{b} \int_{-\pi/2}^{\pi/2} \cos \phi \, d\phi, \tag{14.7}$$

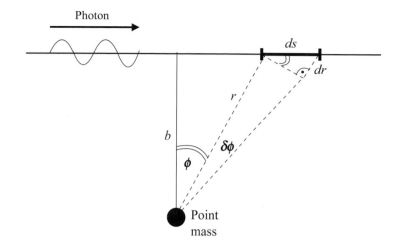

Fig. 14.2 Sketch accompanying the derivation of the deflection angle for the point-mass lens.

or

$$\hat{\alpha} = \frac{4GM}{b} \qquad \text{(deflection angle, point mass)}. \qquad (14.8)$$

Restoring the speed of light, this angle is $\hat{\alpha} = 4GM/(c^2 b)$. For the photon grazing the Sun, the deflection angle is

$$\hat{\alpha}_\odot = \frac{4GM_\odot}{c^2 R_\odot} \simeq 1.7 \,\text{arcsec}. \qquad (14.9)$$

This was just at the limit of observability in Dyson–Eddington's solar eclipse observations.

Given that the impact parameter is related to angle θ via $b = d_L \theta$, the lens equation for the point mass becomes

$$\beta = \theta - \frac{d_{LS}}{d_S d_L} \frac{4GM}{\theta} \qquad \text{(point-mass lens equation)}. \qquad (14.10)$$

From this, one can calculate the angle between the image and the lens θ. Note that θ is quite a bit more useful for comparison with observations than the bending angle $\hat{\alpha}$, since the latter requires an observation of the *unlensed* source position, which is impossible to measure (for cosmological lensing observations where the source–lens–observer configuration does not appreciably change over a period of human lifetime).

14.2.1 Einstein Angle

A particularly instructive case of lensing follows if the source, lens, and observer are *collinear*. Then there will actually be infinitely many images, forming a perfect circle (and observed with the lens at the center); see Fig. 14.3(a). This ring is called the **Einstein ring**, and the radius of the circle is called the Einstein radius. The collinear case occurs when $\beta = 0$ (refer to Fig. 14.1). Then the lens equation for the point mass, Eq. (14.10), can be solved to give the **Einstein angle**

$$\theta_E = \sqrt{\frac{4GMd_{LS}}{d_L d_S}} \qquad \text{(Einstein angle)}. \qquad (14.11)$$

Likewise, the distance that the Einstein angle subtends in the lens plane, $R_E = \theta_E d_L$, is called the **Einstein radius** and is equal to

$$R_E = \sqrt{\frac{4GMd_L d_{LS}}{d_S}}. \qquad (14.12)$$

Equation (14.12) illustrates the fact that gravitational lensing is most effective when the lens is located, very roughly, halfway between the observer and the source. As either d_L or d_{LS} goes to zero, the Einstein radius goes to zero, making the lensing effectively unobservable.

We now plug in some numbers into Eq. (14.11) to compute the Einstein angle for some characteristic cases (derived in Problem 14.1):

- Consider the case where the lens is a cluster[1] of galaxies, while the source is a distant galaxy; in that case the Einstein radius is

$$\theta_E \approx (30\,\text{arcsec}) \left(\frac{M}{10^{14} M_\odot}\right)^{1/2} \left(\frac{d}{1000\,\text{Mpc}}\right)^{-1/2}, \qquad (14.13)$$

  where M is the mass of the (cluster) lens, and d is the distance of the background galaxy (assuming for simplicity the lens is located halfway to the source). This is quite a large angle, and images in a multiply imaged system, provided they are bright enough, will be easily distinguished by telescopes.
- Now consider the case where the lens is a Galactic MACHO (MAssive Compact Halo Object – could be a blob or "planet" of dark matter), while the source is a star in our galaxy. Then

$$\theta_E \approx (4 \times 10^{-4}\,\text{arcsec}) \left(\frac{M}{1M_\odot}\right)^{1/2} \left(\frac{d}{50\,\text{kpc}}\right)^{-1/2}. \qquad (14.14)$$

This angle is more than a thousand times smaller than the typical seeing (resolution limitation mostly due to the atmosphere) from the world's best telescopes, which

---

[1] While a cluster of galaxies is obviously not a point mass, it can be approximately treated as such when the image locations in the lens plane lie outside of most mass enclosed by the cluster.

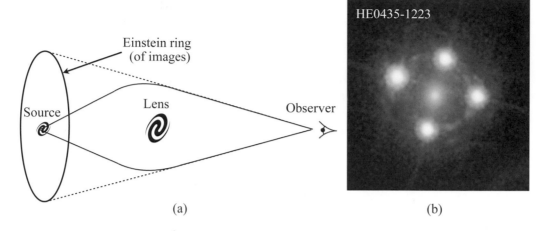

(a)   (b)

Fig. 14.3 (a) Illustration of an Einstein ring, which occurs when the source, lens, and observer are collinear, and the background galaxy can be considered as a point source. In practice, these conditions are not perfectly satisfied, leading to images such as that shown on the right. (b) An example of strongly lensed – that is, multiply imaged – source, obtained by the Hubble Space Telescope. The source is the quasar HE0435-1223, located at $z_s = 1.69$, while the main deflector is a galaxy at $z_l = 0.45$. A faint Einstein ring can also be seen in this image. Credit: NASA, ESA and STScI; Wong *et al.* (2020).

tends to be in the $0.5-1''$ range. Therefore, multiple images from a MACHO, even assuming a sufficiently bright point source, cannot be distinguished.

14.2.2 Image Locations

Let us now consider the lens equation for a non-collinear point-mass case, so that $\beta \neq 0$ in Eq. (14.10). Then there are two solutions to this quadratic equation:

$$\theta_{\pm} = \frac{1}{2}\left(\beta \pm \sqrt{\beta^2 + 4\theta_E^2}\right). \tag{14.15}$$

Note that θ_+ is always greater than the Einstein angle, while θ_- is smaller.

The fact that there are two solutions in Eq. (14.15) may surprise us once we learn that, upon a more detailed mathematical analysis, one can prove something called the **odd number theorem**, which states that the total number of images has to be odd. However, the odd number theorem assumes a continuous, finite distribution of mass, which our point mass does not satisfy (the point mass can be represented by a Dirac delta function in density, so it is formally infinite). Nevertheless, it is also true that we most often observe strongly lensed systems with observed "doubles" (two images) or "quads" (four images); see the example in Fig. 14.3(b). What happened to the odd number theorem in those cases? It turns out that there *does* exist one more, central image (so a third image in a double or a fifth in a quad), but it is very demagnified and sometimes obscured by the lens itself, so that it is usually not observed.

14.3 Beyond the Point Mass

We now perform a more detailed lensing analysis that will allow us to study cases when the lens is not a single point mass, but rather a continuous distribution of mass – the large-scale structure between us (the observer) and the source. We first start with some mathematical background.

14.3.1 Lensing Potential

Gravitational lensing involves observations on the sky, and thus 3D quantities that are projected (integrated) over the radial direction. In particular, we can define the **lensing potential** by integrating the Newtonian potential along the line of sight, with some suitably chosen prefactors:

$$\psi(\boldsymbol{\theta}) = 2 \frac{d_{LS}}{d_L d_S} \int_0^{s_{\mathrm{src}}} \Phi(d_L \boldsymbol{\theta}, s)\, ds, \tag{14.16}$$

where ds is the radial line element. From here on we are assuming a **thin lens approximation**, which says that the lens is thin relative to distances d_L, d_{LS}, and d_S. This allows us to take out the distances in front of the integral, that is, evaluate them at the lens location.

Taking the angular-direction gradient of the projected potential, $\boldsymbol{\nabla}_\theta$, and linking it to the perpendicular-direction gradient, $\boldsymbol{\nabla}_\perp$, via $\boldsymbol{\nabla}_\theta = d_L \boldsymbol{\nabla}_\perp$ (see Fig. 14.1), we get the reduced lensing angle

$$\boldsymbol{\alpha} = \boldsymbol{\nabla}_\theta \psi = 2 \frac{d_{LS}}{d_S} \int_0^{s_{\mathrm{src}}} \boldsymbol{\nabla}_\perp \Phi\, ds. \tag{14.17}$$

Note that this agrees with the equivalent point-mass equation (see the combination of Eqs. (14.4) and (14.6)), except now it is in two dimensions rather than one.

One can now define a quantity called **convergence** κ, defined as

$$\kappa \equiv \frac{1}{2} \nabla_\theta^2 \psi = \frac{d_L d_{LS}}{d_S} \int_0^{s_{\mathrm{src}}} \nabla^2 \Phi\, ds, \tag{14.18}$$

where the latter expression, as before, holds in the thin-lens approximation. Convergence can be thought of as an integrated mass density,[2] which can be seen by recalling that $\nabla^2 \Phi \propto \delta\rho_m$ from the Poisson equation. When specified at every

---

[2] Indeed, convergence can be written as $\Sigma/\Sigma_{\mathrm{crit}}$, where Σ is the surface mass density and Σ_{crit} is its "critical value" for creating multiple images.

point on the sky, convergence contains all the information about gravitational lensing of the system. Because it is a scalar rather than a spin-one-half field, convergence is sometimes easier to handle mathematically than the two-component shear field.

One can invert the expressions for the lensing potential and reduced lensing angle via Green's function formalism to get

$$
\begin{aligned}
\psi(\boldsymbol{\theta}) &= \frac{1}{\pi} \int \kappa(\boldsymbol{\theta}') \ln|\boldsymbol{\theta} - \boldsymbol{\theta}'| d^2\theta' \\
\boldsymbol{\alpha}(\boldsymbol{\theta}) &= \frac{1}{\pi} \int \kappa(\boldsymbol{\theta}') \frac{\boldsymbol{\theta} - \boldsymbol{\theta}'}{|\boldsymbol{\theta} - \boldsymbol{\theta}'|^2} \, d^2\theta'.
\end{aligned}
\tag{14.19}
$$

We will not make use of the above two equations except to note that the convergence field holds the same information as the lensing potential (see the first line in Eq. (14.19)), or the reduced lensing angle (second line). As we will see shortly, convergence is the more convenient quantity to work with.

14.3.2 Magnification

Gravitational lensing does not create or destroy photons, but rather merely bends their trajectories, much like a lens made of glass would. Therefore, the total flux from the source, averaged over a sphere centered at the source, is the same as it would be without the lens. However, the focusing implies that the flux is not distributed isotropically: the source typically appears *magnified* when observed from (roughly) the side of the lens opposite where the source is, and demagnified if observed from other directions. In typical observations where the lens is located between the source and the observer, we therefore expect the source to be magnified – that is, brighter than it would be without the lens.

The object of interest for predicting the magnification is the **distortion matrix** A. Recall first that $\boldsymbol{\beta} = \boldsymbol{x}_{\text{source}} - \boldsymbol{x}_{\text{lens}}$ is the 2D angle between the source and the lens, and $\boldsymbol{\theta} = \boldsymbol{x}_{\text{image}} - \boldsymbol{x}_{\text{lens}}$ is the 2D angle between the image and the lens. Then the magnification matrix A is defined as

$$
A_{ij} \equiv \frac{\partial \beta^i}{\partial \theta^j},
\tag{14.20}
$$

so that A tells us what a small change in the image location, $\boldsymbol{\theta}$, implies for the shift in the source location, $\boldsymbol{\beta}$:

$$
d\boldsymbol{\beta} = A d\boldsymbol{\theta}.
\tag{14.21}
$$

In the case of no lensing, $\boldsymbol{\beta} = \boldsymbol{\theta}$, and A is a unit 2×2 matrix. Since $\boldsymbol{\beta} = \boldsymbol{\theta} - \boldsymbol{\alpha}$, we have

$$
A_{ij} = \delta_{ij} - \frac{\partial \alpha^i}{\partial \theta^j} = \delta_{ij} - \psi_{,ij},
\tag{14.22}
$$

where $\psi_{,ij} \equiv \partial^2 \psi / \partial \theta^i \partial \theta^j$ and where we used $\alpha^i = \psi_{,i}$; see Eq. (14.17).

The inverse of the matrix A is called the **magnification matrix** M:

$$M \equiv \frac{\partial \boldsymbol{\theta}}{\partial \boldsymbol{\beta}} = A^{-1}. \tag{14.23}$$

The magnification matrix lets us map the coordinates of the unlensed image (source position, $\boldsymbol{\beta}$) to the distorted image (lens position, $\boldsymbol{\theta}$).

Lensing distorts the area element described by $\boldsymbol{\beta}$ into one described by $\boldsymbol{\theta}$. The change in the ratio of the fluxes is a scalar quantity called the **magnification**, which is just the Jacobian of this mapping – the determinant of M:

$$\mu \equiv |M| = \left| \frac{\partial \boldsymbol{\theta}}{\partial \boldsymbol{\beta}} \right| = \frac{1}{|A|} \qquad \text{(magnification)}. \tag{14.24}$$

Magnification of one means that the source brightness is unchanged. If the magnification is very small ($\mu \ll 1$), the source image may be unobservable. For typical cases of multiple-imaged lensing, $\mu \gtrsim 1$. Box 14.1 also illustrates a rather extreme example of magnification – the case of gravitational microlensing.

14.3.3 Shear

We already defined convergence in terms of the lensing potential in Eq. (14.18):

$$\kappa = \frac{1}{2} \left(\psi_{,11} + \psi_{,22} \right). \tag{14.25}$$

We now define **shear**, a quantity that has two components:

$$\begin{aligned} \gamma_1 &= \psi_{,11} - \psi_{,22} \\ \gamma_2 &= \psi_{,12} = \psi_{,21}. \end{aligned} \tag{14.26}$$

Shear quantifies the distortion of an image. If an initially circular source is distorted into an ellipse of ellipticity γ and position angle ϕ relative to some fiducial direction, then the components of shear parallel and perpendicular to that direction are given by, respectively

$$\begin{aligned} \gamma_1 &= \gamma \cos(2\phi) \\ \gamma_2 &= \gamma \sin(2\phi), \end{aligned} \tag{14.27}$$

while the total shear is equal to the ellipticity

$$\gamma = \sqrt{\gamma_1^2 + \gamma_2^2}. \tag{14.28}$$

One can rewrite the matrix A as

$$A = \begin{pmatrix} 1 - \kappa - \gamma_1 & -\gamma_2 \\ -\gamma_2 & 1 - \kappa + \gamma_1 \end{pmatrix}. \tag{14.29}$$

| Box 14.1 | MACHO and Gravitational Microlensing |

We now consider the case when the lens is a MACHO – MAssive Compact Halo Object – a hypothetical planet-sized object that can effectively be treated as a point-mass lens. Here, the source is a star in our own galaxy, or else in the nearby Andromeda (M31). Because the lens is typically not very massive (having a ballpark mass of $1M_\odot$), multiple images are not formed. However, the magnification of the source image *can* be observed because it is temporary: the magnification occurs over the period of time when the source–lens–observer system is approximately collinear, which is $O(\text{days})$ for a MACHO-type lens. This type of lensing, when only a temporary amplification is observed, goes under the name of **gravitational microlensing**.

Starting in the 1980s, cosmologists have proposed searching for MACHOs using just this principle: a MACHO would pass across the line of sight between us and a background star, and it would lens the star. While lensing would be weak and multiple images would not form, the MACHO lens would magnify the star for a period of days to months (depending on the MACHO mass and geometry). Several research groups, with lurid acronyms such as MACHO, EROS, and OGLE, have carried out searches for MACHOs by looking for temporary magification of stars. The figure here (reprinted by permission from Springer Nature Customer Service Centre GmbH: Alcock *et al.* (1993), copyright 1993) shows one of the several exciting magnification detections – presumably by a MACHO – obtained by these collaborations.

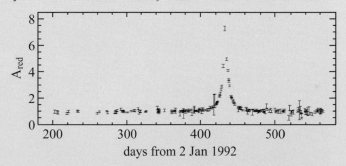

The current results regarding MACHOs are somewhat inconclusive. While the EROS collaboration rules out all objects in the mass range $10^{-7}M_\odot < M < 30M_\odot$, the MACHO collaboration favors ~ 20 percent of the galactic halo is in halos of $M \sim 0.2M_\odot$. In Chapter 11, we discussed how this indicates that the bulk of dark matter is not made up of MACHOs. Instead, the leading candidate for dark matter remains a non-baryonic (microscopic) particle, such as a WIMP.

Therefore, the magnification can be calculated in terms of convergence and shear:

$$\mu = \frac{1}{(1-\kappa)^2 - \gamma^2}. \tag{14.30}$$

We conclude that

> Convergence and shear, when specified at every location in the lens plane,
> separately contain all information about the locations, shapes, and
> magnifications of (source) images.

Of course, one is not *a priori* given the convergence and/or shear; in realistic situations, one *uses* the locations, shapes, and magnifications of the source images in order to infer the convergence and shear across the lens plane. It is interesting, in fact, that we use the aforementioned observations of visible matter in order to infer quantities (shear/convergence) that depend on all matter, visible and dark. In other words, lensing allows us to reconstruct the mass distribution of the lens(es).

An example of the system showing not one but (at least) two sources that are multiply imaged is shown in Fig. 14.4.

We have only scratched the surface of the study of gravitational lensing in the regime where multiple images of the source are observed. The mathematical theory

Fig. 14.4 A quasar at redshift $z_s = 1.734$, SDSS J1004+4112, is split into five images by a galaxy cluster at $z_d = 0.68$ (Sharon *et al.*, 2005). Further observations subsequently showed that a very distant source galaxy, at $z_s = 3.332$, is also multiply imaged. Therefore, there are at least two sources that are multiply imaged, in addition to distortions of numerous other sources (that are not strongly but only weakly lensed), showcasing the complexity of gravitational lensing in realistic situations. Credit: ESA, NASA, K. Sharon (Tel Aviv University) and E. Ofek (Caltech). A black and white version of this figure will appear in some formats. For the color version, please refer to the plate section.

of this subject is very rich, especially once we start studying the realistic cases when the lens is not a point mass, but rather a set of multiple point sources, or else a continuous distribution of mass. Various mathematical properties have been derived about such systems, and sophisticated computer algorithms can be used to solve for the location of images given some distribution of lenses and sources. Here, we move on to discuss situations where multiple images of the sources are *not* observed – the so-called weak gravitational lensing.

14.4 Weak Gravitational Lensing

Weak gravitational lensing occurs in the limit when the shear and convergence are small:

$$\kappa, \gamma \ll 1 \qquad \text{(weak-lensing regime)}. \tag{14.31}$$

In this limit, multiple images of the source are not produced. Rather, the sources are slightly distorted and magnified. The weak-lensing regime occurs *much* more often than strong lensing, which requires a massive lens and favorable geometrical alignment.

Figure 14.5 illustrates how weak lensing works. Light from background source galaxies is focused by the structures between us and those sources, distorting their shapes and typically making them more elliptical by $O(1)$ percent. While very small, this signal is correlated between galaxies that are close on the sky as, roughly speaking, the light from those galaxies crossed the same large-scale structures on its way to us. The ellipticity correlations – known variously as weak lensing or **cosmic shear** – are a statistical signal that can be both predicted and measured. Specifically, the observed ellipticity $\epsilon_{\rm obs}$ can be related to intrinsic (random) ellipticity $\epsilon_{\rm int}$, cosmic shear γ, and ellipticity due to measurement errors $\epsilon_{\rm err}$ as

$$\epsilon_{\rm obs} = \epsilon_{\rm int} + \gamma + \epsilon_{\rm err}. \tag{14.32}$$

The random ellipticities are quite large ($\epsilon_{\rm int} \sim 0.3$) compared to the cosmic-shear signal ($\gamma \sim 0.01$). This illustrates the difficulty of extracting the cosmological signal due to weak lensing. It also points to the only way in which it is possible to measure the statistical properties of γ: through correlations of galaxy shapes. Because the intrinsic shapes are random, they do not contribute to the signal, but only noise. We return to this point in Sec. 14.4.1 when we discuss the convergence/shear power spectrum.

Many results simplify in the weak-lensing limit. For example, the magnification becomes

$$\mu \simeq 1 + 2\kappa \qquad \text{(when } \kappa, \gamma \ll 1\text{)}. \tag{14.33}$$

While weak lensing around individual massive halos was measured in the 1990s, detection of weak lensing by large-scale structures took longer because of the

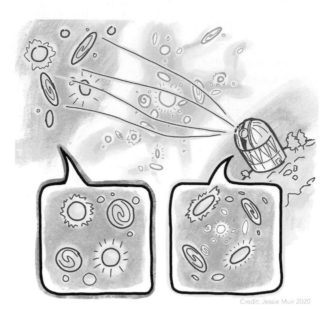

Credit: Jessie Muir 2020

Fig. 14.5 A sketch of how weak gravitational lensing works. The light from background galaxies passes through a large-scale structure on its way to the telescope. In this process, the shape of each background galaxy is slightly sheared (the effect is exaggerated in the sketch for pedagogical purposes). The galaxy shapes also show a tendency to align tangentially around foreground mass overdensities, another effect that is exaggerated here. Drawing made by Jessica Muir.

smallness of its signal. In weak lensing, one needs large statistics in order to separate the lensing effect from the noise represented by random orientations of galaxies. A watershed moment came in the year 2000 when four research groups nearly simultaneously announced the first detection of weak lensing by large-scale structures. Since that time, weak lensing has matured into an increasingly accurate and powerful probe of dark matter and dark energy.

The principal power of weak lensing – like with any flavor of lensing – comes from the fact that it responds to all matter: dark matter and visible (or, more generally, baryonic) matter. Therefore, modeling of the visible-to-dark matter bias, a thorny and complicated subject, is altogether avoided when using weak lensing. As we have seen in Chapter 9, matter clustering statistics (and, hence, the weak-lensing signal) can be modeled exactly in linear theory and, with the help of numerical simulations, to a good accuracy even in the mildly to moderately nonlinear regime. Because of our ability to model its signal accurately, weak lensing has great intrinsic power to probe dark matter and dark energy in the universe.

The other principal reason why weak lensing is powerful comes from the nature of its observable quantity. Galaxy shear is nowadays measured relatively straightforwardly, and comes from millions of galaxies typically observed in current surveys

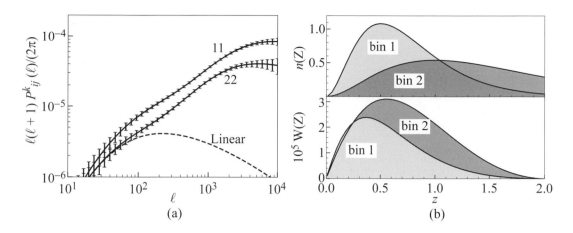

Fig. 14.6 (a) Convergence power spectrum P_{ij}^κ (multiplied, as is convention, by $\ell(\ell+1)/(2\pi)$), as a function of multipole ℓ. We assume a two-bin tomography, and plot the auto-correlations in the two bins, but (for clarity) not the cross-correlation P_{12}^κ. (b) The number density of source galaxies $n_i(z)$ assumed in the analysis, as well as the corresponding weights $W_i(z)$, for the two bins. We assumed the fiducial ΛCDM cosmological model from Table 3.1.

(100 million galaxy shapes to be precise, in the case of the DES Year-3 shear catalog). While *individual* galaxy shear measurements do not provide cosmological information, correlation of galaxy shear across the sky does. One can also split source galaxies into multiple redshift bins, thus revealing more information about the distribution of the matter along the line of sight in this so-called **weak-lensing tomography**. Such tomographic measurements are illustrated in Figs. 14.6 and 14.7.

We now work out the theory behind weak gravitational lensing.

14.4.1 Shear/Convergence Power Spectrum

We now turn to the theoretical modeling of a shear (or convergence) field on the sky, with immediate emphasis on the statistical structure of its correlations – or, its power spectrum.

Let us assume a lens mass characterized by the convergence field $\kappa(\mathbf{n})$. Transforming the convergence into multipole space

$$\kappa_{\ell m} = \int d\hat{\mathbf{n}}\, \kappa(\hat{\mathbf{n}})\, Y_{\ell m}^*(\hat{\mathbf{n}}), \tag{14.34}$$

and assuming statistical isotropy, the power spectrum of convergence $P^\kappa(\ell)$ is then defined as the harmonic transform of the two-point correlation function of the convergence

$$\langle \kappa_{\ell m} \kappa_{\ell' m'} \rangle = \delta_{\ell\ell'}\, \delta_{mm'}\, P^\kappa(\ell). \tag{14.35}$$

Note that this parallels almost exactly the case of transforming the CMB temperature fluctuations to harmonic space, covered in Chapter 13.

One can show that the convergence power spectrum is identical to the shear power spectrum in the limit of weak distortions (i.e., in the weak-lensing limit); $P^\gamma(\ell) \simeq P^\kappa(\ell)$. For a Gaussian field, either P^γ or P^κ would contain all the information. However, since the weak-lensing field is non-Gaussian on small scales, higher-order correlations contain additional information and may be useful to exploit.

We would like to relate convergence to theory, so that we can have a from-scratch theoretical prediction for the weak-lensing power spectrum in either real or harmonic space. To do this, we start from Eq. (14.18) for the convergence in terms of the projected potential, and use the Poisson equation. Note a few things:

- We cannot use the thin-lens approximation any more as the foreground mass distribution – the large-scale structure itself – is distributed in three dimensions, so the distances in the convergence formula move into the line-of-sight integral.

- It is easiest to do the calculation in comoving coordinates, where $\nabla_{\mathrm{comov}} = a\nabla$. In comoving coordinates, the usual Poisson equation $\nabla^2\Phi = 4\pi G\,\rho_M\delta = 4\pi G\rho_{\mathrm{crit},0}\,\Omega_M a^{-3}\delta$ becomes $\nabla^2_{\mathrm{comov}}\Phi = 4\pi G\rho_{\mathrm{crit},0}\,\Omega_M a^{-1}\delta$. Moreover, the angular diameter distances (e.g., $d_S \equiv d_A(z_S)$) become the comoving distances (e.g., $r(z_S) \equiv r(\chi_S)$), where χ_S is the coordinate distance evaluated at the source.

- We are starting out assuming all source galaxies are at a single redshift, but will drop that assumption shortly below.

Then, employing Eq. (14.18) along with the Poisson equation, we have

$$
\begin{aligned}
\kappa(\hat{\mathbf{n}}) \equiv \frac{1}{2}\nabla^2_\theta\psi(\hat{\mathbf{n}}) &= \int_0^{\chi_S} \frac{r_L r_{LS}}{r_S}\,\nabla^2_{\mathrm{comov}}\Phi(\hat{\mathbf{n}},\chi)\,d\chi \\
&= \int_0^{\chi_S} \frac{r(\chi)r(\chi_S - \chi)}{r(\chi_S)}\,4\pi G\rho_{\mathrm{crit},0}\,\Omega_M a^{-1}\delta(\hat{\mathbf{n}},\chi)\,d\chi \\
&= \int_0^{\chi_S} \frac{r(\chi)r(\chi_S - \chi)}{r(\chi_S)}\,\frac{3}{2}H_0^2\Omega_M(1+z)\,\delta(\hat{\mathbf{n}},\chi)\,d\chi \\
&\equiv \int_0^{\chi_S} W(\chi)\delta(\hat{\mathbf{n}},\chi)\,d\chi.
\end{aligned}
\tag{14.36}
$$

Here χ is the coordinate distance (recall, $d\chi = dz/H(z)$ and, if the universe is flat, $r(\chi) = \chi$), and we have defined a weight function W as

$$
W(\chi) \equiv \frac{3}{2}H_0^2\Omega_M(1+z)\frac{r(\chi)r(\chi_S - \chi)}{r(\chi_S)}.
\tag{14.37}
$$

If we drop the artificial simplifying assumption that source galaxies lie at a single redshift, and assume instead they are distributed along the radial direction with some distribution $n(\chi)$, then the weight function generalizes to

$$
W(\chi) \to \frac{3}{2}H_0^2\Omega_M(1+z)r(\chi)\int_\chi^\infty d\chi_S n(\chi_S)\frac{r(\chi_S - \chi)}{r(\chi_S)},
\tag{14.38}
$$

where $\chi = \chi(z)$, We then produce the angular power spectrum of convergence, P^κ, by using something called the **Limber approximation** (see Problem 14.6 for details). The Limber approximation assumes that only the modes *transverse* to the line-of-sight modes contribute, and that the modes parallel to the line of sight can be ignored. This is an excellent approximation everywhere except on the largest scales (lowest multipoles), and tremendously simplifies the numerical computation of the convergence power spectrum by turning two highly oscillatory integrals into a single smooth integral. The resulting Limber expression for the weak-lensing convergence power spectrum is given by (Problem 14.6)

$$P^\kappa(\ell) = \int_0^\infty dz \, \frac{W^2(z)}{r(z)^2 \, H(z)} \, P\left(\frac{\ell}{r(z)}, z\right). \tag{14.39}$$

The integral is conservatively set to go out to infinity, although in practice it contributes only up to the redshift of most distant source galaxies (typically $z \sim 2$). Most of the integral is contributed where the lensing weight is maximal, roughly halfway out to sources or typically $z \sim 0.2-0.5$.

Recall again that, in the weak-lensing limit, the convergence and shear power spectrum are identical, $P^\kappa(\ell) = P^\gamma(\ell)$. Moreover, note the fact that the convergence/shear power spectrum, P^κ/P^γ, is an integral of the matter power spectrum $P(k)$ along the line of sight, which intuitively fits the picture that the weak-lensing signal is a sum of the contributions provided by the matter field between us and the source galaxies.

According to Eq. (14.39), for lenses at redshift z and at a fixed multipole ℓ, the 3D matter power spectrum $P(k)$ contributes at the wavenumber

$$k = \frac{\ell}{r(z)}. \tag{14.40}$$

This correspondence comes from the Limber approximation, which says that only the modes transverse to the line of sight contribute, and hence $\lambda \simeq r\theta$, or $k^{-1} \simeq r\ell^{-1}$, where λ is the wavenumber of the perturbation, r its distance to us, and θ an angle at which it is observed. The k-to-ℓ mapping is not sharp; a fixed ℓ roughly corresponds to about a factor of two in k around $k = \ell/r(z)$. And vice versa, a fixed k corresponds to a range of ℓ values centered at $\ell \simeq kr(z)$.

14.4.2 Weak-Lensing Tomography

Weak-lensing tomography – slicing of the shear signal in redshift bins – enables extraction of additional information from the weak-lensing shear, as it makes use of the radial information. Consider correlating shears in some redshift bin i to those in the redshift bin j. The tomographic cross-power spectrum for these two redshift bins, at a given multipole ℓ, is defined by

$$\langle (\kappa_{\ell m})_i (\kappa_{\ell' m'})_j \rangle = \delta_{\ell\ell'} \, \delta_{mm'} \, P^\kappa_{ij}(\ell), \tag{14.41}$$

and can be related to theory via a generalization of Eq. (14.39):

$$P_{ij}^{\kappa}(\ell) = \int_0^{\infty} dz \, \frac{W_i(z) \, W_j(z)}{r(z)^2 \, H(z)} \, P\left(\frac{\ell}{r(z)}, z\right), \tag{14.42}$$

where $r(z)$ is the comoving angular diameter distance and $H(z)$ is the Hubble parameter. The weights W_i are given by

$$W_i(\chi) = \frac{3}{2} \, \Omega_M H_0^2 \, q_i(\chi) \, (1+z), \tag{14.43}$$

where

$$q_i(\chi) = r(\chi) \int_{\chi}^{\infty} d\chi_S n_i(\chi_S) \frac{r(\chi_S - \chi)}{r(\chi_S)}, \tag{14.44}$$

and n_i is the normalized ($\int n(z)dz = 1$) comoving density of galaxies if χ_S falls in the distance range bounded by the ith redshift bin, and zero otherwise. The weight in Eq. (14.43) is therefore a straightforward generalization of the no-tomography expression in Eq. (14.38).

Because a tomographic measurement is sensitive to signals from multiple source planes, tomography is particularly useful in measuring how the growth of structure evolves through cosmic time. This, in turn, helps better pin down the dark energy and other cosmological parameters.

The convergence power spectrum for our fiducial cosmological model, as a function of multipole ℓ, is shown in Fig. 14.6(a). Here we show the case of two-bin tomography. Panel (b) of the same figure shows the number densities of sources that we assumed, as well as the corresponding weights $W_i(z)$ that enter the weak-lensing kernel in Eq. (14.42). The error bars in panel (a) were obtained using Eq. (14.47) discussed just below, assuming sky coverage of $f_{\rm sky} = 0.1$, total observed ellipticity of $\langle \epsilon_{\rm obs}^2 \rangle^{1/2} = 0.2$, and number density of galaxies n_i corresponding to 10 galaxies per square arcminute in each bin.

The convergence power is smooth because it is a radial projection of the matter power spectrum, that is, it encodes the canceling effect of the superposition of overdensities and underdensities along the line of sight. Note also that the full, nonlinear convergence power spectrum has more power than what we would expect from purely linear clustering, for the same reasons as the matter power spectrum in Fig. 9.6. Finally, note that the statistical error increases going to larger scales (lower ℓ) due to cosmic variance, and also increases going to smaller scales due to shot noise. For our choice of the survey parameters, the error is minimal at scales of $\ell \sim 1000$.

14.4.3 Uncertainty in P^{κ} Measurements

The observed convergence power spectrum has an additional contribution from the shot noise given by random galaxy shapes. The total, actually observed convergence power spectrum is then

$$\underbrace{C_{ij}^{\kappa}(\ell)}_{\text{observed power}} = \underbrace{P_{ij}^{\kappa}(\ell)}_{\text{"signal"}} + \underbrace{\delta_{ij}\frac{\langle \epsilon_{\text{obs}}^2 \rangle}{\bar{n}_i}}_{\text{shot noise}}, \tag{14.45}$$

where $\langle \epsilon_{\text{obs}}^2 \rangle^{1/2}$ is the rms observed shear in each of the two shear components (and is typically around 0.2), and $\bar{n}_i$ is the average number of galaxies (per steradian) with well-measured shapes in the ith redshift bin. The first term on the right-hand side, which dominates on large scales, comes from cosmic variance of the mass distribution. The second term is called the **shot noise**, and it results from a limited number of observed galaxies. This shot-noise term is familiar from everyday applications of statistics, where the Poisson (random) error in counting N events goes as $N^{-1/2}$; the only novelty here is that this factor is multiplied by the rms measured ellipticity $\langle \epsilon_{\text{obs}}^2 \rangle^{1/2}$, and then squared to get the contribution to the variance. Moreover, the shot noise is always uncorrelated between different tomographic bins, hence the Kronecker delta function.

Going from Eq. (14.32) to Eq. (14.45), we see that, once we take the two-point correlation function of the observed ellipticities, only the cosmic shear γ (or, equivalently, the convergence κ) contributes to the signal, but *all* sources of ellipticity that make up ϵ_{obs} contribute to shot noise. The latter are dominated by randomness of galaxy shapes ($\langle \epsilon^2 \rangle^{1/2} \simeq 0.2$), because the cosmic signal ($\langle \epsilon^2 \rangle^{1/2} \simeq 0.01$) is much smaller.

The statistical uncertainty in measuring the shear power spectrum is, by Wick's theorem (which we covered in Box 13.3 in Chapter 13)

$$\text{Cov}\left[C_{ij}^{\kappa}(\ell), C_{kl}^{\kappa}(\ell')\right] = \frac{\delta_{\ell\ell'}}{(2\ell+1)f_{\text{sky}}}\left[C_{ik}^{\kappa}(\ell)C_{jl}^{\kappa}(\ell) + C_{il}^{\kappa}(\ell)C_{jk}^{\kappa}(\ell)\right], \tag{14.46}$$

where f_{sky} is the fraction of sky area covered by the survey. The above expression, as usual, holds only for the Gaussian kappa-field, which is a pretty good approximation at large scales (and with no primordial non-Gaussianity, which appears to hold to an excellent approximation). At small scales, nonlinearities will introduce non-Gaussianities, and additional terms in Eq. (14.46) will appear. For the no-tomography case, Eqs. (14.45) and (14.46) show that the error in the observed power spectrum simplifies to a form familiar from our study of the CMB:

$$\sigma_{C^{\kappa}(\ell)} \equiv \text{Cov}\left[C^{\kappa}(\ell), C^{\kappa}(\ell')\right]^{1/2} = \sqrt{\frac{2\delta_{\ell\ell'}}{(2\ell+1)f_{\text{sky}}}}\left[P^{\kappa}(\ell) + \frac{\langle \epsilon_{\text{obs}}^2 \rangle}{\bar{n}}\right]. \tag{14.47}$$

Modern measurements of the tomographic power spectrum of weak-lensing shear are shown in Fig. 14.7. Note that the figure shows the real-space shear correlation functions $\xi_{\pm}^{ij}(\theta)$ (multiplied by θ for readability) for any two tomographic bins i and j. These real-space correlation functions are just harmonic transforms of the convergence power spectrum $P^{\kappa}(\ell)$, and are used because the weak-lensing

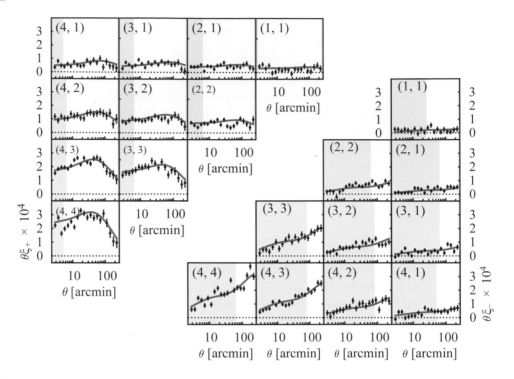

Tomographic shear measurements of the two-point correlation function of shear in real space, $\xi^{ij}_\pm(\theta)$, from the Year-3 data analysis of the Dark Energy Survey. The pair of numbers in ξ (written in the top left corner of each panel) refers to the pair of redshift bins from which the measurement is made (e.g., "2,4" means spatial correlation of galaxies in bin 2 to those in bin 4). The plots above the diagonal show $\theta\xi^{ij}_+(\theta)$ as a function of θ, while those below show $\theta\xi^{ij}_-(\theta)$. The curvy solid line shows the best-fit ΛCDM model. The grey regions show scales *not* used in the final parameter constraints due to concerns about modeling of the nonlinear clustering at these small scales. Adapted and reprinted with permission from Secco *et al.* (2022), copyright (2022) by the American Physical Society.

measurements are often performed in real space (in θ), rather than in multipole space (in ℓ). The functions are obtained from the convergence power spectrum via

$$\xi^{ij}_\pm(\theta) = \frac{1}{2\pi} \int_0^\infty P^\kappa_{ij}(\ell) J_{0,4}(\ell\theta)\ell d\ell, \tag{14.48}$$

where $J_{0,4}(x)$ is the Bessel function of zeroth and fourth order, respectively. Note the impressive agreement between the shear measurements and the best-fit ΛCDM model in Fig. 14.7, extending even to the scales that were conservatively not used in the cosmological analysis (shaded in gray).

14.4.4 Galaxy–Galaxy Lensing

In addition to measuring shapes of distant galaxies across the sky and correlating them, there are other ways to leverage weak-lensing observations to learn about the structure in the universe. One option is to measure the correlation of the *shear* of background galaxies with the *positions* of the foreground galaxies. Because the foreground object is a galaxy (and not, e.g., a cluster of galaxies), there is typically not enough mass to generate multiple images of the background objects. Rather, the shapes of background galaxies are sheared in the direction tangential to the radius-vector pointing from them to the foreground galaxy.

These measurements are commonly called "**galaxy–galaxy lensing**," which is surely one of the most confusing names in cosmology. The method should perhaps be called galaxy–galax*ies* lensing, as it correlates the position of one foreground galaxy with shapes of a number of background galaxies that are near it on the sky. Another better name is **galaxy–shear cross-correlation**, as galaxy positions are correlated with (other galaxies') shear.

The amount of shearing of background galaxies depends, as usual, on the lensing geometry as well as the total (baryonic plus dark-matter) mass of the lens. In this aspect galaxy–galaxy lensing is similar to weak lensing by large-scale structure discussed earlier. The key difference is that galaxy–galaxy lensing correlates the shears of the source galaxies not with each other as in weak lensing, but with known positions of foreground galaxies. Because the lenses are "guaranteed" massive galaxies along the line of sight, galaxy–galaxy lensing is also typically easier to detect than weak lensing by large-scale structure.

To connect the galaxy–galaxy lensing measurements to theory, we make use of the surface mass density contrast $\Delta\Sigma(R)$, which is proportional to the mass density integrated along the line of sight, and evaluated within radius R away from the lens galaxy:[3]

$$\Delta\Sigma(R) \equiv \overline{\Sigma}(<R) - \overline{\Sigma}(R) = \Sigma_{\mathrm{crit}} \times \gamma_t(R). \tag{14.49}$$

Here $\overline{\Sigma}(<R)$ is the mean surface density within radius R, $\overline{\Sigma}(R)$ is the azimuthally averaged surface density at R, γ_t is the tangential shear of background galaxies, and the critical surface density Σ_{crit} is a known function of the distances to the source and the lens. The surface density contrast $\Delta\Sigma(R)$ can be further inverted to obtain radial density profiles of dark-matter halos.

Galaxy–galaxy lensing measurements constrain the density profiles and bias of dark-matter halos, and the relation between their masses and luminosities. Ongoing surveys have started to use galaxy–galaxy lensing as a probe of dark matter and dark energy in its own right, as well as combining it with measurements of cosmic shear (shear–shear correlations) and galaxy clustering in a combined analysis –

---

[3] Radius R is defined in the plane of the galaxy, perpendicular to the line-of-sight. One typically measures the *angles* rather than physical radii in this plane, but cosmologists often prefer to use the measured redshift to the lens, along with a fiducial cosmological model, to convert to distance; $R = d_A(z)\theta$.

sometimes called the "**3×2-pt analysis**" (three two-point correlation functions: galaxy clustering, cosmic shear, galaxy–galaxy lensing) – which benefits from the complementarity of these probes.

14.4.5 Calibrating Masses of Galaxy Clusters with Weak Lensing

Another application of weak gravitational lensing is to help determine the mass of galaxy clusters. Clusters are typically detected in the optical, X-ray wavebands, or else via their SZ signatures[4] (see Chapter 13). In a classic cosmological test, one counts the number of clusters per solid angle and per redshift, above some minimal mass $M_{\min}$:

$$\frac{d^2 N}{d\Omega dz} = \frac{r^2(z)}{H(z)} \int_{M_{\min}}^{\infty} \frac{dn(M, z)}{dM} \, dM, \tag{14.50}$$

where $dV/(d\Omega dz) = r^2/H$ is the volume element and $n(z)$ is the number density. This test therefore requires the redshifts and masses of clusters. The former can typically be determined quite accurately from the redshifts of clusters' member galaxies. In contrast, cluster masses are notoriously difficult to pin down – they are typically inferred from the clusters' luminosities, and this determination is subject to significant systematic biases. This is where weak lensing comes in: by measuring tangential shear profiles of background galaxies near a foreground cluster, one can determine the mass of the cluster. While the statistical error in this method – currently about 30 percent error in each cluster's mass – is fairly substantial, its future prospects are good.

14.4.6 Bullet Cluster and Evidence for Dark Matter

Much important information about the dark matter and gas content of galaxy clusters can be inferred with combined lensing, X-ray, and optical observations. In 2006, this was demonstrated spectacularly with observations of the "Bullet" cluster. In this system, a smaller subcluster has passed through a larger cluster, and the shock from that interaction can be seen as heated gas – baryons – that glows in X-rays. The total mass distribution, on the other hand, is inferred from a combination of strong-lensing and weak-lensing measurements, and is clearly *offset* from the hot gas; see Fig. 14.8(a).

The Bullet cluster is often used as direct, visual evidence for the existence of dark matter. And justifiably so, as it provides a spectacular literal image of the system where the mass and the (X-ray) light clearly do not appreciably overlap. However, as far as the overall evidence for dark matter goes, the Bullet cluster's $\lesssim 10\sigma$ evidence is very solid but takes a back seat to the bumps and wiggles of

[4] The weak-lensing signal can in principle also be used to *detect* galaxy clusters. However, such detections are challenging because of the contamination with other mass along the line of sight to the cluster (or beyond the cluster). Hence, in most practical situations thus far, weak lensing is used to calibrate masses of clusters already detected by other means.

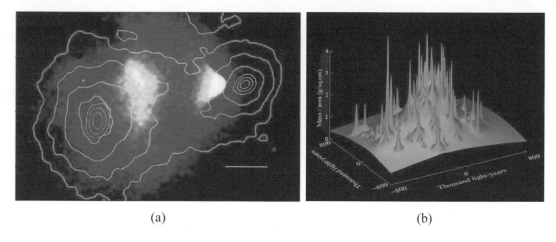

(a)　　　　　　　　　　　　　　　(b)

Fig. 14.8 (a) X-ray emission from the "Bullet" cluster of galaxies observed by ground and space telescopes. Colored features correspond to X-ray emission observed by Chandra Space Telescope, while the green contours correspond to the mass reconstruction from weak-lensing observations. Note that the X-ray features, which are dominated by baryons, are clearly separated from the green contours where most of the mass is located. Reprinted with permission from Clowe *et al.* (2006), © AAS. (b) Reconstruction of the density profile of galaxy cluster CL 0024+1654 at $z = 0.39$ using strong gravitational lensing alone. The density peaks correspond to individual galaxies in the cluster, while the broad central feature is largely due to dark matter. Credit: Tony Tyson; see also Tyson *et al.* (1998). A black and white version of this figure will appear in some formats. For the color version, please refer to the plate section.

the CMB angular power spectrum which would be off from the measurements by some $\sim 100\sigma$ in cosmological models with no dark matter; see the discussion around Eq. (11.19) in Chapter 11.

Bibliographical Notes

Gravitational lensing is covered in a number of reviews and several textbooks. Very clear introductions to the general-relativistic foundations of the mathematics of lensing are given in Carroll (2019) and Hartle (2021). Gravitational lensing as a whole is covered at a fairly technical level in Dodelson (2017) and Congdon and Keeton (2018). Weak lensing in particular is reviewed at a basic level in Hoekstra and Jain (2008) and Mandelbaum (2018), and in more mathematical detail in Bartelmann and Schneider (2001). A very nice popular book that focuses on lensing (calling it "Einstein's telescope") is the one by Gates (2009).

Problems

14.1 **Einstein angle.** Let us get some practice evaluating the Einstein angle for typical cosmological lenses.

(a) As a warm-up, evaluate the quantity $\sqrt{4GM_\odot/(1\,\text{Mpc})}$, which is dimensionless (in natural units).

(b) With that in hand, verify Eqs. (14.13) and (14.14).

14.2 **Peculiar velocity and gravitational lensing.** In this problem you will estimate how long it will take for peculiar velocities of the lens to make the lensed system "out of alignment" (so that we don't see lensing any more).

(a) Consider first the case where the lens is a cluster of galaxies, while the source is a distant galaxy; in that case the Einstein radius is (assuming for simplicity that the lens is located halfway to the source) given by Eq. (14.13):

$$\theta_E \approx (30\,\text{arcsec})\left(\frac{M}{10^{14}M_\odot}\right)^{1/2}\left(\frac{d}{1000\,\text{Mpc}}\right)^{-1/2},$$

where M is the mass of the (cluster) lens and d is the distance of the background galaxy. Assume the galaxy at a truly cosmological distance with $d = 5000\,\text{Mpc}$ and the cluster lens has mass $10^{15}M_\odot$. How long does it take for the peculiar velocity to move the lens out of alignment? To make the estimate, assume that the characteristic peculiar velocity of the cluster lens is $\sigma_{\text{cluster}} \approx 1000\text{km/s}$.

(b) Now consider the case where the lens is a galactic MACHO (MAssive Compact Halo Object – could be a blob or "planet" of dark matter), while the source is a star either in our galaxy or in the nearby Andromeda. Then the Einstein angle is given by Eq. (14.14):

$$\theta_E \approx (4 \times 10^{-4}\,\text{arcsec})\left(\frac{M}{1M_\odot}\right)^{1/2}\left(\frac{d}{50\,\text{kpc}}\right)^{-1/2},$$

where M is the mass of the (MACHO) lens and d is the distance of the background star, and assuming again that the lens is located halfway to the source. Assume that MACHOs are solar mass and reside in the Andromeda galaxy (M31), that is, $M \approx 1M_\odot$ and $d \approx 780\,\text{kpc}$. The characteristic velocity of the MACHO, however, is comparable to that of stars in a galaxy, so that $\sigma_{\text{MACHO}} \approx 200\text{km/s}$. How long does it take for the peculiar velocity to move the lens out of alignment?

14.3   **Singular isothermal sphere.** [Adapted from Sherry Suyu and Peter Schneider.] The singular isothermal sphere (SIS), though simplistic, is one of the best-known models for the mass distribution in a lens. Here, the density distribution is modeled with

$$\rho(r) = \frac{\sigma_v^2}{2\pi G r^2},$$

where r is the (3D) distance from the center of the lens, σ_v is the velocity dispersion, and G is Newton's gravitational constant. In what follows, assume all angles are 1D for simplicity rather than 2D (e.g., $\boldsymbol{\theta} \to \theta$). Then

(a) Integrate the density profile in the radial direction s to obtain the surface density Σ of an SIS lens:

$$\Sigma(R) = \int_{-\infty}^{\infty} \rho(\sqrt{R^2 + s^2})ds,$$

where R is the coordinate in the transverse direction, that is, in the plane of the sky.

(b) Find the convergence $\kappa(\theta)$. To do so, make use of Eq. (14.18), and the Poisson equation which relates the Laplacian of the 3D potential to density:

$$\nabla^2 \Phi = 4\pi G \rho(r).$$

You should get $\kappa(\theta) = C_1 \theta_E / |\theta|$, where C_1 is some numerical constant you need to determine, and where the Einstein angle is defined as

$$\theta_E \equiv 4\pi \sigma_v^2 \frac{d_{LS}}{d_S}.$$

(c) It turns out that, for a spherically symmetric lens, the reduced lensing angle in the second line of Eq. (14.19) simplifies to

$$\alpha(\theta) = \frac{1}{\pi\theta} \int_0^\theta \kappa(\theta') \, d^2\theta',$$

where $d^2\theta' = 2\pi\theta'd\theta'$. Use this equation, along with the expression for κ from part (b), to obtain the reduced lensing angle $\alpha(\theta)$ for the SIS.

(d) Now you have all the ingredients to solve the lens equation, $\beta = \theta - \alpha(\theta)$, for the SIS case. Sketch this lens equation by plotting θ on the x-axis, and two alternate quantities on the y-axis: $\alpha(\theta)$ that you found above and $\theta - \beta$ for a few values of β. From this sketch, deduce the condition for the source location β to create multiple images. [Your condition should feature the Einstein angle θ_E.]

14.4   **[Computational] Optical depth for lensing.** The probability of lensing of a given object in the universe, P, is given by the optical depth for lensing, τ, as $P = 1 - e^{-\tau} \simeq \tau$ (when $\tau \ll 1$ which will be the case here). The lensing optical depth depends on both the source and the lens redshift, as well as the mass of the lens.

Here, you will numerically estimate the lensing optical depth. From its definition, the optical depth is equal to the number density (of potential lenses) times the cross-section for lensing, times the distance between the observer and the source: $\tau \simeq n\sigma_{\text{lens}}d$. In more detail, the lensing optical depth is equal to

$$\tau = \int_0^{z_S} dz \, \frac{dD}{dz}(1+z)^3 \sigma_{\text{lens}}(z, M) \int_{M_{\text{min}}}^{M_{\text{max}}} \frac{dn}{d\ln M}(z, M) d\ln M,$$

where z_S is the source redshift, M_{min} and M_{max} are the minimal and maximal mass of the lenses considered, and D stands for the angular diameter distance. The term $(1+z)^3$ serves to convert the comoving number density $dn/d\ln M$ into the physical quantity, and thus match the distance D and the cross-section σ_{lens} which are also physical. To simplify the evaluation of the expression above, you can use the fitting function for the mass function from Evrard *et al.* (2014) evaluated at $z = 0.23$:

$$\frac{dn}{d\ln M} \simeq A_{\text{MF}} \, e^{-\beta_1\mu_1 - \beta_2\mu_2 - \beta_3\mu_3},$$

where $\mu \equiv \ln(M/M_P)$ with $M_P = 2 \times 10^{14} M_\odot$. The constants in this fitting function are $A_{\text{MF}} = 1.944 \times 10^{-6}$ (Mpc)$^{-3}$, $\beta_1 = 1.97$, $\beta_2 = 0.70$, and $\beta_3 = 0.40$. Even though we are utilizing this fitting function at a fixed redshift (here, $z = 0.23$), rather than varying z, it will still give us a useful estimate of τ. Finally, model the lensing cross-section as $\sigma_{\text{lens}} = \pi R_E^2$, where R_E is the Einstein radius from Eq. (14.12). Note that, with our redshift-independent approximation for the mass function, and with our definition of σ_{lens}, the redshift and mass integrals separate completely. Note also that the quantity $\sqrt{4GM_\odot/(1\,\text{Mpc})}$, evaluated in Problem 14.1 (a), could be useful here.

Evaluate the optical depth for a source galaxy at $z_S = 1$ and lenses above $M_{\text{min}} = 10^{14} M_\odot$. The result you find will give you a *very* rough idea for the probability of lensing of a source by galaxy clusters. [In reality, this optical depth is likely quite a bit smaller mainly because our modeling of the lensing cross-section was too simplistic.]

14.5 **[Computational] Weak-lensing angular power spectrum.** You will now numerically evaluate the angular power spectrum of convergence (nearly equivalently, shear), $P^\kappa(\ell)$.

(a) Evaluate and plot the shear power spectrum in the Limber approximation, which is given by Eq. (14.39):

$$P^\kappa(\ell) = \int_0^\infty dz \, \frac{W^2(z)}{r(z)^2 H(z)} P\left(\frac{\ell}{r(z)}, z\right).$$

Adopt the cosmological parameters of our fiducial model. As per usual convention, plot the quantity $\ell(\ell+1)P^\kappa(\ell)/(2\pi)$ vs. ℓ and make both axes uniform in the log. Assume a single-source population at $z_S = 1$ – so $n(z)$ of source galaxies is a delta function at this redshift. Assume that $P(k,z)$ is the linear power spectrum. While this is not a particularly good approximation at most scales of interest (see Fig. 14.6), it is much simpler than incorporating the full nonlinear matter power spectrum using one of the prescriptions.

For a smooth plot, you don't have to evaluate at every multipole, but at, say, a few tens of multipoles spread uniformly in $\log \ell$, in the range $10 \le \ell \le 10,000$. In a flat universe, $r(\chi) = \chi$, so that $r(\chi_S - \chi) = r(z_S) - r(z)$. Finally, think about what redshift you really need to integrate the above equation out to.

(b) Describe qualitatively how the angular power spectrum changes if you make the following parameter variations, one at a time:

- increase matter density Ω_M;
- increase amplitude of mass fluctuations A_s;
- increase spectral index n;
- increase the redshift of source galaxies z_S.

In each case say very briefly (in one sentence) why the change makes sense physically. You can then test your predicted trends numerically, re-plotting the convergence power spectrum with these changes implemented.

(c) Measuring shear is hard; the tiny cosmological-lensing signal is swamped by systematic errors, including "rounding" of the galaxy shapes by the atmosphere and local tidal gravitational (and not lensing) interactions between galaxies. But weak-lensing shear measurements offer a fundamental advantage over the (much easier to perform!) measurements of galaxy clustering in terms of its power to probe cosmology, specifically dark matter, dark energy, and so on. Briefly, what is this key advantage?

14.6   **Deriving the weak-lensing convergence power spectrum formula.**
Derive Eq. (14.39), helped with some hints along the way. First, assume a flat universe so that you don't need to worry about any differences between χ and r (we will adopt r). Then:

- Start with the combination of Eqs. (14.34) and (14.36):

$$\kappa_{\ell m} = \int dr \, d\hat{\mathbf{n}} \, W(r)\delta(r,\hat{\mathbf{n}})Y^*_{\ell m}(\hat{\mathbf{n}}).$$

- Write down the correlation $\langle \kappa_{\ell m} \kappa^*_{\ell' m'} \rangle$. In it, implement the relation $\langle \delta \delta^* \rangle \equiv \xi(r) = 1/(2\pi)^3 \int P(k)e^{i\mathbf{k}\mathbf{x}}d^3k$. You will get a long, unwieldy looking expression.

- Break up into transverse and parallel to line-of-sight modes, à la $e^{i\mathbf{k}\mathbf{x}} = e^{ik_\parallel (x-x')}e^{ik_\perp |\hat{\mathbf{n}} - \hat{\mathbf{n}}'|r}$ (see the figure below). Do the $k_\parallel$ integral. The result will be proportional to a Dirac delta function.

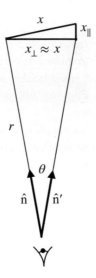

- Adopt the Limber approximation, where $k_\perp r \simeq kr \equiv \ell \gg 1$, and argue that $P(k)$ can be pulled out of the $k_\perp$ integral. The integral can then be done analytically.

- Use orthonormality of spherical harmonics.

And you should be able to get the desired expression!

Appendix A: Natural Units

Physicists are good at saving trees: they often take a massive shortcut in dropping several fundamental constants of nature from their equations, thus surely decreasing paper usage around the world. In other words, they use *natural units*. This fundamentally corresponds to a simple change in the choice of units — for example, speed is no longer in units of meters per second, but rather in units of the speed of light c.

Natural units were covered in Chapter 1 (Sec. 1.7) as well as Box 1.3. Here we explain them in a little more detail, and give some more examples in order to build familiarity with this system.

Most of us are familiar with using natural units in the sense of expressing a physical quantity in a different set of units. For example, one expresses the height of a building in terms of how many stories it goes up, with each story (floor) counting for about five meters. Thus, saying that the building is 10 stories high is equivalent to saying that it is 50 meters high. An amusing further example is the length of Harvard Bridge (near MIT, in Cambridge, MA) which, according to markings on it, spans[1] "364.4 smoots": the length of the bridge is quoted in units of the height of a certain Oliver R. Smoot, who was an MIT student whose body his fraternity brothers used to measure the length of the bridge (in 1958).

Natural units go well beyond distance unit conversions. When using natural units, we set the following familiar physical quantities to unity:

$$c = 1 \qquad \text{(speed of light)} \tag{A.1}$$

$$\hbar = 1 \qquad \text{(reduced Planck constant)} \tag{A.2}$$

$$k_B = 1 \qquad \text{(Boltzmann constant)} \tag{A.3}$$

It turns out that, with these choices, we can dramatically simplify calculations in typical problems in cosmology.

Let us start with distance and time. Because a distance L and time t are related by the speed of light which is now simply one, we have $[L] = [t]$, where the square parentheses stand for "units of". Quantitatively, the conversion is of course

$$\frac{L}{1\,\text{m}} = 2.9979 \times 10^8 \frac{t}{1\,\text{s}}. \tag{A.4}$$

For example, the closest star to our Sun, Proxima Centauri, is about 4×10^{16} m away from us, which converts to about 1.33×10^8 s, or 4.22 years. We could thus justifiably

---

[1] Actually "364.4 smoots $\pm$ 1 ear"; it includes the measurement error!

say that this star is 4.22 years away though, in this specific case, astronomers usually refer to *light-years* rather than years as convenient units of distance.

As explained in Chapter 1, energy E and mass m are not only mutually equivalent (in natural units), but can be related to distance/time. Specifically, the correspondence is

$$[E] = [m] = [T] = [L]^{-1} = [t]^{-1}. \qquad (A.5)$$

where T stands for temperature.

A particularly relevant example for cosmology is that of the Hubble constant, $H_0 = 100h\,\mathrm{km/s/Mpc}$, with $h \simeq 0.7$. The Hubble distance is (Eq. (2.29))

$$\frac{1}{H_0} \equiv \frac{c}{H_0} = 2997.9\,h^{-1}\mathrm{Mpc} = 9.251 \times 10^{25}\,h^{-1}\,\mathrm{m}$$

$$\equiv 3.086 \times 10^{17}\,h^{-1}\,\mathrm{s} = 9.784 \times 10^9\,h^{-1}\,\mathrm{yr}, \qquad (A.6)$$

where the second row lists the Hubble *time*, which is equivalent to Hubble distance in natural units. Because $1\,\mathrm{GeV}^{-1} = 1.98 \times 10^{-16}\,\mathrm{m}$ (see Box 1.3), one can also express the Hubble constant in mass/energy units

$$H_0 = 1.081 \times 10^{-26} h\,\mathrm{m}^{-1}$$

$$= 2.133 \times 10^{-33}\,h\,\mathrm{eV} \qquad (A.7)$$

$$\equiv 2.475 \times 10^{-29}\,h\,\mathrm{K},$$

where the last equality follows from the conversion between energy and temperature, $1\,\mathrm{GeV} = 1.21 \times 10^{13}\,\mathrm{K}$ (see again Box 1.3). The Hubble parameter can then be used very flexibly. Consider, for example, the Klein–Gordon equation for the evolution of inflationary scalar field, Eq. (8.32)

$$\ddot{\phi} + 3H\dot{\phi} + \frac{dV}{d\phi} = 0, \qquad (A.8)$$

where ϕ is the field value, $V(\phi)$ is the potential, and dots are the derivatives with respect to time. Given that the field has units of mass, and in fact that all of the following quantities have units of mass

$$[\phi] = [V]^{1/4} = [H] = [t]^{-1} \equiv [m], \qquad (A.9)$$

it follows that each term in the ϕ evolution equation (A.8) scales as mass cubed. Similarly, adopting Eq. (3.11)

$$\frac{d}{dt} = H\frac{d}{d\ln a}, \qquad (A.10)$$

we can rewrite the ϕ evolution equation as (Problem 8.9)

$$\phi'' + \left(3 + \frac{H'}{H}\right)\phi' + \frac{1}{H^2}\frac{dV}{d\phi} = 0, \qquad (A.11)$$

where $' \equiv d/d\phi$. This again contains terms that all manifestly have the same dimensionality — linear in mass. Further, if one works with the field value in units of

Planck mass (for example) — so, defining $\tilde{\phi} \equiv \phi/m_{\rm Pl}$ — each term in Eq. (A.11) becomes dimensionless, making it particularly convenient for numerical implementations. All of this gymnastics with units and easy dimensionality checks is enabled by our use of natural units.

As a final example, let us consider Big Bang nucleosynthesis, the subject of Chapter 7, and let us assume we know next to nothing about its physics. We could still estimate, *very* roughly, the time at which the BBN takes place as follows. We might guess that the temperature (or energy) at which the BBN takes place is comparable to the typical binding energy per nucleon of nuclei, or $T_{\rm BBN} \sim 1\,{\rm MeV}$ (see Fig. 7.1). What time after the Big Bang does this correspond to? From the Friedmann I equation, it follows that

$$H^2 \sim \frac{\rho}{m_{\rm Pl}{}^2} \simeq \frac{T^4}{m_{\rm Pl}{}^2} \tag{A.12}$$

(assuming the radiation energy density that goes as T^4), from which it follows that

$$H_{\rm BBN} \sim \frac{T_{\rm BBN}^2}{m_{\rm Pl}} \sim \frac{(1\,{\rm MeV})^2}{10^{19}\,{\rm GeV}} \simeq 10^{-16}\,{\rm eV}. \tag{A.13}$$

It is also true that the Hubble parameter, having units of inverse time, is *roughly* equal to the inverse age of the universe[2]. Then

$$t_{\rm BBN} \simeq H_{\rm BBN}^{-1} \simeq 10^{16}\,{\rm eV}^{-1} \simeq 1\,{\rm s}. \tag{A.14}$$

[A more accurate estimate, laid out in Chapter 7, gives the BBN temperature around $0.1\,{\rm MeV}$, and the corresponding time of around 100 seconds after the Big Bang.] Again, the quick conversions between different physical quantities are enabled by the use of natural-unit reasoning.

---

[2] This is discussed around Eq. (2.30) while a much more precise relation between time and temperature in the BBN era is given in Eq. (4.37).

Appendix B: Useful Constants

Fundamental constants (in MKS units)

Reduced Planck constant $\hbar$ $\qquad\qquad$ $6.6261 \times 10^{-34}\,\mathrm{m^2\,kg/\,s}$

Boltzmann constant k_B $\qquad\qquad$ $1.3807 \times 10^{-23}\,\mathrm{m^2\,kg\,J/\,s^{-2}\,K^{-1}}$

Speed of light c $\qquad\qquad$ $2.9979 \times 10^8\,\mathrm{m/s}$

Newton's constant G $\qquad\qquad$ $6.67 \times 10^{-11}\,\mathrm{m^3/kg/m^2}$

Planck mass $m_{\mathrm{Pl}} \equiv \sqrt{\dfrac{\hbar c}{G}}$ $\qquad\qquad$ $1.221 \times 10^{19}\,\mathrm{GeV}$

Planck length $l_{\mathrm{Pl}} \equiv \sqrt{\dfrac{\hbar G}{c^3}}$ $\qquad\qquad$ $1.616 \times 10^{-35}\,\mathrm{m}$

proton mass m_p $\qquad\qquad$ $938.3\,\mathrm{MeV}$

neutron mass m_n $\qquad\qquad$ $939.6\,\mathrm{MeV}$

electron mass m_e $\qquad\qquad$ $511\,\mathrm{MeV}$

Sun's mass $1 M_\odot$ $\qquad\qquad$ $1.989 \times 10^{30}\,\mathrm{kg}$

Cosmological quantities
[with $h \equiv H_0/(100\,\mathrm{km/s/Mpc})$]

H_0^{-1} (Hubble time) $\qquad\qquad$ $2997.9\,h^{-1}\mathrm{Mpc}$

H_0^{-1} $\qquad\qquad$ $9.251 \times 10^{25}\,h^{-1}\,\mathrm{m}$

H_0^{-1} $\qquad\qquad$ $3.086 \times 10^{17}\,h^{-1}\,\mathrm{s}$

H_0^{-1} $\qquad\qquad$ $9.784 \times 10^9\,h^{-1}\,\mathrm{yr}$

H_0 (Hubble constant) $\qquad\qquad$ $2.133 \times 10^{-33}\,h\,\mathrm{eV}$

H_0 $\qquad\qquad$ $2.475 \times 10^{-29}\,h\,\mathrm{K}$

t_0 (age of the universe) $\qquad\qquad$ $4.352 \times 10^{17} \times \left(\dfrac{t_0}{13.8\,\mathrm{Gyr}}\right)\,\mathrm{s}$

t_0	$1.304 \times 10^{29} \times \left(\dfrac{t_0}{13.8\,\mathrm{Gyr}} \right)$ m
t_0	$4.228 \times \left(\dfrac{t_0}{13.8\,\mathrm{Gyr}} \right)$ Gpc
$\rho_{\mathrm{crit},0}$ (critical density today)	$8.096 \times 10^{-11}\, h^2\,\mathrm{eV}^4$
$\rho_{\mathrm{crit},0}$	$1.877 \times 10^{-26}\, h^2\,\mathrm{kg/m}^3$
$\rho_{\mathrm{crit},0}$	$1.053 \times 10^{10}\, h^2\,\mathrm{eV/m}^3$
$\rho_{\mathrm{crit},0}$	$2.775 \times 10^{11}\, h^2\, M_\odot\,\mathrm{Mpc}^{-3}$
T_0 (temperature today)	2.349×10^{-4} eV
$n_{\gamma,0}$ (photon density today)	4.107×10^8 m^{-3}
$n_{B,0}$ (baryon density today)	$0.252\,(\Omega_B h^2/0.022396)$ m^{-3}

Unit conversions

1 K	$436.96\,\mathrm{m}^{-1}$
1 K	1.536×10^{-40} kg
1 K	8.619×10^{-5} eV
1 kg	5.612×10^{35} eV
1 eV	$5.067 \times 10^6\,\mathrm{m}^{-1}$
1 eV	$1.519 \times 10^{15}\,\mathrm{s}^{-1}$
1 eV	1.782×10^{-36} kg
1 eV^4	$2.3192 \times 10^{-16}\,\mathrm{kg/m}^3$
1 eV^4	$3.428 \times 10^{21}\, M_\odot\,\mathrm{Mpc}^{-3}$
1 Mpc	$1.563 \times 10^{29}\,\mathrm{eV}^{-1}$
1 m	3.240×10^{-23} Mpc
1 s	9.714×10^{-15} Mpc
1 AU	1.496×10^{11} m
1 pc	3.086×10^{16} m

C   Appendix C: Fundamental Particles

The table below shows particles of the Standard Model of particle physics. The columns show: particle type; particle symbol; its mass; its spin; and its relativistic factor g (see Chapter 4). All masses are given to three significant digits, except for some quarks whose masses are less precisely known.

Table C.1

Type	Particle(s)	Mass	Spin	$g_{\text{particles}} \cdot g_{\text{spin}} = g$
quarks	$t, \bar{t}$	173 GeV	1/2	$2 \cdot 2 \, (\cdot 3 \, \text{colors}) = 12$
	$b, \bar{b}$	4.2 GeV	1/2	12
	$c, \bar{c}$	1.3 GeV	1/2	12
	$s, \bar{s}$	93 MeV	1/2	12
	$d, \bar{d}$	5 MeV	1/2	12
	$u, \bar{u}$	2 MeV	1/2	12
gluons	$g_i; \; i = \{1, \ldots, 8\}$	0	1	$8 \cdot 2 = 16$
leptons	$\tau^{\pm}$	1.78 GeV	1/2	$2 \cdot 2 = 4$
	$\mu^{\pm}$	106 MeV	1/2	4
	$e^{\pm}$	511 keV	1/2	4
	$\nu_\tau, \bar{\nu}_\tau$	< 0.1 eV	1/2	$2 \cdot 1 = 2$
	$\nu_\mu, \bar{\nu}_\mu$	< 0.1 eV	1/2	2
	$\nu_e, \bar{\nu}_e$	< 0.1 eV	1/2	2
gauge bosons	$W^{\pm}$	80.4 GeV	1	$2 \cdot 3 = 6$
	Z	91.2 GeV	1	$1 \cdot 3 = 3$
photon	γ	0	1	$1 \cdot 2 = 2$
Higgs	H	125 GeV	0	$1 \cdot 1 = 1$

D   Appendix D: Symbol Definitions

Symbol	Meaning	Introduced in
a	Scale factor	Eq. (2.5)
$a_{\ell m}$	Harmonic coefficient of temperature fluctuation	Eq. (13.64)
A	Area	Eqs. (2.11), (3.66)
A_{ij}	Distortion matrix (Sec. 14.3.2 only)	Eq. (14.20)
α_s	Running of the scalar spectral index	Eq. (8.62)
b	Galaxy bias	Eq. (9.93)
B	B-mode of polarization (Sec. 13.5 only)	Eq. (13.83)
B	Binding energy (Chapter 7 only)	Box 7.1
c_s	Speed of sound	Eq. (9.20)
$\mathbf{C}$	Covariance matrix	Eq. (10.14)
$C(\theta)$	Angular two-point correlation function	Eq. (13.10)
C_ℓ	Harmonic two-point correlation function	Eq. (13.10)
d_L	Luminosity distance	Eq. (3.65)
d_A	Angular diameter distance	Eq. (3.70)
d_p	Proper distance	Eq. (2.24)
d_H	Hubble distance	Eq. (2.27)
$D(a)$	Linear growth function	Eq. (9.46)
$D \equiv \mathbf{d}$	Data (Chapter 10 only)	Eq. (10.18)
D	Convective derivative (Sec. 9.3.1 only)	Eq. (9.14)
δ	Matter overdensity	Eq. (9.1)
δ_c	Critical overdensity for collapse	Eq. (9.86)
$\Delta^2(k)$	Dimensionless matter power spectrum	Eq. (9.60)
$\Delta_s^2(k)$	Scalar (curvature) power spectrum	Eq. (8.54)
$\Delta_t^2(k)$	Tensor power spectrum	Eq. (8.59)
E	Energy	N/A
E	E-mode of polarization (Sec. 13.5 only)	Eq. (13.83)
ϵ	First slow-roll parameter	Eq. (8.39)
ϵ	Ellipticity (Chapter 14 only)	Eq. (14.32)

η	Second slow-roll parameter	Eq. (8.41)
η	Baryon-to-photon ratio	Eq. (6.19)
η	Conformal time (Sec. 2.5.3 only)	Eq. (2.32)
$f(\mathbf{x}, \mathbf{p}, t)$	Phase-space distribution function	Eqs. (4.2), (6.1)
f	Flux	Eqs. (3.65), (12.1)
f_{NL}	Non-Gaussianity parameter	Eq. (13.88)
F_{ij}	Fisher matrix	Eq. (10.37)
$g(a)$	Linear growth suppression factor	Eq. (9.46)
g	Multiplicity factor of states	Eq. (4.1)
g_*	Effective number of rel. degrees of freedom	Eq. (4.27)
g_{*S}	Similar to g_*; see definition	Eq. (4.33)
$\gamma_{1,2}$	Shear (Chapter 14 only)	Eq. (14.26)
Γ	Annihilation rate	Eq. (4.15)
H	Hubble parameter	Eq. (2.7)
H_0	Hubble constant	Eq. (2.3)
k	Wavenumber	Eq. (9.47)
κ	Dimensionless curvature	Eq. (2.14)
κ	Convergence (Chapter 14 only)	Eq. (14.18)
ℓ	Multipole	Eqs. (13.10), (13.65)
L	Luminosity	Eqs. (3.65), (12.1)
$\mathcal{L}$	Likelihood	Eq. (10.18)
m, M	Mass	N/A
m	Apparent magnitude (Chapter 12 and Sec. 10.6 only)	Eq. (12.2)
M	Model (Chapter 10 only)	Eq. (10.18)
M_{ij}	Magnification matrix (Sec. 14.3.2 only)	Eq. (14.23)
μ	Chemical potential (Chapter 4 only)	Eq. (4.9)
μ	Mean (Chapter 10 only)	Eq. (10.5)
μ	Magnification (Chapter 14 only)	Eq. (14.30)
n	Number density	N/A
n_s	Scalar spectral index	Eqs. (8.60), (9.53)
n_t	Tensor spectral index	Eq. (8.64)
N	Number (dimensionless)	N/A
N_{eff}	Effective number of relativistic species	Eq. (5.26)
$\mathcal{N}$	Gaussian normal distribution	Eq. (10.13)
ν	Frequency	N/A

ν_c	Peak height corresponding to δ_c (Sec. 9.6 only)	Eq. (9.88)
$\xi(r)$	Two-point correlation function	Eq. (9.8)
Ω	Solid angle	N/A
Ω_i	Density relative to critical of component i	Eq. (2.50)
Ω_k	"Omega in curvature"	Eq. (3.57)
P	Pressure	Eq. (3.1)
P	Probability (Chapter 10 only)	Eq. (10.1)
$P(k)$	Matter power spectrum	Eq. (9.52)
$P^\kappa(\ell)$	Convergence power spectrum (Chapter 14 only)	Eq. (14.39)
$\mathbf{p}$	Momentum	Eq. (4.1)
$\mathbf{p} = \{p_j\}$	Cosmological parameters (Chapter 10 only)	Eq. (10.18)
q_0	Deceleration parameter	Eq. (3.45)
Q	Neutron–proton mass difference	Eq. (7.3)
r	Comoving distance	Eq. (3.61)
r	Tensor-to-scalar ratio	Eq. (8.67)
r_s	Sound horizon	Eq. (13.52)
R	Rate of interactions (Sec. 11.4 only)	Eq. (11.30)
R	Baryon-to-photon perturbation ratio (Chapter 13 only)	Eq. (13.51)
R_0	Radius of curvature	Eqs. (2.45), (3.58)
$\mathcal{R}$	Curvature perturbation	Eq. (8.50)
ρ	Energy density	Eq. (1.4)
ρ_{crit}	Critical energy density	Eq. (2.47)
s	Entropy density	Eq. (4.32)
S	Entropy	Eqs. (4.8), (9.20)
σ	Rms amplitude of linear mass fluctuations	Eq. (9.83)
σ_8	The above evaluated at $z = 0$, $r = 8\,h^{-1}\mathrm{Mpc}$	Eq. (9.84)
σ	Standard deviation (Chapter 10 only)	Eq. (10.9)
t	Age of the universe (or just time)	Eq. (3.43)
T	Temperature	N/A
$T(k)$	Transfer function	Eqs. (9.61), (9.63)
τ	Optical depth	Eq. (13.38)
τ_n	Neutron lifetime (Chapter 7 only)	Eq. (7.4)
Θ	Temperature fluctuation (Sec. 13.3.3 only)	Sec. 13.3.3
v	Velocity	N/A
V	Volume	N/A

$V(\phi)$	Potential of scalar field	Eq. (8.30)
w	Equation of state parameter	Eq. (3.9)
$w(\theta)$	Angular two-point correlation function	Eq. (9.13)
x_e	Ionization fraction	Eq. (13.24)
X	Random variable (Chapter 10 only)	Eq. (10.1)
y	SZ y-parameter (Sec. 13.6 only)	Eq. (13.86)
Y	Number-density-to-entropy ratio	Eqs. (4.34), (6.10)
Y_p	Helium mass fraction	Eq. (7.21)
z	Redshift	Eq. (2.1)
z_{eq}	Redshift of matter–radiation equality	Eq. (3.44)
$\zeta(r, s, |\mathbf{r}-\mathbf{s}|)$	Three-point correlation function	Eq. (9.11)
χ	Radial coordinate	Eq. (2.19)
Φ	Gravitational (metric) potential	Eq. (13.39)
Ψ	"The other" metric potential	Eq. (13.39)
ψ	Projected gravitational potential (Chapter 14 only)	Eq. (14.16)

References

Abbott, T. M. C., Abdalla, F. B., Alarcon, A., *et al.* [DES Collaboration]. 2018. Dark Energy Survey year 1 results: Cosmological constraints from galaxy clustering and weak lensing. *Phys. Rev. D*, **98**(4), 043526.

Abbott, T. M. C., Aguena, M., Alarcon, A., *et al.* [DES Collaboration]. 2022. Dark Energy Survey year 3 results: Cosmological constraints from galaxy clustering and weak lensing. *Phys. Rev. D*, **105**(2), 023520.

Ade, P. A. R., Aikin, R. W., Barkats, D., *et al.* [BICEP 2 Collaboration]. 2014a. Detection of *B*-mode polarization at degree angular scales by BICEP2. *Phys. Rev. Lett.*, **112**(24), 241101.

Ade, P. A. R., Aghanim, N., Alves, M. I. R., *et al.* [Planck Collaboration]. 2014b. Planck 2013 results. I. Overview of products and scientific results. *Astron. Astrophys.*, **571**, A1.

Ade, P. A. R., Ahmed, Z., Amiri, M., *et al.* [BICEP/Keck Collaboration]. 2021. Improved constraints on primordial gravitational waves using Planck, WMAP, and BICEP/Keck observations through the 2018 observing season. *Phys. Rev. Lett.*, **127**(15), 151301.

Aghanim, N., Akrami, Y., Arroja, F., *et al.* [Planck Collaboration]. 2020a. Planck 2018 results. I. Overview and the cosmological legacy of Planck. *Astron. Astrophys.*, **641**, A1.

Aghanim, N., Akrami, Y., Ashdown, M., *et al.* [Planck Collaboration]. 2020b. Planck 2018 results. VI. Cosmological parameters. *Astron. Astrophys.*, **641**, A6. [Erratum: *Astron. Astrophys.*, **652**, C4 (2021).]

Akrami, Y., Ashdown, M., Aumont, J., *et al.* [Planck Collaboration]. 2020a. Planck 2018 results. IV. Diffuse component separation. *Astron. Astrophys.*, **641**, A4.

Akrami, Y., Arroja, F., Ashdown, M., *et al.* [Planck Collaboration]. 2020b. Planck 2018 results. X. Constraints on inflation. *Astron. Astrophys.*, **641**, A10.

Albrecht, A., and Steinhardt, P. J. 1982. Cosmology for grand unified theories with radiatively induced symmetry breaking. *Phys. Rev. Lett.*, **48**, 1220–1223.

Alcock, C., and Paczynski, B. 1979. An evolution free test for non-zero cosmological constant. *Nature*, **281**, 358–359.

Alcock, C., Akerlof, C. W., Allsman, R. A., *et al.* 1993. Possible gravitational microlensing of a star in the Large Magellanic Cloud. *Nature*, **365**, 621–623.

Ali-Haimoud, Y., and Hirata, C. M. 2011. HyRec: A fast and highly accurate primordial hydrogen and helium recombination code. *Phys. Rev. D*, **83**, 043513.

Allen, S. W., Evrard, A. E., and Mantz, A. B. 2011. Cosmological parameters from observations of galaxy clusters. *Ann. Rev. Astron. Astrophys.*, **49**, 409–470.

Arbey, A., Auffinger, J., Hickerson, K. P., and Jenssen, E. S. 2020. AlterBBN v2: A public code for calculating Big-Bang nucleosynthesis constraints in alternative cosmologies. *Comput. Phys. Commun.*, **248**, 106982.

Bahcall, N. A., and Soneira, R. M. 1983. The spatial correlation function of rich clusters of galaxies. *Astrophys. J.*, **270**(July), 20–38.

Bardeen, J. M. 1980. Gauge invariant cosmological perturbations. *Phys. Rev. D*, **22**, 1882–1905.

Bardeen, J. M., Bond, J. R., Kaiser, N., and Szalay, A. S. 1986. The statistics of peaks of Gaussian random fields. *Astrophys. J.*, **304**, 15–61.

Bartelmann, M., and Schneider, P. 2001. Weak gravitational lensing. *Phys. Rept.*, **340**, 291–472.

Bassett, B. A., and Hlozek, R. 2009. Baryon acoustic oscillations. arXiv:0910.5224.

Baumann, D. 2011. Inflation. In *Elementary Particle Physics: Physics of the Large and the Small*, pp. 523–686. Boulder, CO: Theoretical Advanced Study Institute.

Baumann, D. 2013. *Cosmology Lecture Notes*. Available at `http://cosmology.amsterdam/education/cosmology/`. [accessed 1 June 2021].

Bennett, C. L., Banday, A., Gorski, K. M., *et al.* 1996. Four-year COBE DMR cosmic microwave background observations: Maps and basic results. *Astrophys. J. Lett.*, **464**, L1–L4.

Bennett, C. L., Larson, D., Weiland, J. L., *et al.* 2013. Nine-year Wilkinson Microwave Anisotropy Probe (WMAP) observations: Final maps and results. *Astrophys. J. Suppl.*, **208**, 20.

Bennett, C. L., Turner, M. S., and White, M. 1997. The cosmic Rosetta Stone. *Phys. Today*, **50**(11), 32–38.

Bernal, N., Heikinheimo, M., Tenkanen, T., Tuominen, K., and Vaskonen, V. 2017. The dawn of FIMP dark matter: A review of models and constraints. *Int. J. Mod. Phys. A*, **32**(27), 1730023.

Bertone, G., and Hooper, D. 2018. History of dark matter. *Rev. Mod. Phys.*, **90**(4), 045002.

Bertschinger, E. 1993. Cosmological dynamics: Course 1. In *Les Houches Summer School on Cosmology and Large Scale Structure* (Session 60), pp. 273–348.

Blas, D., Lesgourgues, J., and Tram, T. 2011. The Cosmic Linear Anisotropy Solving System (CLASS) II: Approximation schemes. *JCAP*, **07**, 034.

Burles, S., Nollett, K. M., and Turner, M. S. 2000. Deuterium and Big Bang nucleosynthesis. *Nucl. Phys. A*, **663**, 861c–864c.

Caldwell, R. R., Kamionkowski, M., and Weinberg, N. N. 2003. Phantom energy and cosmic doomsday. *Phys. Rev. Lett.*, **91**, 071301.

Carroll, S. M. 2001. The cosmological constant. *Living Rev. Rel.*, **4**, 1.

Carroll, S. M. 2019. *Spacetime and Geometry*. Cambridge: Cambridge University Press.

Carroll, S. M., Press, W. H., and Turner, E. L. 1992. The cosmological constant. *Ann. Rev. Astron. Astrophys.*, **30**, 499–542.

Clowe, D., Bradac, M., Gonzalez, A. H., *et al.* 2006. A direct empirical proof of the existence of dark matter. *Astrophys. J. Lett.*, **648**, L109–L113.

Cole, S., Percival, W. J., Peacock, J. A., *et al.* 2005. The 2dF Galaxy Redshift Survey: Power-spectrum analysis of the final dataset and cosmological implications. *Mon. Not. Roy. Astron. Soc.*, **362**, 505–534.

Congdon, A. B, and Keeton, C. R. 2018. *Principles of Gravitational Lensing*. Cham: Springer Nature.

Cooke, R. J., Pettini, M., and Steidel, C. C. 2018. One percent determination of the primordial deuterium abundance. *Astrophys. J.*, **855**(2), 102.

Copeland, E. J., Sami, M., and Tsujikawa, S. 2006. Dynamics of dark energy. *Int. J. Mod. Phys. D*, **15**, 1753–1936.

Copi, C. J., Huterer, D., Schwarz, D. J., and Starkman, G. D. 2015. Lack of large-angle TT correlations persists in WMAP and Planck. *Mon. Not. Roy. Astron. Soc.*, **451**(3), 2978–2985.

Cyburt, R. H., Fields, B. D., Olive, K. A., and Yeh, T.-H. 2016. Big Bang nucleosynthesis: 2015. *Rev. Mod. Phys.*, **88**, 015004.

Dalal, N., Dore, O., Huterer, D., and Shirokov, A. 2008. Imprints of primordial non-Gaussianities on large-scale structure: Scale-dependent bias and abundance of virialized objects. *Phys. Rev. D*, **77**, 123514.

Davis, M., Efstathiou, G., Frenk, C. S., and White, S. D. M. 1992. The end of cold dark matter? *Nature*, **356**(6369), 489–494.

Davis, T. M., and Lineweaver, C. H. 2004. Expanding confusion: Common misconceptions of cosmological horizons and the superluminal expansion of the universe. *Publ. Astron. Soc. Austral.*, **21**, 97.

de Lapparent, V., Geller, M. J., and Huchra, J. P. 1986. A slice of the universe. *Astrophys. J. Lett.*, **302**, L1–L5.

Dodelson, S. 2017. *Gravitational Lensing*. Cambridge: Cambridge University Press.

Dodelson, S., and Schmidt, F. 2020. *Modern Cosmology*. Amsterdam: Academic Press.

Doran, M., and Robbers, G. 2006. Early dark energy cosmologies. *JCAP*, **06**, 026.

Duncan, T., and Tyler, C. 2008. *Your Cosmic Context*. Upper Saddle River, NJ: Pearson.

Durrer, R. 2020. *The Cosmic Microwave Background*. Cambridge: Cambridge University Press.

Efstathiou, G. 2003. The statistical significance of the low CMB multipoles. *Mon. Not. Roy. Astron. Soc.*, **346**, L26.

Eisenstein, D. J., Zehavi, I., Hogg, D. W., *et al.* 2005. Detection of the baryon acoustic peak in the large-scale correlation function of SDSS luminous red galaxies. *Astrophys. J.*, **633**, 560–574.

Evrard, A. E., Arnault, P., Huterer, D., and Farahi, A. 2014. A model for multiproperty galaxy cluster statistics. *Mon. Not. Roy. Astron. Soc.*, **441**(4), 3562–3569.

Feldman, G. J., and Cousins, R. D. 1998. A unified approach to the classical statistical analysis of small signals. *Phys. Rev. D*, **57**, 3873–3889.

Ferreira, E. G. M. 2021. Ultra-light dark matter. *Astron. Astrophys. Rev.*, **29**(1), 7.

Frieman, J., Turner, M., and Huterer, D. 2008. Dark energy and the accelerating universe. *Ann. Rev. Astron. Astrophys.*, **46**, 385–432.

Gates, E. 2009. *Einstein's Telescope.* New York: WW Norton.

Gil-Marín, H., Percival, W. J., Brownstein, J. R., *et al.* 2016. The clustering of galaxies in the SDSS-III Baryon Oscillation Spectroscopic Survey: RSD measurement from the LOS-dependent power spectrum of DR12 BOSS galaxies. *Mon. Not. Roy. Astron. Soc.,* **460**(4), 4188–4209.

Górski, K. M., Hivon, E., Banday, A. J., *et al.* 2005. HEALPix – a framework for high resolution discretization, and fast analysis of data distributed on the sphere. *Astrophys. J.,* **622**, 759–771.

Guth, A H. 1981. The inflationary universe: A possible solution to the horizon and flatness problems. *Phys. Rev. D,* **23**, 347–356.

Guth, A. H. 1998. *The Inflationary Universe.* Boulder, CO: Perseus Books.

Guth, A. H. 2013. *8.286 The Early Universe.* MIT OpenCourseWare. `https://ocw.mit.edu`.

Hand, N., Addison, G. E., Auborg, E., *et al.* 2012. Evidence of galaxy cluster motions with the kinematic Sunyaev–Zel'dovich effect. *Phys. Rev. Lett.,* **109**, 041101.

Hartle, J. B. 2021. *Gravity.* Cambridge: Cambridge University Press.

Hauser, M. G., and Peebles, P. J. E. 1973. Statistical analysis of catalogs of extragalactic objects. II. The Abell catalog of rich clusters. *Astrophys. J.,* **185**(Nov.), 757–786.

Heath, D. J. 1977. The growth of density perturbations in zero pressure Friedmann–Lemaître universes. *Mon. Not. R. Astron. Soc.,* **179**(May), 351–358.

Hinshaw, G., Spergel, D. N., Verde, L., *et al.* 2003. First-year Wilkinson Microwave Anisotropy Probe (WMAP) observations: The angular power spectrum. *Astrophys. J. Suppl.,* **148**, 135.

Hlozek, R., Grin, D., Marsh, D. J. E., and Ferreira, P. G. 2015. A search for ultralight axions using precision cosmological data. *Phys. Rev. D,* **91**(10), 103512.

Hoekstra, H., and Jain, B. 2008. Weak gravitational lensing and its cosmological applications. *Ann. Rev. Nucl. Part. Sci.,* **58**, 99–123.

Hou, Z., Keisler, R., Knox, L., Millea, M., and Reichardt, C. 2013. How massless neutrinos affect the cosmic microwave background damping tail. *Phys. Rev. D,* **87**, 083008.

Hu, W. 2008. Lecture notes on CMB theory: From nucleosynthesis to recombination. arXiv:0802.3688.

Hu, W., and Dodelson, S. 2002. Cosmic microwave background anisotropies. *Ann. Rev. Astron. Astrophys.,* **40**, 171–216.

Hubble, E. 1929. A relation between distance and radial velocity among extragalactic nebulae. *Proc. Nat. Acad. Sci.,* **15**, 168–173.

Huterer, D., and Shafer, D. L. 2018. Dark energy two decades after: Observables, probes, consistency tests. *Rept. Prog. Phys.,* **81**(1), 016901.

Huterer, D., and Turner, M. S. 1999. Prospects for probing the dark energy via supernova distance measurements. *Phys. Rev. D,* **60**, 081301.

Huterer, D., and Turner, M. S. 2001. Probing the dark energy: Methods and strategies. *Phys. Rev. D,* **64**, 123527.

Kaiser, N. 1984. On the spatial correlations of Abell clusters. *Astrophys. J. Lett.*, **284**, L9–L12.

Kaiser, N. 1987. Clustering in real space and in redshift space. *Mon. Not. Roy. Astron. Soc.*, **227**, 1–27.

Kamionkowski, M., and Kovetz, E. D. 2016. The quest for B modes from inflationary gravitational waves. *Ann. Rev. Astron. Astrophys.*, **54**, 227–269.

Kardar, M. 2007. *Statistical Physics of Particles*. Cambridge: Cambridge University Press.

Kolb, E. W., and Turner, M. S. 1994. *The Early Universe*. Philadelphia, PA: Westview Press.

Komatsu, E., Dunkley, J., Nolta, M. R., *et al.* 2009. Five-year Wilkinson Microwave Anisotropy Probe (WMAP) observations: Cosmological interpretation. *Astrophys. J. Suppl.*, **180**, 330–376.

Kovac, J., Leitch, E. M., Pryke, C., *et al.* 2002. Detection of polarization in the cosmic microwave background using DASI. *Nature*, **420**, 772–787.

Kragh, H. 1999. *Cosmology and Controversy*. Princeton, NJ: Princeton University Press.

Kragh, H., and Longair, M. (eds). 2019. *The Oxford Handbook of the History of Modern Cosmology*. Oxford: Oxford University Press.

Landy, S. D., and Szalay, A. S. 1993. Bias and variance of angular correlation functions. *Astrophys. J.*, **412**, 64.

Lang, K. R. 1999. *Astrophysical Formulae*, Vol. 1: *Radiation, Gas Processes and High-Energy Astrophysics*; Vol. 2: *Space, Time, Matter and Cosmology*. Berlin: Springer.

Lee, N., and Ali-Haïmoud, Y. 2020. HYREC-2: A highly accurate submillisecond recombination code. *Phys. Rev. D*, **102**(8), 083517.

Lesgourgues, J. 2011. The Cosmic Linear Anisotropy Solving System (CLASS) I: Overview. arxiv:1104.2932.

Lesgourgues, J., Mangano, G., Miele, G., and Pastor, S. 2013. *Neutrino Cosmology*. Cambridge: Cambridge University Press.

Lewis, A., Challinor, A., and Lasenby, A. 2000. Efficient computation of CMB anisotropies in closed FRW models. *Astrophys. J.*, **538**, 473–476.

Liddle, A. 2015. *An Introduction to Modern Cosmology*, 3 ed. Hoboken, NJ: Wiley-Blackwell.

Liddle, A. R., and Lyth, D. H. 2000. *Cosmological Inflation and Large-Scale Structure*. Cambridge: Cambridge University Press.

Linde, A. D. 1982. A new inflationary universe scenario: A possible solution of the horizon, flatness, homogeneity, isotropy and primordial monopole problems. *Phys. Lett. B*, **108**, 389–393.

Linder, E. V. 1997. *First Principles of Cosmology*. Boston, MA: Addison Wesley.

Linder, E. V. 2008. The dynamics of quintessence, the quintessence of dynamics. *Gen. Rel. Grav.*, **40**, 329–356.

Lisanti, M. 2017. Lectures on dark matter physics. In *Elementary Particle Physics: New Frontiers in Fields and Strings*, pp. 399–446. Boulder, CO: Theoretical Advanced Study Institute.

Long, A. J., Lunardini, C., and Sabancilar, E. 2014. Detecting nonrelativistic cosmic neutrinos by capture on tritium: Phenomenology and physics potential. *JCAP*, **08**, 038.

Longair, M. S. 2006. *The Cosmic Century*. Cambridge: Cambridge University Press.

Lupton, R. 1993. *Statistics in Theory and Practice*. Princeton, NJ: Princeton University Press.

Lyth, D. H. 1997. What would we learn by detecting a gravitational wave signal in the cosmic microwave background anisotropy? *Phys. Rev. Lett.*, **78**, 1861–1863.

Mandelbaum, R. 2018. Weak lensing for precision cosmology. *Ann. Rev. Astron. Astrophys.*, **56**, 393–433.

Mo, H. J., and White, S. D. M. 1996. An analytic model for the spatial clustering of dark matter halos. *Mon. Not. Roy. Astron. Soc.*, **282**, 347.

Mo, H. J., van den Bosch, F. C., and White, S. 2010. *Galaxy Formation and Evolution*. Cambridge: Cambridge University Press.

Mohr, J. J., Carlstrom, J. E., Holder, G. P., *et al.* 2000. A Sunyaev–Zel'dovich effect survey for high redshift clusters. In *VLT Opening Symposium: ESO Astrophysics Symposia European Southern Observatory*, pp. 150–156.

Mukhanov, V. 2005. *Physical Foundations of Cosmology*. Cambridge: Cambridge University Press.

Navarro, J. F., Frenk, C. S., and White, S. D. M. 1997. A universal density profile from hierarchical clustering. *Astrophys. J.*, **490**, 493–508.

Ostriker, J. P., and Peebles, P. J. E. 1973. A numerical study of the stability of flattened galaxies: Or, can cold galaxies survive? *Astrophys. J.*, **186**, 467–480.

Padilla, L. E., Tellez, L. O., Escamilla, L. A., and Vazquez, J. A. 2021. Cosmological parameter inference with Bayesian statistics. *Universe*, **7**(7), 213.

Peacock, J. A. 1998. *Cosmological Physics*. Cambridge: Cambridge University Press.

Peebles, P. J. E. 1968. Recombination of the primeval plasma. *Astrophys. J.*, **153**, 1.

Peebles, P. J. E. 1973. Statistical analysis of catalogs of extragalactic objects. I. Theory. *Astrophys. J.*, **185**(Oct), 413–440.

Peebles, P. J. E. 1980. *Large-Scale Structure of the Universe*. Princeton, NJ: Princeton University Press.

Peebles, P. J. E. 1982. Large-scale background temperature and mass fluctuations due to scale-invariant primeval perturbations. *Astrophys. J. Lett.*, **263**, L1–L5.

Peebles, P. J. E. 2020. *Cosmology's Century: An Inside History of Our Modern Understanding of the Universe*. Princeton, NJ: Princeton University Press.

Peebles, P. J. E., and Yu, J. T. 1970. Primeval adiabatic perturbation in an expanding universe. *Astrophys. J.*, **162**, 815–836.

Peebles, P. J. E., Page Jr, L. A., and Partridge, R. B. 2009. *Finding the Big Bang*. Cambridge: Cambridge University Press.

Perlmutter, S., Aldering, G., Goldhaber, G., *et al.* 1999. Measurements of Ω and Λ from 42 high redshift supernovae. *Astrophys. J.*, **517**, 565–586.

Phillips, M. M. 1993. The absolute magnitudes of Type IA supernovae. *Astrophys. J. Lett.*, **413**, L105–L108.

Press, W. H., and Schechter, P. 1974. Formation of galaxies and clusters of galaxies by self-similar gravitational condensation. *Astrophys. J.*, **187**, 425–438.

Press, W. H., Teukolsky, S. A, Vetterling, W. T, and Flannery, B. P. 2007. *Numerical Recipes*, 3rd ed. Cambridge: Cambridge University Press.

Raine, D. J., and Thomas, T. E. G. 2001. *An Introduction to the Science of Cosmology*. Boca Raton, FL: CRC Press.

Reeves, H., Andouze, J., Fowler, W. A., and Schramm, D. N. 1973. On the origin of light elements. *Astrophys. J.*, **179**, 909–930.

Riess, A. G., Filippenko, A., Challis, P., *et al.* 1998. Observational evidence from supernovae for an accelerating universe and a cosmological constant. *Astron. J.*, **116**, 1009–1038.

Ringeval, C. 2010. Cosmic strings and their induced non-Gaussianities in the cosmic microwave background. *Adv. Astron.*, **2010**, 380507.

Rood, H. J. 1981. Clusters of galaxies. *Rep. Prog. Phys.*, **44**(10), 1077–1122.

Ryden, B. 2016. *Introduction to Cosmology*, 2 ed. Cambridge: Cambridge University Press.

Samtleben, D., Staggs, S., and Winstein, B. 2007. The cosmic microwave background for pedestrians: A review for particle and nuclear physicists. *Ann. Rev. Nucl. Part. Sci.*, **57**, 245–283.

Sayre, J. T., Reichardt, C. L., Henning, J. W., *et al.* 2020. Measurements of B-mode polarization of the cosmic microwave background from 500 square degrees of SPTpol data. *Phys. Rev. D*, **101**(12), 122003.

Schael, S., Barate, R., Bruneliere, R., *et al.* 2006. Precision electroweak measurements on the Z resonance. *Phys. Rept.*, **427**, 257–454.

Schneider, P. 2010. *Extragalactic Astronomy and Cosmology*. Berlin: Germany.

Scolnic, D. M., Jones, D. O., Rest, A., *et al.* 2018. The complete light-curve sample of spectroscopically confirmed SNe Ia from Pan-STARRS1 and cosmological constraints from the combined Pantheon Sample. *Astrophys. J.*, **859**(2), 101.

Secco, L. F., Samuroff, S., Krause, E., *et al.* 2022. Dark Energy Survey Year 3 results: Cosmology from cosmic shear and robustness to modeling uncertainty. *Phys. Rev. D*, **105**(2), 023515.

Sharon, K., Ofek, E. O., Smith, G. P., *et al.* 2005. Discovery of multiply imaged galaxies behind the cluster and lensed quasar SDSS J1004+4112. *Astrophys. J. Lett.*, **629**, L73–L76.

Slatyer, T. R. 2018. Indirect detection of dark matter. In *Elementary Particle Physics: Anticipating the Next Discoveries in Particle Physics*, pp. 297–353. Boulder, CO: Theoretical Advanced Study Institute.

Springel, V., White, S. D., Jenkins, A., *et al.* 2005. Simulating the joint evolution of quasars, galaxies and their large-scale distribution. *Nature*, **435**, 629–636.

Steigman, G., Schramm, D. N., and Gunn, J. E. 1977. Cosmological limits to the number of massive leptons. *Phys. Lett. B*, **66**, 202–204.

Strigari, L. E. 2013. Galactic searches for dark matter. *Phys. Rept.*, **531**, 1–88.

Sunyaev, R. A., and Zeldovich, Ya. B. 1970. Small scale fluctuations of relic radiation. *Astrophys. Space Sci.*, **7**, 3–19.

Sunyaev, R. A., and Zeldovich, Ya. B. 1972. The observations of relic radiation as a test of the nature of X-ray radiation from the clusters of galaxies. *Comm. Astrophys. Space Phys.*, **4**, 173–178.

Suzuki, N., Rubin, D., Lidman, C., *et al.* 2012. The Hubble Space Telescope Cluster Supernova Survey: V. Improving the dark energy constraints above $z > 1$ and building an early-type-hosted supernova sample. *Astrophys. J.*, **746**, 85.

Tegmark, M. 2003. Cosmology: Parallel universes. *Sci. Am.*, **288**(5), 40–51.

Tegmark, M., Taylor, A., and Heavens, A. 1997. Karhunen–Loeve eigenvalue problems in cosmology: How should we tackle large data sets? *Astrophys. J.*, **480**, 22.

Thorne, K. S., and Blandford, R. D. 2017. *Modern Classical Physics*. Princeton, NJ: Princeton University Press.

Tremaine, S., and Gunn, J. E. 1979. Dynamical role of light neutral leptons in cosmology. *Phys. Rev. Lett.*, **42**, 407–410.

Trotta, R. 2008. Bayes in the sky: Bayesian inference and model selection in cosmology. *Contemp. Phys.*, **49**, 71–104.

Turner, M. S. 2022. The road to precision cosmology. *Ann. Rev. Nucl. Part. Sci.*, **72**, 1–33.

Tyson, J. A., Kochanski, G. P., and Dell'Antonio, I. P. 1998. Detailed mass map of CL0024+1654 from strong lensing. *Astrophys. J. Lett.*, **498**, L107.

van Albada, T. S., Bahcall, J. N., Begeman, K., and Sancisi, R. 1985. The distribution of dark matter in the spiral galaxy NGC-3198. *Astrophys. J.*, **295**, 305–313.

Vikhlinin, A., Kravtsov, A. V., Burenin, R. A., *et al.* 2009. Chandra Cluster Cosmology Project III: Cosmological parameter constraints. *Astrophys. J.*, **692**, 1060–1074.

Walsh, D., Carswell, R. F., and Weymann, R. J. 1979. 0957 + 561 A, B – Twin quasistellar objects or gravitational lens. *Nature*, **279**, 381–384.

Warren, M. S., Abazajian, K., Holz, D. E., and Teodoro, L. 2006. Precision determination of the mass function of dark matter halos. *Astrophys. J.*, **646**, 881–885.

Wechsler, R. H., and Tinker, J. L. 2018. The connection between galaxies and their dark matter halos. *Ann. Rev. Astron. Astrophys.*, **56**, 435–487.

Weinberg, D. H., Mortonson, M. J., Eisenstein, D. J., *et al.* 2013. Observational probes of cosmic acceleration. *Phys. Rept.*, **530**, 87–255.

Weinberg, S. 1989. The cosmological constant problem. *Rev. Mod. Phys.*, **61**, 1–23.

Weinberg, S. 2008. *Cosmology*. Oxford: Oxford University Press.

White, M. J., and Hu, W. 1997. The Sachs–Wolfe effect. *Astron. Astrophys.*, **321**, 8–9.

Wong, K. C., Suyu, S. H., Chen, G. C.-F., *et al.* 2020. H0LiCOW – XIII. A 2.4 per cent measurement of H_0 from lensed quasars: 5.3σ tension between early- and late-Universe probes. *Mon. Not. Roy. Astron. Soc.*, **498**(1), 1420–1439.

Zyla, P. A., Barnett, R. M., Beringer, J., *et al.* [Particle Data Group]. 2020. Review of particle physics. *PTEP*, **2020**(8), 083C01.

Index